THE MECHANISMS OF METALLURGICAL FAILURE

THE MECHANISMS OF METALLURGICAL FAILURE

The Origin of Fracture

JOHN CAMPBELL

ELSEVIER

Butterworth-Heinemann
An imprint of Elsevier

Butterworth-Heinemann is an imprint of Elsevier
The Boulevard, Langford Lane, Kidlington, Oxford OX5 1GB, United Kingdom
50 Hampshire Street, 5th Floor, Cambridge, MA 02139, United States

Library of Congress Cataloging-in-Publication Data
A catalog record for this book is available from the Library of Congress

British Library Cataloguing-in-Publication Data
A catalogue record for this book is available from the British Library

ISBN: 978-0-12-822411-3

For information on all Butterworth-Heinemann publications visit our website at https://www.elsevier.com/books-and-journals

Publisher: Matthew Deans
Acquisitions Editor: Dennis McGonagle
Editorial Project Manager: Charlotte Rowley
Production Project Manager: Prasanna Kalyanaraman
Cover Designer: Mark Rogers

Typeset by TNQ Technologies

To Zoe

Contents

Preface

When writing this book on fracture, I did not wish readers to think me presumptuous by appearing to consider that I was competent to enter a field so complex and so full of experts. However, I sought expert collaboration in vain. I believed I had an important contribution to make and so had no choice but to go at it alone. It is my hope, therefore, that fracture experts will be generous and tolerant, identifying the valuable offerings contained here and not damning the whole for the failings of its parts.

The book is written for the fracture, engineering, and metallurgical communities. They will, in general, not be familiar with the technicalities of melting and casting. Conversely, the early parts of the book will, or should, be largely familiar to melters and casters, being based on my earlier casting texts.

Be it for melters, casters, or fracture experts, the central issue of this work will be a shock to many. Fig. 1 represents a cracked carbide inclusion as might commonly be seen under the microscope in a grain boundary of a steel. The crack may or may not extend somewhat into the matrix. The traditional common-sense explanation for this structure is that the inclusion was originally crack-free but weak and brittle, and the application of stress caused it to crack. Naturally, the crack may now propagate into the matrix. Thus, the logic proceeds from (i) inclusion to (ii) stress to (iii) crack.

It is hoped the reader will forgive a non-too-serious diversion which illustrates the improbability of this explanation. Consider a clean, rounded, smooth pebble, perhaps selected from the beach, which is enclosed in the center of a block of soft butter. If the block of butter is then sheared, the question is "Will the pebble break?" The answer is "Don't be ridiculous!"

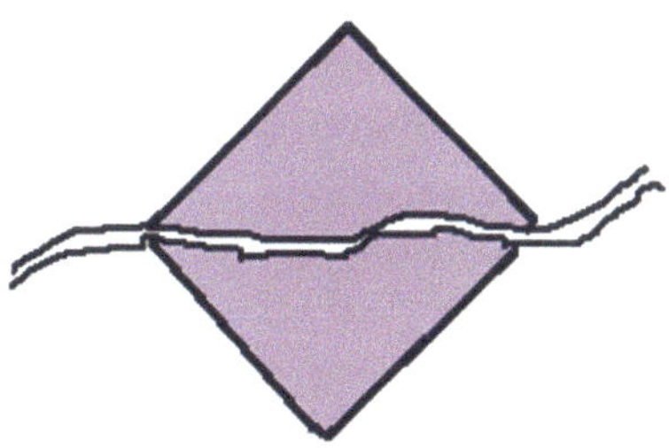

Figure 1 An apparently cracked inclusion in a steel.

Clearly, there may be something disquietingly incorrect with the conventional approach which assumes that the cracked inclusion has been fractured by stress. More seriously, we shall find the fundamental logic to be quite wrong.

Instead of the inclusion being present first, then stress, and finally the crack, the logic emphasized repeatedly throughout this book is reversed; it is proposed that the crack existed first, the inclusion second, and at no stage is stress necessary to form the original crack. (I am sure that Griffith would have been interested to discover his *preexisting cracks* as bifilms in the metal, and even more interested to note that inclusions nucleated and grew on bifilm cracks.) Pursuing the logic ruthlessly, we shall conclude that both the stress and inclusion are, in general, redundant to the cracking mechanism. The cracked inclusion merely indicates the presence of a preexisting bifilm crack.

If this is true, how could we have got it so wrong? I find myself concluding that traditional reversed thinking has been not merely unfortunate, but has been a disaster for our understanding of metallurgical failure and a disaster for metallurgical understanding in general. It has provided an inverted paradigm to mislead our thinking, research, and teaching for nearly a century. Griffith would have been disappointed with us all.

The proposed reversal of our normal metallurgical thinking is so alarmingly counterintuitive that a reasoned explanation could hardly be achieved in a paper. It seemed to me to require a book—a book sufficiently large to gather and collate the wide-ranging evidence, but not so large as to deter a reader. Clearly, we need good evidence to encourage and enable us to reverse our mindset, although the treatment in this book cannot be in any great depth and will be justly criticized by many.

For the first time, fracture is here considered in three parts, corresponding to the Sections of the book: (I) the liquid phase, (II) partly solidified phases, and (III) solid phases. This counterintuitive layout was essential for the logic and coherence of the approach to be clear. In particular, the existence of cracks in the liquid state is not an easy concept for those of us who have received merely an education and training in physical metallurgy. Many of us know no process metallurgy.

The conditions in the liquid which influence fracture in the solid are essentially simple in nature, but give rise to numerous outcomes demonstrating the ubiquity of good physics, which have yet to be included in the syllabus of our teaching institutions. Because of the paucity of knowledge in this area, the phenomena in the liquid state, Sections I and II of the book, will occupy much of this work. Melters and casters will have to bear with the repetition of much that I have already presented elsewhere. I hope other readers will be intrigued by the intricacy and importance of the numerous contributions of the liquid phenomena which, for the first time, explain many well-known aspects of solid state metallurgy.

When, finally, we arrive at Section III, fracture in the solid state, I found myself shocked to realize the widespread tacit acknowledgment in our metallurgical community that, in fact, none of the currently proposed mechanisms for crack initiation appear to be valid. Despite this amazing fact that there seems to be no viable mechanism for the metallic lattice to initiate a crack, it seems researchers and teachers have acquiesced unperturbed. It is astonishing how this outstanding and overwhelmingly important issue has remained ignored. In due course, I hope the reader will become convinced that the bifilm not only clearly fulfills this role but also is the only contender to fill this vacuum.

The other feature of Section III is what some will distain as "speculative." In a not-too-serious moment, Einstein is said to have commented, "Prediction is difficult; especially of the future." From a viewpoint of Relativity Theory, most would agree. In contrast, the central feature of this book, Entrainment Theory, is amazing because it is so simple. Perhaps too simple to be believable? Nevertheless, this simplicity has encouraged many predictions, all of which appear to have so far proven correct. These include the double nature of the bifilm, bubble trails, their role as favored substrates for graphite in cast iron, carbides in steels, topologically close-packed phases in Ni superalloys, and many other phases in different alloys.

Speculation? Perhaps. But the concepts of entrainment and the bifilm have opened a new, rich, and fertile field, in which simple and commonsense "speculation" has been rewarded, finding itself, again and again, to be accurate prediction.

At this time, Section III naturally relies heavily on incidental evidence from the metallurgical literature because, of course, the targeted research has not yet been carried out. However, it seems to this writer that the amount of circumstantial evidence, all pointing unerringly to the fundamental and necessary involvement of the bifilm, cannot be ignored. Another widespread reported admission is that the experimenters are unable to explain their findings, which is understandable when they are unaware of the major controlling factor in their work. The few isolated direct proofs are rare and, but once again, are supportive. We look forward to the luxury of more targeted critical experiments.

We need to remind ourselves therefore that our well-established theories of brittle fracture require a *preexisting* crack to initiate failure, as elegantly proposed by Griffith as early as 1921. Griffith proposed "the general conclusion may be drawn that the weakness of isotropic solids, as ordinarily met with, is due to the presence of discontinuities, or flaws, as they may be more correctly called, whose ruling dimensions are large compared with molecular distances. The effective strength of technical materials might be increased 10 or 20 times at least if the flaws could be eliminated."

Griffith's farseeing suggestion coincides with the thesis presented in this book. He sensed the reason underlying the improbability of initiating a crack in a metal: the bonds between atoms in the lattice are far too strong to permit the initiation of a crack; atoms are effectively inseparable.

There is a sense in which this book constitutes a reappraisal of the enormity of the implications of Griffith's theory. It takes Griffith further because Griffith never questioned how the population of cracks could have been formed. He did not question their structure or how they might be avoided. He was also unaware of how the precipitation of second phases could prize open the cracks, leading to entirely new failure mechanisms.

We are at the point of the development of physical metallurgy in which the presence of a dense population of *preexisting* cracks in our metals, as suggested by Griffith, has been consistently overlooked. This has been a serious oversight. Our preoccupation with physical metallurgy, ignoring the stark realities of process metallurgy, has been a regrettable diversion. We need the humility to start repairing this momentous error in our collective metallurgical understanding. We all need to find out how the "cracks" originate and how they might be controlled.

Our ignorance of the dense populations of bifilm cracks often present in our metals has been costly. Nearly all our metallurgical researches, published as papers in our respected international journals, start with words such as "We took a slab of steel..." or "Ingots were provided by..." or "Our starting point was 2 mm sheet...." None of the authors of these researches thinks to specify the casting conditions which define the defect types and density, and which will most probably be the major factor affecting the outcome of their research. Most of our vast library of metallurgical research carried out so far will therefore have to be repeated. We may then achieve a logical and consistent understanding of metallurgy.

I hope at last we shall find ourselves reevaluating many of our standard metallurgical clichés as "brittle grain boundary carbides" and "weak grain boundaries" finding these concepts to have been quite wrong and require to be discarded. We shall find ourselves admitting that carbides are not only hard, strong, and crack-resistant but also ductile. In addition, we shall find that grain boundaries are also immensely strong and resistant to decohesion.

Overall, we shall come to the realization that our current metals cannot be relied upon. They break, sometimes at stresses well below their expected strength. The gas pipeline explodes because of hydrogen embrittlement; the stainless fittings fail in the seawater; the turbine blade fatigues. Clearly, at this time, our metals are not reliable. Action is required to correct what appears to be their natural disposition to failure. Furthermore, such action is possible; a disposition to fail is not to be expected and correct processing can eliminate the disposition to fail.

All of this is counter to traditional metallurgical teaching. Clearly, I have had to unlearn my metallurgy and relearn it. I found this hard. Hence this book. I hope the reader will find the book, with all its undoubted faults, to make the fundamental transition in our thinking easier. There is much to be gained.

Acknowledgments

I am grateful to my numerous students whose work I find myself quoting all the time. In addition, of course, there are those foundries who have taken on board the vital importance of reducing oxide film entrainment during casting, and the concept of the bifilm, and whom are proving the benefits every day. My ex-son-in-law, Professor Robert Matthews, drew my attention to the work of Barenblatt whose crack initiation model fundamentally underpins the bifilm model. Finally, the women in my life: there has been tolerance and encouragement from my wife, Sheila, who, without complaint, well, not too many complaints, has endured my obsession with the writing of this book. Meanwhile, daughter, Zoe, stands by to look after her dad and his books.

Introduction

We are to make a new start on an attempt to reveal the mysteries of metallurgical science, attempting to unravel the origin of fracture, the failure mechanism central to the concern of all metallurgy and engineering. It seems to me as fundamental as Darwin's unleashing of his Origin of Species. These are, once again, new times requiring new thinking.

The fact that most metals fracture when subjected to sufficient tensile stress seems so obvious and so undeniable as to constitute a fundamental axiom of metallurgy. However, this book presents the evidence that if initiators for fracture are excluded from the metal, the metal cannot fracture. Thus, if one of our high strength steels was subjected to tension in a tensile testing machine, the sample would not break; after exceeding the yield point it would continue to extend plastically, despite displaying immensely high strength. Final failure would occur by the test piece necking down, eventually coming apart at the last pair of atoms, thus achieving 100% reduction in area. This behavior should be expected for practically all our engineering metals and alloys if processed correctly, as discussed at length throughout this work.

Similarly, creep or superplastic specimens would never fail by cavitation from inclusions or the apparent decoherence of grain boundaries. The performance of polycrystalline, directionally solidified, and single crystal superalloys would all exhibit the same large strains to failure. Fatigue test specimens should never fail by the propagation of a fatigue crack. Only occasionally would the highest strength alloys fail by fracture, and such fracture would occur at the extremely high theoretical tensile strength of the solid. Such very high stress fractures will be dangerously explosive, and not recommended, as will also be discussed.

Naturally, these predictions may appear as wild flights of fancy. However, in this work we shall lay out the case for transforming the performance of our future alloys. Traditionally the metallurgist laboring to develop alloys has found that strength is relatively easily achieved, but toughness is elusive, not knowing that toughness lay not in his hands, but in the hands of the foundryman, who, unfortunately, was also unaware that he had this power. For this reason, many strong alloys have remained essentially unusable as a result of their unnecessarily brittle nature. This work illustrates how, in general, it is possible to avoid the introduction of crack initiators, allowing apparently brittle metals and alloys to acquire ductility.

One of my main purposes in writing this book has been to draw attention to the parallels between fracture initiation in the liquid and in the solid states. These conditions tend to be widely separated within the academic disciplines of (i) melting and casting and those of (ii) physical metallurgy and fracture mechanics.

But the parallels are compelling and valuable. In the pure metallic liquid, and in the solid lattice, the interatomic distances are closely similar, so that both states of matter have similar resistance to fracture. More importantly, it will become clear that fracture in the solid state cannot be understood without an understanding of the prior liquid processing. Furthermore, as a result of the extremely high interatomic bonding, we shall conclude during the course of this short study that the resistance to fracture is so high that fracture in our engineering metals should be generally impossible.

The reader's patience is requested: the book presents the evidence, eventually reaching the surprising conclusion that metals, in both the liquid and solid states, even under extreme engineering stresses, should never fail by cracking. The fact that we live with cracking, fatigue, creep, and other common modes of failure of critical engineering applications such as engines, bearings for wind turbines, helicopter drive trains, bridges, etc., does not appear to be explicable. We shall find our current metallurgical knowledge wanting.

In other words, the book attempts to present the little-known information that metals do not and cannot fail from natural, internal, and intrinsic mechanisms; they fail from externally introduced features which are common to most metals (although there are interesting exceptions which we shall discuss).

These externally introduced features are bifilms; the surface oxides on liquid metals which are entrained during the casting process. It is important to realize that these features are not associated with solidification. Although vast amount of research continues to be carried out on the details of solidification, this simple phase transition, in comparison, is largely irrelevant to the strength and fracture resistance of metals. All those features we thought of for years as being important to properties, such as dendrite arm spacing (DAS), are seen to be merely secondary features, in which the primary control of properties rests firmly with bifilms. In fact, we shall find that without the presence of bifilms, DAS appears to have only insignificant influence over properties. Similarly, the huge and continuing efforts to reduce grain size result in only minor benefits, and, as we shall see, even much of this benefit is an accidental issue resulting from the unwitting reduction of the bifilm population.

The huge and continuing research into failure by hot tearing in the liquid plus solid state will also be concluded to have little to do with solidification, but it is once again powerfully controlled by bifilms formed during casting of the liquid metal for the hot tear experiments. We can surmise that, perhaps apart from segregation, *solidification* defects probably do not exist, and if they do exist, they do not matter; only *casting* defects are important.

The casting defects which arise as a result of entrainment mechanisms are bifilms and bubbles. (Occasionally, foreign inclusions may also be entrained; although apart from sand inclusions from badly cast sand molds, entrained foreign inclusions in metals are rare. We shall not spend time on this comparatively trivial issue.) Bifilms are excellent initiators for cracks because they possess an unbonded interface, like a crack. Furthermore, the crack tip (the edge of the bifilm) is sharp, being usually only a few nanometers in radius.

We shall examine the evidence that the tensile failure of metals occurs probably *exclusively* from *unbonded* interfaces *entrained* during manufacture.

Put another way, all pores and cracks observed in the solid originated as entrainment defects in the liquid state.

The new and overwhelmingly important outcome of this work is that failure of metals might be eliminated simply by the elimination of entrainment defects, which we can choose to do if we wish.

Until now, we appear to have chosen to use metallurgical processes which introduce generous quantities of entrainment defects into our metals. We need not choose such damaging processes. Bifilms need not be present in our metals. If, therefore, for the first time in history, we take the decision to avoid entraining defects into metals, it follows that we should obtain metals which never crack. Current developments in foundries using new casting technology to reduce bifilms are already yielding astonishing results, although it is clear the processes have much potential for additional improvement. The future for metals has never looked so exciting.

CHAPTER 1

The fracture of liquids

Contents

The Mechanisms of Metallurgical Failure
ISBN: 978-0-12-822411-3
https://doi.org/10.1016/B978-0-12-822411-3.00001-0

1.1 Theoretical strength of liquids

1.1.1 Classical continuum theory

When a liquid is subjected to an increasing hydrostatic tension, equivalent to a negative pressure, it will eventually "fracture." The fracture process for liquids is the formation of a pore. Above a critical size it will expand explosively, relieving the elastic stress in the liquid. The critical stress in the liquid at which this occurs is called "the fracture pressure, P_f."

It was probably Temperley in 1953 who first proposed that the tensile strength of liquids might be estimated from the minimum in the pressure P versus volume V relation predicted by the van der Waals's equation. As the van der Waals description derives from the kinetic theory of gases, being a first approximation of the behavior of a highly compressed gas, this is definitely a counterintuitive approach, but illustrates the power and delight of physics. The well-known general gas equation is

$$PV = \mathrm{NRT}$$

N is Avogadro's number, being approximately 6×10^{23}; the number of atoms per mole. If we modify this relation introducing the constants "a" and "b" to take approximate account of the forces and volumes, respectively, of the atoms, we have the van der Waals equation

$$\left(P + \frac{aN^2}{V^2}\right)(V - Nb) = \mathrm{NRT}$$

It is noted that when $Nb = V$, the whole volume is occupied by atoms so that the pressure approaches infinity, but when V is large, the equation approaches the general gas equation. This equation is a cubic in terms of the volume V (as is clear if the terms of the equation are all multiplied by V^2) and so has either 1, 2, or 3 solutions for V for a given value of P, as seen in Fig. 1.1.

The left-hand branch of the curve shows how the volume of the liquid will increase slightly as the pressure falls, even when pressure continues to fall, becoming negative, as a tensile stress (but a hydrostatic tensile stress of course means the stress has the same value in all directions). However, at P_f, pressure becomes a minimum, corresponding to the ultimate tensile strength of the liquid. Beyond this point, the pressure is predicted to jump unstably to the maximum vapor pressure. This upward-sloping region of

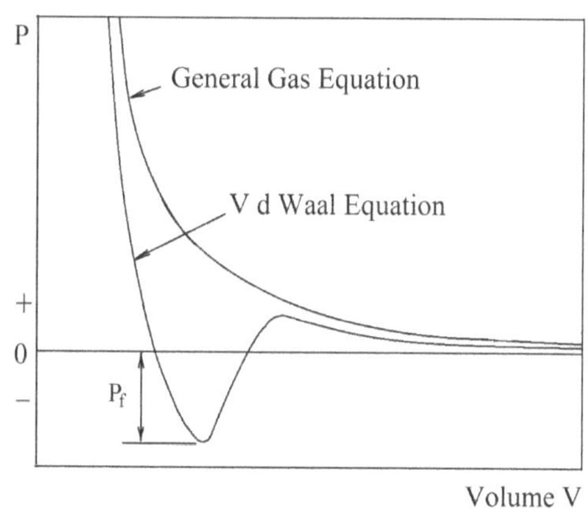

Figure 1.1 The van der Waals relation between pressure P and volume V of a liquid, indicating its theoretical strength.

the curve is unstable because the volume is predicted to increase at the same time as the pressure is increasing, which is clearly not realistic. For the right-hand portion of the van der Waals curve, the metal has become a vapor, which, with increasing pressure, reduces its volume as would be expected.

The equation has to be solved iteratively to find P_f. We shall not attempt that here. Solutions are posted on the internet for interested readers. In addition, Fletcher (1993) gives an excellent account. We simply note that the equation predicts values for P_f approximately in agreement with those of simpler approaches described below.

1.1.2 Classical bubble nucleation

Using an alternative approach of classical physics, it is easy and quick to demonstrate that a pore cannot be nucleated in a liquid metal: the well-known equation for the mechanical stability of a spherical pore is

$$P = 2\gamma / r \tag{1.1}$$

where P is the internal pressure (or external hydrostatic tensile stress), γ is the surface tension, and r is the radius of the pore. For liquid aluminum and liquid iron, we can take γ as approximately 1 and 2 N/m, respectively.

We do not know the radius r to find out P_f. However, we can guess it is only a very few atomic diameters. To obtain an estimate, if we assume Fisher's values (see below) for P_f are between approximately 3 and 4 GPa for Al and approximately 7.0 GPa for Fe, Eq. (1.1) indicates the critical bubble sizes are close to 2–2.5 atom diameters (assuming atom radii 0.28 and 0.25 nm, respectively) corresponding to a pore size of approximately 8–14 vacancies in each case. These values seem reasonable and are taken as approximately correct from now on.

1.1.3 Homogeneous nucleation

The above estimates for the ultimate strength of liquid metals are confirmed to within a factor of about 2 by a number of elegant, classical theoretical studies to compute the theoretical strength of liquid metals. The impressive compilation of data by Tiryakioglu for aluminum confirms the reliability of the strength data (Fig. 1.2). In the short review of this topic by the author (Campbell, 1968), the treatment by Fisher (1948) stands out because of its simplicity and elegant logic.

Fisher quantifies the conditions required for the formation of porosity in liquid metals. From energy considerations, using macroscopic concepts such

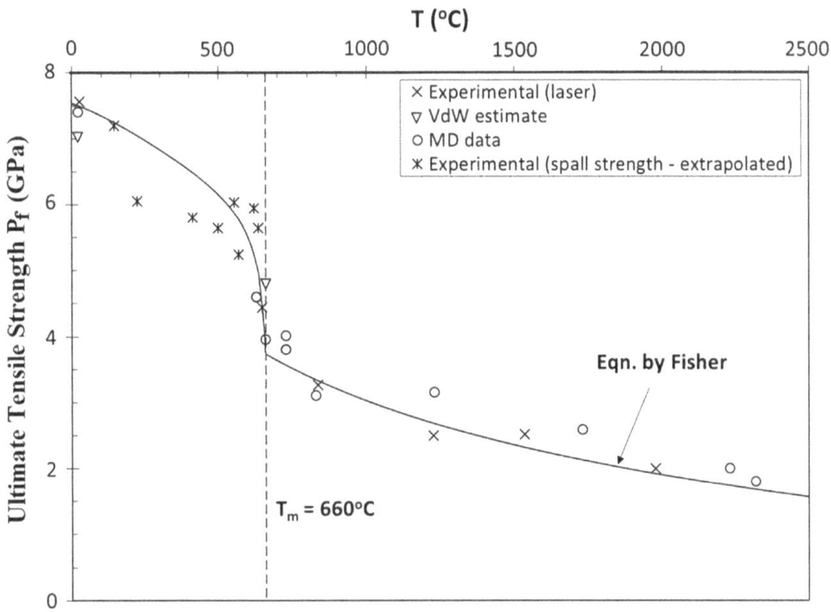

Figure 1.2 The ultimate tensile strength of solid and liquid aluminum as a function of temperature (Tiryakioglu, 2018).

as surface tension, he finds the quantity of work associated with the reversible formation of a bubble in a liquid. If the local pressure in the liquid is P_e, we need to carry out an amount of work P_eV to push back the liquid far enough to create a bubble of volume V.

The formation and stretching out of the new liquid–gas interface of area A requires additional work γA, where γ is the interfacial energy per unit area.

The work required to fill the bubble with vapor or gas at pressure P_i is negative and equal to $-P_iV$. (The negative sign arises because the pressure inside the bubble clearly helps the formation of the bubble, as opposed to the other work requirements, which tend to oppose bubble formation.) Thus, the total work is

$$\begin{aligned}\Delta G &= \gamma A + P_eV - P_iV \\ &= 4\gamma\pi r^2 + (4/3)\pi r^3(P_e - P_i)\end{aligned}$$

where clearly $(P_e - P_i)$ is the pressure difference between the exterior and the interior of the bubble, which we may write as ΔP for convenience. Similar to dense-phase nucleation, a plot of ΔG versus bubble radius r shows a maximum that constitutes an energy barrier to nucleation, as in Fig. 1.3. The critical radius r^* in this case is

$$r^* = 2\gamma/\Delta P^* \tag{1.2}$$

As bubbles growing from the bulk liquid will grow an atom at a time as the result of statistical thermal fluctuations, it is evident that small bubbles with radii less than r^* will tend to disappear. Only exceptionally will a long chain of favorable energy fluctuations produce a bubble exceeding the critical radius r^*. When this rare event does happen, the embryonic bubble will then have the potential to grow to an observable size.

Fisher goes on to apply some delightfully elegant rate theory to derive values for the critical pressure difference ΔP^* at which nucleation will occur. The reader is strongly recommended to consult Fisher's original paper. His formula for the tensile strength of liquid gives values of 3.1 GPa for liquid Al (Tiryakioglu's smoothed interpolation indicates nearer 4.0 GPa at the melting point) and 7.0 GPa for liquid Fe.

Erzi and Tiryakioglu (2019) present results for the fracture strengths of six metals, Pb, Al, Ti, Cu, Fe, and Ni, calculated using a number of completely different methods: Fisher (classical physics), van der Waals ("mechanistic thermodynamics"), and molecular dynamics (computer simulation). The results are all in impressive agreement.

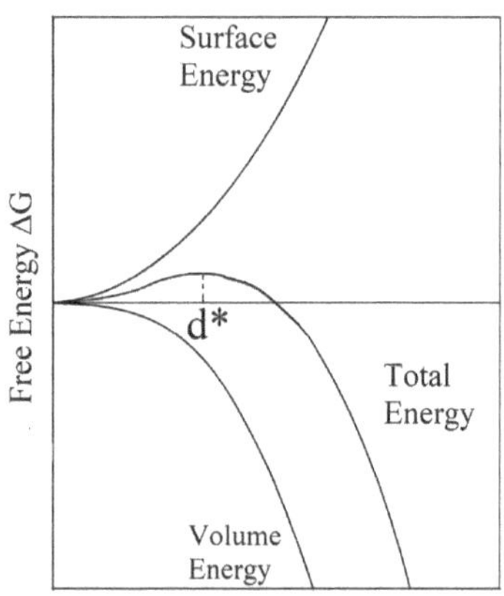

Figure 1.3 Surface and volume energies of an embryo pore growing in a liquid give the total energy as shown. Below the critical size d* any embryos will tend to shrink and disappear. Above d* increasing size reduces the total energy, so growth will be increasingly favored, becoming a runaway process.

The agreement is reassuring. However, the important message is that the pressures required for nucleation are all *extremely* high and reflect the impossibility of homogeneous nucleation of pores in liquid metals. It is clear that the strengths of liquid metals are almost as high as those of solid metals. As noted previously, this is hardly surprising because the atomic structure is similar, liquid metals being close-packed random structures, compared with solid metals being close-packed regular structures. In either case, the atoms are about the same distance apart, and it is similarly difficult to separate them; they are resistant to be forced apart to create a void.

The fracture pressure is reduced by the presence of surface-active impurities in the liquid. The nonmetals oxygen, sulfur, and phosphorus are particularly active in iron melts: the presence of only 0.2 weight percent of oxygen reduces the surface tension of liquid iron from 1.9 to approximately 1.0/Nm. This approximately halves the estimates of the pressure required for nucleation as shown by Eq. (1.2). Similar reductions in surface tension (and therefore in fracture pressures) are to be found in liquid copper when contaminated with high levels of the nonmetals, O, S, and P.

Such high concentrations of oxygen (and the other surface-active nonmetals) are probably often found on solidification because of the concentration of solutes ahead of the freezing front. For the case of liquid iron, the partition coefficient for oxygen is approximately 0.05, giving a factor of 20 increase in concentration at the advancing front. Thus, an average of only 0.01 weight percent oxygen in the bulk melt can produce 0.2 weight percent ahead of the front. If the levels of oxygen rise sufficiently to precipitate FeO as a liquid inclusion at the front, then the nucleation problem is reduced yet further because FeO has a surface tension of between 0.6 and 0.5/Nm, depending on its oxygen content (Popel and Esin, 1956). A pore will therefore preferentially nucleate in such a liquid, where the critical pressure is easily shown to be reduced to around 1.7 GPa. Effectively, this is still homogeneous nucleation in a pure liquid where in this case the pure liquid in the liquid Fe is in the form of regions, possibly minute droplets, of FeO.

Even this pressure is still so high as to be probably unattainable. What other possibilities are there?

1.1.3.1 Heterogeneous nucleation

Fisher considers the case of the nucleation of a bubble against the surface of a solid substrate. Conceptually, the liquid is considered to be in perfect atomic contact with the substrate, but as the pore grows, vacancy by vacancy, the pore forms a cap shape with an angle θ against the solid. This contact angle defines the extent of wetting; θ = 0 degree means complete wetting (the continuation of perfect atomic contact), whereas θ = 180° is complete nonwetting (the complete detachment—the decohesion—of the liquid from the substrate). The geometry is shown in Fig. 1.4. Fisher shows that nucleation is easier by a factor

$$P^*_{het}/P^*_{hom} = 1.12\{(2 - \cos\theta)(1 + \cos\theta)^2/4\}^{1/2} \quad (1.3)$$

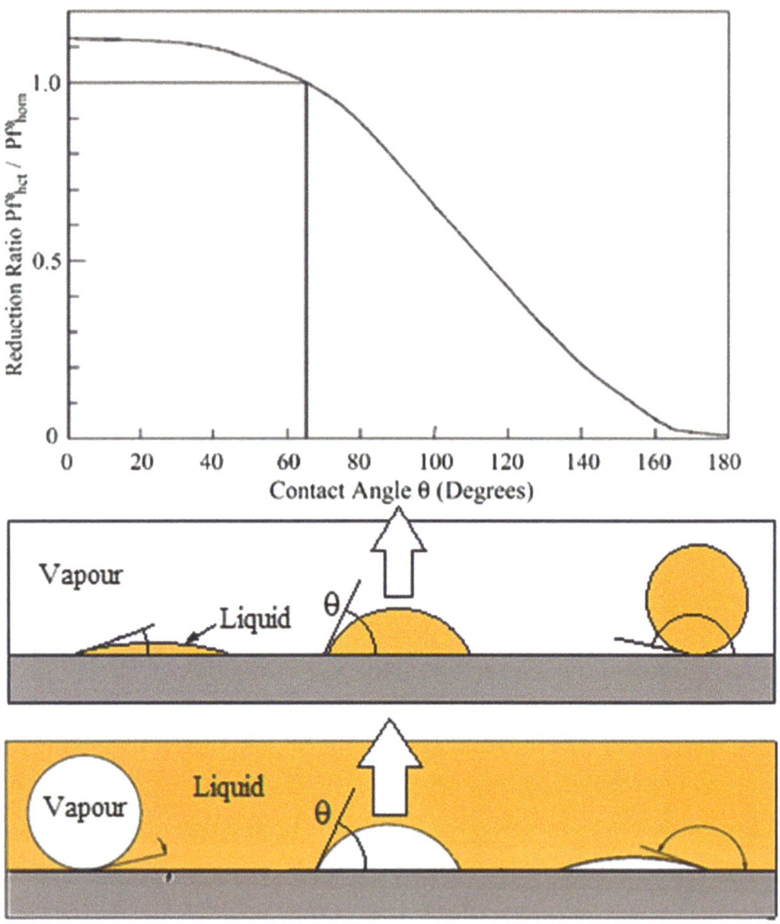

Figure 1.4 Spreading of the liquid on a substrate (good wetting), whereas nucleating a pore (equivalent to decohering from the substrate) is favored by poor wetting (high contact angle).

The factor 1.12 arises because of the fewer sites for nucleation on a plane surface of atoms, compared to the greatly increased number of possibilities in the bulk liquid. This factor alone makes the nucleation of pores on wetted surfaces unfavorable. Interestingly, of course, this includes the solidification front and surfaces of dendrites (although the segregation of gases, potential nuclei, and solutes which reduce surface tension are all concentrated ahead of the front, in practice reversing this prediction). In fact, nucleation on clean solid surfaces does not become favorable until the contact angle exceeds 60 or 70 degrees as shown in Fig. 1.4.

It is also important to note that not all inclusions are good nucleation sites for porosity. Those that are well-wetted (small contact angle) will not be favored. In steels, these include the rather more metallic inclusions such as borides, carbides, and nitrides. The wetting requirements for the nucleation of pores are completely opposite: good nuclei exhibit poor wetting characterized by high contact angle. Such substrates include the nonmetals such as oxides.

Fig. 1.4 indicates that as the contact angle increases to 180 degrees, any difficulty of heterogeneous nucleation should fall to zero. In fact, there are good reasons to believe that such perfect nonwetting is probably not possible and that the maximum contact angle attainable in practice is perhaps close to 160 degrees. No contact angle greater than 160 degrees appears ever to have been observed (see, for instance, the work by Livingston and Swingley (1971)). Assuming this to be true, then from Fig. 1.4, we see that heterogeneous nucleation on the most nonwetted solid known, having a contact angle of 160 degrees, reduces the pressure required for nucleation in the bulk liquid by a factor of 20.

Even after being reduced by a factor of 20, the fracture stresses for the liquid remain impressively high, in the region of 155 MPa for Al and 350 MPa for Fe. Even the most weakly bonded interfaces on the most nonwetted inclusions would not be expected to decohere during normal stresses occurring during solidification. It follows that even the most favorable heterogeneous substrates are probably incapable of nucleating a pore. Therefore, both homogeneous and heterogeneous nucleation seem impossible. We conclude, in agreement with the study by Yousefian and Tiryakioglu (2018), that classical nucleation of bubbles in liquid metals, by either homogeneous or heterogeneous mechanism, does not occur.

If porosity cannot be nucleated in cast alloys, how is it that pores in cast materials are the norm rather than the exception? Clearly, there is a major profound mismatch between theory and fact. The failure of classical nucleation theory to account for porosity is not widely appreciated. In this work, the fundamental inability of the classical theory has driven the evaluation of other pore initiation processes as outlined below.

This conclusion is so important and so surprising that it is worth emphasizing by stating it another way. The student will usually have been persuaded that because classical *homogeneous* nucleation of pores in liquid metals is difficult, classical *heterogeneous* nucleation on a solid particle therefore *must* occur. The key conclusion from our above discussion is that this is almost certainly not true.

The classical theory for homogeneous and heterogeneous nucleation is so well known that the reader may wonder why it has been included here in this fairly detailed presentation. This has been carried out to emphasize the elegant atomic credentials of these theories, particularly heterogeneous nucleation. In this very specific process, a bubble cap grows on a substrate vacancy by vacancy until a critical size, at which point the bubble can grow explosively, releasing all the stored elastic energy in the liquid.

The extremely rigorous classical conditions of this atomic mechanism are rarely understood. It has to be set apart from much sloppy science which claims to be heterogeneous nucleation. Other mechanisms which have nothing to do with nucleation are often called heterogeneous nucleation. This loose and inaccurate thinking is discussed later under Section 1.1.6.

1.1.4 Nucleation conditions for shrinkage pores

The problem of the nucleation of shrinkage cavities has also been widely overlooked. Somehow it has been assumed that they are fundamentally different to gas pores and that they "just arrive." After all, it is argued that they *must* occur in an unfed isolated volume of liquid because the concept of shrinkage means that there is a volume deficit. It is assumed that this volume deficit *must* result in a cavity. However, this is far from true.

As we have seen already, there really is a difficulty in the initiation of a cavity in a liquid. The various analogous systems described later in Section 1.3 demonstrate that the liquid can withstand hydrostatic tension and may never fail. If we accept this, then it follows that the tensioned liquid in the center of a solidifying casting is stretched elastically, and the surrounding solidified shell must be drawn inward, first elastically and then plastically as the stress in the liquid increases, provided no pore initiates to dispel the stress (Campbell, 1967). The surrounding solidified regions of the casting deform plastically, collapsing inward to relieve the internal tension. For this reason, the volume deficit does not result in a cavity in the cast material; the liquid never reaches the critical fracture pressure P_f. The conditions for classical nucleation of volume defects such as pores or cracks are never met in the defect-free liquid (by defect-free we mean a liquid in which neither bifilm nor bubble is present).

The inability of a casting to withstand any high level of internal tensile stress is very limited as every foundry person knows; a poorly fed casting can collapse, its external surfaces dishing inward under only atmospheric pressure (0.1 MPa) when operating over macroscopic areas. For microscopic volumes of unfed liquid, as might be trapped between dendrite arms, having a radius of the order of 100 μm, higher tensile stress in the liquid can be generated. The plastic collapse model (Campbell, 1967) predicts tensile stress up to 700 atm (70 MPa). Where the outside of the casting is cooling rapidly and contracting onto the interior of the cast metal, these tensile stresses can be reduced and even reversed as shown by Oki (1970). It seems the internal stress will always be far too low to generate pores by homogeneous or heterogeneous nucleation. In summary, therefore, the tensile stresses sufficient to create volume defects in castings cannot be generated; interatomic forces are too high to allow pores to open; and hot cast metal is in general too weak to support such stresses.

1.1.5 Joint gas and shrinkage conditions

The cooperative action of gas and shrinkage to create pores in liquid metals was quantified by Whittenberger and Rhines (1952). They considered this cooperative action in a simple but insightful model. The model is extremely helpful to the understanding of nucleation conditions for pores during the solidification of metals as (i) the effective gas pressure in solution is raised by segregation and (ii) tensile stress in the liquid is raised by shrinkage effects. A full discussion of these interesting details is not within the scope of this book. The interested reader is directed to my Complete Casting Handbook 2015.

1.1.6 Pseudo-heterogeneous nucleation of pores

At this point in the quest for a pore-forming mechanism in liquid metals, many texts now conclude that some preexisting pore must now be postulated, such as a pocket of gas trapped in a recess in an inclusion (Fig. 1.5). Such assumptions presuppose the very problem we are attempting

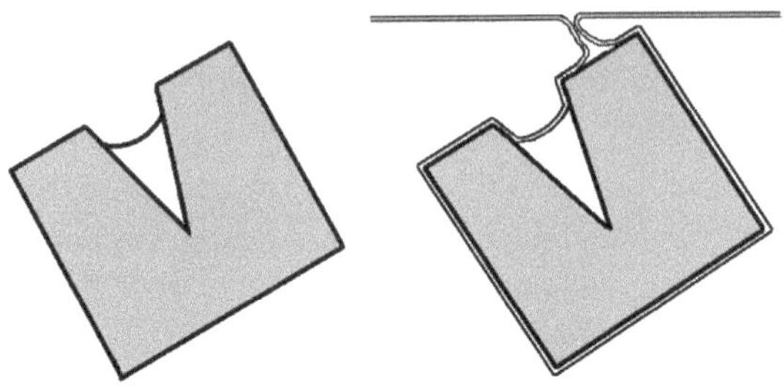

Figure 1.5 A crevice of entrained refractory containing a co-entrained pocket of air. (Because it is entrained, the right hand figure may be more accurate, showing the accompanying entrained oxide film.)

to explain. How could a void or gas pocket occur in a solidification process? The atomic movements during the reorganization of the metal from liquid to solid are only small fractions of an atomic diameter; the structure of the liquid is that of a randomly close packed solid, and the structure of the solid is, of course, also close packed but regular; both are rather similar, with quite similar interatomic spacings. The high forces that keep the atoms together effectively forbid the opening of a void in either phase, as suggested by Eq. (1.1).

The pore sitting at the base of the crevice in the solid inclusion cannot therefore be nucleated or formed by solidification processes, but must have originated outside the liquid metal and be introduced or "entrained." Once entrained, it can easily expand to create new pores at almost negligible applied stress or gas content in solution. However, once again, if such an entrained inclusion were somehow avoided or eliminated, we return to the situation that a pore becomes impossibly difficult to form.

The pore arriving in the liquid as part of an entrained inclusion as shown in Fig. 1.5 is clearly not the result of a heterogeneous nucleation event. Entrainment is a mesoscopic phenomenon in complete contrast to nucleation, which is an atomic process.

1.2 Experimental demonstration of hydrostatic tensions in liquids

It is important for the reader to be at ease with the concept of a liquid having strength; when somehow subjected to tension, the liquid will stretch elastically, and if the stress becomes sufficiently high, it will finally "break." The nature of the "break" in the case of a liquid is the nucleation of a bubble. As in a solid, the break can occur with a bang. For the liquid, the bubble expands explosively, practically at the speed of sound in the liquid, so as to dissipate the built-up elastic stress in the liquid.

A number of examples illustrating the tensile behavior of liquids are described in this section, although, of course, few are metals. Most relate to water or other organic systems. Even so, it will be clear that the principles apply perfectly to liquid metals.

Liquids have been known to be able to withstand tension for many years. Mercury is an interesting case. In a clean glass barometer freshly filled with clean mercury, the liquid would adhere to the glass, "hanging up," supporting the weight of the long length of the heavy liquid like a rope in tension. If the glass was tapped sharply, breaking the liquid away from the glass, the liquid would fall, opening up a vacuum above it. It would settle at the height at which it could be supported by the pressure of 1 atm acting at its base. This height is approximately 760 mm, the barometric height. The so-called vacuum above the mercury is of course not perfect. A tiny pressure due to the vapor pressure of mercury is present, pushing the upper surface down about 1 μm at room temperature, which can normally be ignored. Actually, if the glass and the metal had been really clean, in principle, the huge weight of a mercury column could have hung from a height somewhere in the stratosphere, as the tensile strength of the liquid is extremely high.

The very high tensile strengths of liquid metals should not be a surprise. The interatomic distances in liquids, as random close-packed atomic structures, are only a few percent larger than the distances in solid metals, which have more regular close-packed structures. Thus the strengths would be expected to be of the same order, as is shown to be true in the case of aluminum in the interesting compilation of data by Tiryakioglu (Fig. 1.2).

As long ago as 1850, Berthelot investigated the strength of various organic liquids. He cooled the liquids in sealed glass tubes, watching until cavitation occurred as a shower of small bubbles. The number of bubbles was subsequently explained by Lewis (1961), who observed that the rupture was always associated with the sudden appearance and collapse of a single bubble, which was immediately followed by the formation of many tiny bubbles along the length of the tube. The asymmetric collapse of the primary bubble would form an impacting jet and create a series of shock waves, triggering the formation of additional bubbles. This simple experiment is of special interest because it is analogous to the solidifying casting, with the internal solidifying liquid progressively occupying less volume and with the surrounding solid gradually being pulled inward, its elastic and plastic resistance to deformation steadily increasing, building up the tensile stress in the liquid, to the point at which the liquid fractures.

Berthelot's experiment has been repeated many times in different ways. For instance, Vincent and Simmons (1943) sealed their tube by freezing the liquid in the mouth of the capillary. They found that water would withstand −15.7 MPa and a mineral oil −11.9 MPa. This is a fascinating observation because it indicates that a solidification front will withstand high tensions; it is not necessarily a good nucleation site for pores, as predicted in Section 1.1. The facts reveal the error of those assumptions, widely presumed in the solidification literature, of ease of pore nucleation at the solidification front. (Perversely, as a result of microsegregation at an advancing front, the surface tension can be low and the gas in solution in the liquid can be high. In addition, potential nuclei can be pushed ahead. Thus, despite its theoretically unfavored state, a freezing front tends to be the location for bubble formation for quite separate reasons.)

For a liquid sealed in a tube, Scott et al. (1948) calculated the build-up of tensile stress ΔP as the liquid cools an amount ΔT, using the coefficient of thermal expansion of the liquid α and the glass σ, together with the modulus of rigidity G and the coefficient of compressibility β of the glass tube:

$$\Delta P = [(\alpha - \sigma) / (\beta - (1/G))]\Delta T$$

They also checked their results by measuring the length of the liquid columns after fracture. This work emphasizes the elastic strains built up in the liquid and the surrounding solid during the experiment.

Lewis (1961) found tensile strengths in water and carbon tetrachloride in excess of 6.0 MPa. Also, when water and carbon tetrachloride were introduced into the same tube, the two liquids separated at a clear interface. Fracture was not observed to occur at the interface. The liquid–liquid interface seems as poor a nucleation site as the liquid–solid interface above.

Tyndall (1872) carried out experiments on the internal melting of ice by shining light through the ice. The light was absorbed by particles frozen inside. The heated particles then melted a small surrounding region of the shape of a snowflake (known nowadays as a Tyndall figure or negative crystal). However, because ice reduces in volume on melting, the liquid volume created in this way was under tension. As melting progressed, the tensile stress became so great that the liquid fractured with an audible click. The melted region was then observed to contain a small bubble. Kaiser (1966) demonstrated that the critical pressure for fracture was in excess of −15.0 MPa. Tyndall proved that the bubble was a vapor cavity containing practically no air by carefully melting the ice toward the figure, using warm water. As soon as a link was established with the bubble, it collapsed and disappeared; no air bubble rose to the surface. The Tyndall figures are analogous to the isolated regions of metal castings, where vapor cavities are expected to be formed at critical stresses during solidification if suitable points of initiation are present.

Briggs (1950) centrifuged water in a capillary tube, spinning it at increasing speed to stretch the water until it was observed to snap at a tensile stress of 28.0 MPa.

As will have been clear, all this early work on the determination of the tensile strengths of liquids held in various containers such as Berthelot tubes gave notoriously scattered results. This was probably to be expected from the different degrees of cleanness of the container and any microscopic imperfections it may have had on its walls. Experiments to test the liquid away from the influence of walls have been carried out using ultrasonic waves focused in the center of a volume of liquid. The rarefaction (tensile) half of the vibration cycle can cause cavitation in the liquid. These experiments have been moderately successful in increasing the limits, but impurities in the liquid have tended to obscure results.

More recently, Carlson (1975) has reflected shock waves through a thin film of mercury and recorded strengths of up to 3 GPa. This result is one of the first to give a strength value of the same order as that predicted by various theoretical approaches. Despite the dynamic nature of the test method, the result is not particularly sensitive to the length of the duration of the stress (varying only from 2.5 to 3.0 GPa as the time decreased from 1 s to 10^{-7} s). This again is to be expected because nucleation is a process involving the rearrangement of atoms, where the atoms are vibrating about their mean positions at a rate of approximately 10^{13} times per second.

The data gathered by Tiryakioglu for Fig. 1.1, based on femtosecond laser pulses and extrapolated spalling test data, found the tensile strength of liquid aluminum to be close to 4.0 GPa. The spalling test seems to be a fruitful new approach in addition to MD calculation or experiment using laser or electron irradiation of the surface of a liquid or solid. The surface impact generates a compressive wave, followed by a tensile wave which travels through the metal. The compressive wave is reflected as a tensile wave from the far surface and meets the oncoming tensile wave. If the addition of the two stresses exceeds the tensile strength of the metal, the far surface spalls off. The technique is used to evaluate the strengths of a wide variety of liquid metals (Mayer et al., 2014, 2015).

In general, therefore, it seems that the various early attempts to measure the strengths of liquids have achieved results intermediate between very low and scattered results and high results close to those predicted for homogeneous nucleation, but recent spalling tests appear to be achieving consistently high values (Table 1.1). This is probably as good an outcome as can be expected. The reassuringly consistent high strengths suggest that the physics seems basically correct, confirming that the atoms of liquid metals strongly resist being pulled apart.

Because of the evident difficulty of reaching high enough stresses in a liquid to promote fracture by a classical nucleation processes, a number of "easier" nonclassical mechanisms have been suggested from time to time. Some of these are described in the following section.

1.3 Nonclassical pore formation mechanisms

1.3.1 High energy radiation

The radioactive decay of naturally occurring isotopes and radioactive substances occurs around us all at every point of our lives. Naturally occurring radioactive isotopes are relatively common in a number of metals and alloys. In addition, general radioactive contamination is presently increasing annually, arising from industrial and medical sources, and general fall out since the first nuclear explosions in 1945. This is a sad outcome, but even if all future contamination of the environment were to be prevented, we will still have to come to terms with accepting the historical legacy of a radioactive environment as a fact of life.

Liquid metals are, in common with all other present-day materials, subjected to a constant barrage of high-energy particles from these internal radioactive decay processes. The passage of the high-energy particles

Table 1.1 Tensile strengths of selected liquids.

			P_f*	
Liquid	**Surface tension N/m**	**Atom diameter nm**	**From Eq. (1.1) GPa**	**From Fisher GPa**
Water	0.072	-	-	0.13
Mercury	0.5	0.30	1.67	2.23
Aluminum	0.9	0.29	3.10	3.00
Copper	1.3	0.26	5.00	5.00
Iron	1.9	0.25	7.6	7.00

through the liquid causes thermal or displacement spikes, the name given to regions of intense heating, or actual displacement of atoms, effectively raising the local temperature of the liquid to well above its boiling point. It is possible, therefore, that these transient heated regions might become vapor bubbles sufficiently large to satisfy Eq. (1.2) and thus constituting effective nucleation sites for gas or shrinkage pores.

Johnson and Orlov in their review (1986) describe defect regions in solid metals of up to 100 atoms in diameter. Energy can be channeled away from such events along crystallographic directions in ordered solids, reducing the local damage, but the lack of any long-range order in liquids means that no such safety valve is possible, so that energy deposition is intensely localized. It follows that the production of a bubble of 100 atoms diameter in liquid iron should be easy, giving an equivalent fracture pressure of approximately 1500 atm, much lower than that required for classical homogeneous nucleation. In liquid FeO, the fracture pressure would be 400 atm. If the event occurred close to a nonwetted surface, then the fracture pressure would be only 20 atm. Such pressures are much more easily met in castings.

Analogous events are actually observed directly in the bubble chamber, a device full of a transparent liquid that can be vaporized by a high-energy particle, and so thus define its path. It is sobering to note that a bubble chamber can only be constructed using steel made before 1945, commonly sourced from the battleships from the seabed at Scapa Flow. Later steel introduces too much spurious background radiation.

Claxton (1967) has carried out a detailed study of the nucleation of vapor bubbles in liquid sodium subjected to a wide variety of different high-energy particles, including photons, electrons, protons, neutrons, alpha particles, xenon and strontium as fission fragments, and alpha recoils. His preliminary analysis suggests that the only interactions capable of initiating nucleation are "knocked-on" atoms of the liquid produced by fast neutrons. Claxton (1969) suggests that for heavy recoils arising from alpha-particle decay, the rate of energy transfer in liquid aluminum would be about three times higher than that in sodium, and in liquid iron should be about 10 times, giving rise to the possibility of nucleation, depending on the isotope responsible.

Furthermore, if a particularly troublesome radioactive isotope happened to be present in the melt, as an alloy in solution, it would be fundamentally different in character from suspensions of bubbles or inclusions, in that it could not simply be eliminated by filtration. These uncertainties have never been investigated.

Differences in the levels of contamination by trace isotopes may be just one of the many possible reasons for the occasional different behavior from one batch of metal to the next that is often experienced in the foundry and that often seems inexplicable. Kato (1999) is one of the first to record a check for the presence of alpha emitters in his high purity copper castings. It is a concern that one day such checks may have to become a routine.

Accepting that bubble nucleation events can occur spontaneously in liquid metals as a result of such energetic particle interactions, it remains to ascertain how important this may be. Overall, in view of the importance of bifilms, to be described later, for which bubble formation can occur at nearly zero stress, the atomic mechanisms are clearly not favored.

In the longer term, however, when casting processes are under proper control such that bifilm-free metals can be produced, the background radioactively induced pores may present a limit to the attainment of perfection in metals. It is an irony that we should consider ourselves fortunate in having a long way to go to experience such a limit.

1.3.2 Preexisting suspension of bubbles

Numerous studies have indicated that there may be a microbubble spectrum in most liquids.

For instance, Hammitt (1973) has carried out work that implies distributions of bubbles in liquid sodium circulating as coolant in atomic reactors, and high densities of bubbles have been measured by Outlaw et al. (1981) in vacuum-cast pure aluminum. In addition, Chen and Engler (1994) examine in some detail the clearly related and very old proposal that pores exist in irregular crevices of solid inclusions (Fig. 1.5).

All these approaches seem to be bifilm phenomena in different guises. Bifilms are to be expected in most liquid metals and alloys. They are not envisaged as a distribution of spherical pores, but as a wide spectrum of sizes and shapes of films of air trapped between folded entrained surface films (the crack-like pores illustrated by Chen and Engler are good examples of bifilms in the early stages of opening). They are, of course, a kind of crevice, so that, to some extent, the theory elaborated by Chen and Engler remains appropriate. Even so, we shall see how the bifilm model as a somewhat flexible folded film, exhibiting some rigidity, and enclosing films and bubbles of entrained gas, in general, fits the facts more closely than that of a model of a solid particle with a gas-filled crevice.

In contrast with all other pore initiation mechanisms, the bifilm is seen to possess the potential to initiate pores with negligible difficulty. It simply opens by the separation of its unbonded halves. Surface tension is not involved (as is usually assumed when nucleating a pore in a liquid). This is simply a mechanical action between parallel films of (effectively) vanishingly thin oxide separated by a vanishingly thin layer of gas. The acid test of any theory of pore initiation is its capability to explain the experimental data described below. Prior to the concept of bifilms, these findings have so far proved inexplicable.

The definitive research on the growth of porosity in aluminum alloy castings was that carried out at Alcan Kingston Laboratories by a team led by Fred Major (Tynelius, 1993). In this exemplary work, small tapered plates and end-chilled plates of Al 7Si 0.4 Mg alloy were cast under varying conditions to separate the effects of gas content, alloy composition, freezing time, and solidus velocity on the growth of porosity in the castings. This impressively logical piece of work has been analyzed in detail in a previous study ("Castings second Edition, 2003"). The results of this work were stunningly counterintuitive but were found completely explicable by the bifilm approach. The interested reader is recommended to study the explanation presented in 2003. It is not presented here because our focus is on the initiation of volume defects such as pores and cracks in metals, rather than their growth. The formation of pores and cracks from bifilms is, in every case, simply a matter of growth. Nucleation is circumvented.

Interestingly, there remains a case for a relatively pure spherical bubble population in some high-temperature and reactive liquid metals such as steel and titanium. If an oxide bifilm is entrained, the oxygen in the air entrapped in the folded film will be quickly consumed by reaction with the metal to form more oxide, and the nitrogen subsequently consumed to form nitrides (Raiszadeh and Griffiths, 2008). Ultimately, however, there will be a tiny residue of only about 1% of the original volume of entrapped air, consisting mainly of argon. The inert gases are practically insoluble in liquid metals (Boom et al., 2000). Thus a spectrum of very fine volumes of inert gas, trapped within oxide and nitride fragments (if these elements have not already completely dissolved as expected in titanium), will be expected to be rather stable over long periods of time.

1.4 Entrainment processes

If perfectly clean water is poured or is subject to a breaking wave, the newly created liquid surfaces fall back together again and so impinge and mutually assimilate. The body of the liquid reforms seamlessly. We do not normally even think to question such an apparently self-evident process.

However, the same is not true for many common liquids, the surface of which is not a liquid but a solid, often an invisible film of extreme thinness. Aqueous liquids often exhibit films of surface-active compounds including proteins or other large molecular compounds.

Metals in their liquid state are a special case. The surface of most liquid metals comprises an oxide film. If the surface happens to fold by the action of a breaking wave or suffers impingement by droplets or splashes, the surface oxide and even pockets of air can become entrapped and submerged in the bulk liquid (Fig. 1.6). The mechanism of entrainment is a key process with very specific outcomes. The surface oxide on the liquid always folds or impinges so as to result in the impingement of the dry-side-to-dry-side contact of two films, forming a *double film*, which I call a *bifilm*. The interior dry-side-to-dry-side interface between the two films has practically zero bonding and so acts as a crack in the liquid. Turbulent casting processes are common, so that most liquid metals are full of cracks.

Conversely, the two outer interfaces of the bifilm are formed by the undersides of the original oxide film from the surface of the liquid metal. The bifilm exterior interfaces have grown, atom by atom from the liquid, and are therefore in perfect atomic contact with the matrix. We shall see that these interfaces form preferred substrates for the precipitation and growth of many intermetallics and second phases. The interaction of the bifilm with second phases in this way is of major importance for the development of metallurgical structure and the control of properties as will become clear in this text.

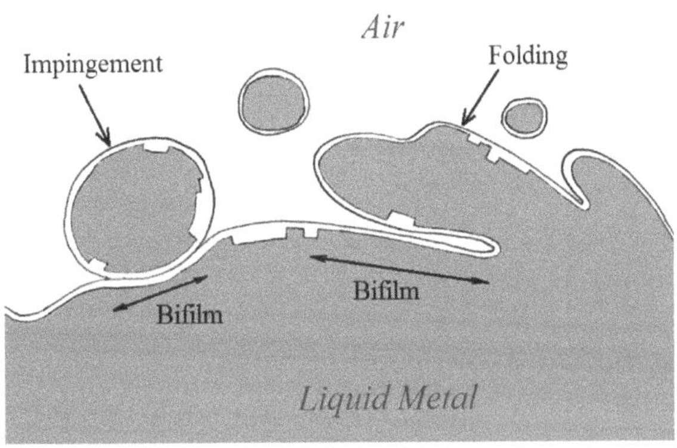

Figure 1.6 Folding and impingement during surface turbulence creating bifilms which often contain microbubbles in folds and creases, and sometimes macrobubbles sufficiently large to be actively buoyant.

The *nonbonding inside*, between the two halves of the double film, and the *perfect bonding outside*, in contact with the surrounding matrix, give the bifilm its unique features.

This entrapment and submergence of the surface oxide film and air pockets to form bifilm and bubbles we shall call "*entrainment.*" Both bifilms and bubbles are major defects in metals which have been overlooked throughout the history of metallurgy. We shall call them both "*entrainment defects.*"

Accidental entrainment of the surface during pouring is, unfortunately, only to be expected. This phenomenon of the degradation of liquid metals is, unfortunately, perfectly natural and fundamental to the quality and reliability issues for metals. Because these defects are inherited by wrought metals, nearly all our engineering metals are degraded. It is amazing that such a simple mechanism could have arrived at the 21st century having escaped notice for thousands of years and more recently of thousands of workers, researchers, and teachers.

Anyway, it is now clear that the entrained bifilm has the potential to become one of the most severely damaging defects in cast products (and, as we shall see, in wrought products too). It is essential, therefore, to understand film formation and the way in which films can become incorporated into metals so as to 'form bifilms and thereby' damage properties. These are vitally important issues.

It is worth repeating that a surface film is not harmful while it continues to stay on the surface, as illustrated in Fig. 1.7. In fact, in the case of the oxide on liquid aluminum in air, it is doing a valuable service in protecting the melt from catastrophic oxidation. This is clear when comparing with liquid magnesium in air. Because the magnesium oxide is not protective, the liquid magnesium can burn, generating its characteristic brilliant flame until the whole melt is converted to oxide. In the meantime, so much heat is evolved that the liquid melts its way through the bottom of the crucible, through the base of the furnace, and will continue down through a concrete floor, taking oxygen from the concrete to sustain the oxidation process until all the metal is consumed. This is the incendiary bomb effect. Oxidation reactions can be impressively energetic!

As we have seen, the surface film grows from the surface of the liquid, atom by atom, as each metal atom combines with newly arriving atoms or molecules of the surrounding gas. Thus, for an alumina film on the surface of liquid aluminum, the underside of the film is in perfect atomic contact with the melt and can be considered to be "well-wetted" by the liquid.

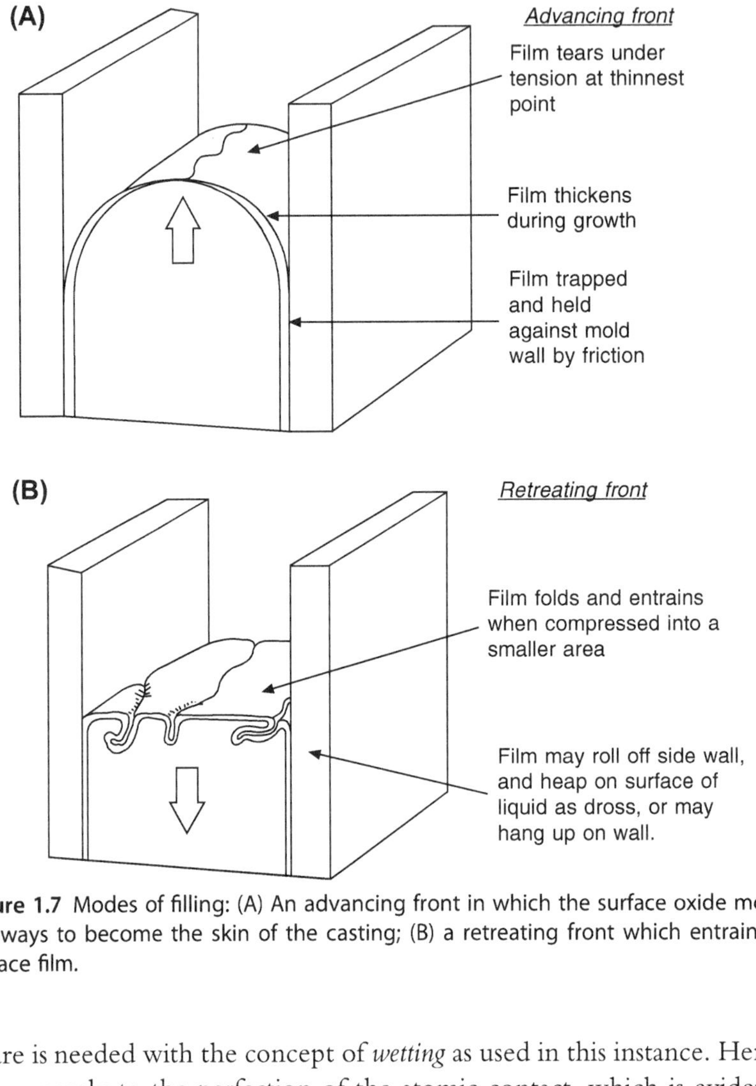

Figure 1.7 Modes of filling: (A) An advancing front in which the surface oxide moves sideways to become the skin of the casting; (B) a retreating front which entrains its surface film.

(Care is needed with the concept of *wetting* as used in this instance. Here it refers merely to the perfection of the atomic contact, which is evidently automatic when the film is grown in this way. The concept contrasts with the use of the term *wetting* for the case where a sessile drop is placed on an alumina substrate. Perfect atomic contact is now unlikely to exist where the liquid covers the substrate so that at its edges the liquid will form a large contact angle with the substrate, indicating, in effect, that it does not wish to be in contact with such a surface. Technically, the creation of the liquid/solid interface raises the total energy of the system. The *wetting* in this case is said to be poor.)

The problem with the surface film only occurs when it becomes *entrained* and thus submerged in the bulk liquid. The entrainment mechanism converts the single film to a double film, thereby forming a crack.

When considering submerged oxide films, it is important to emphasize that the side of the film which was originally in contact with the melt will continue to enjoy its perfect atomic contact with the liquid. As such it will adhere well and be an unfavorable nucleation site for volume defects such as cracks, gas bubbles, or shrinkage cavities. In other words, the oxide/liquid interface is strong, being expected to decohere only when subjected to stresses corresponding to the ultimate tensile strength of the interface, probably in the tens of GPa. When the metal solidifies, the metal oxide bond will be expected to continue to be strong, as in the perfect example of the oxide on the surface of all solid aluminum products, especially noticeable in the macroscopic example of anodized aluminum.

Turning now to consider the upper surface of the solid oxide on the liquid. This is, of course, dry. On a microscale, the upper surfaces of oxide films can be extremely rough (Zeng et al., 2013). When folded into a bifilm, the central unbonded interface therefore touches only rarely, in isolated points, and occludes a significant amount of air.

The other key feature of oxide surface films is the great speed at which they grow. Thus, in the fraction of a second (probably between 10 and 100 ms) that it takes to cause a splash or to enfold the surface, the expanding surface, newly creating additional area of liquid, will react with its environment to cover itself in new film. The reaction is so fast to be effectively instantaneous for the formation of oxides. The film can only grow to a few nanometers thickness in this short time, explaining the extreme thinness of bifilms, and their ability to have hidden in plain sight for 6000 years.

Other types of surface films occur on liquid metals. One important variety is the liquid oxide film. The entrainment of liquid oxides is a totally different behavior which is of extreme value in avoiding bifilm formation. This important phenomenon is discussed in a separate section. In contrast, solid graphitic films seem to be common when liquid metals are cast in hydrocarbon-rich environments and are common in the casting of cast iron. In addition, there is some evidence that other films such as sulfides and oxychlorides are important in some conditions. Fredriksson (1996) describes TiN films on alloys of Fe containing Ti, Cr, and C when melted in a nitrogen atmosphere. Nitride films are known to occur in aluminum in the absence of oxygen and may lead to the well-known rock candy fractures in steels the rock candy fracture may be an entrainment defect.

1.4.1 Entrainment of bifilms

The *entrainment process* can be a folding action as seen in Fig. 1.6. Alternatively, also shown in the figure, volumes of the liquid can impinge, as droplets or splashes falling back into the main reservoir of liquid. In both cases, the films on both bodies of liquid necessarily come together dry-side to dry-side. The submerged surface films are therefore *necessarily always double* and *necessarily exhibit negligible bonding* between the two dry ceramic films and which now *necessarily resemble* and *necessarily act as cracks*. The underlines and italics are there because, for some, coming to the concepts for the first time, the fundamental concepts of bifilms are curiously difficult to convey.

Turbulent pouring of liquid metals can therefore quickly fill a batch of liquid metal with cracks. The cracks have a relatively long life and in many alloys can survive long enough to be frozen into the cast metal. We shall see how they have a key role in the creation of other defects during the process of freezing and how they degrade the properties of the final cast metal.

For convenience, I call these double oxide films "bifilms." This convenient short hand denotes the *double* nature of the film defect, as in the word *bicycle*. A better example is the marine shellfish; the *bivalve*, whose two leaves of its shell are hinged, allowing it to open and close.

It seems also worth underlining the fact that there are fundamental reasons to assume the bifilm will never become separated once again into single films. This is because where the two halves are in contact, interfacial forces will be important, and interfilm friction will ensure that both halves of the bifilm will move as one. In addition, the continued consumption of oxygen and nitrogen by the films will reduce the volume of air in the interfilm gap. The exterior atmospheric and metallostatic pressures will thereby pressurize the two halves into greater contact, ensuring that the films continue their association as a pseudo-single entity, despite the absence, or near-absence, of bonding forces between the halves.

Entrainment does not necessarily occur only by the dramatic action of a breaking wave as seen in Fig. 1.6. It can occur simply by the contraction of a "free liquid" surface. In the case of a liquid surface that contracts in area, its area of oxide, being a solid, is not able to contract. Thus, the excess area is forced to buckle, folding in a concertina-like fashion. Considerations of buoyancy (in all but the most rigid and thick films) confirm that the fold will be inward and so entrained (Fig. 1.7). Such loss of surface is common during rather gentle undulations of the surface, the slopping and surging that can occur during the filling of molds. Such gentle folding might be available to unfold again during a subsequent expansion, so that the entrained surface might almost immediately detrain once again. This potential for reversible entrainment may not be important compared to the probability that much enfolded material will remain enfolded and entrained. Masses of entrained oxides will entangle and adhere to the sides of containers: launders (the refractory channels for transferring liquid metals) and molds, but more severe bulk turbulence may tear it away and transport it elsewhere.

In passing, in the usual case of an alloy with a solid oxide film, it is of interest to examine whether the presence of an oxide film in a melt necessarily implies that the film is double. For instance, why cannot a single piece of oxide be simply taken and immersed in a melt to give a single (i.e., nondouble) interface with the melt? The reason is that as the piece of oxide is pushed through the surface of the liquid, the surface film on the liquid is automatically pulled down either side of the introduced oxide, coating both sides with a double film, as illustrated in Fig. 1.8. The entrainment mechanism therefore *necessarily* results in a submerged film that is at least double and therefore always has the nature of a crack.

During the melting of a charge of ingots or returned castings in a furnace, as the metal melts, it gradually moves up around the solid pieces of charge material, the meniscus on the liquid gently rolling up against the pieces of charge as the melted metal rises and the melting charge materials descend (Fig. 1.8A). The charge materials, however, have a thick oxide film as a result of being exposed to the heat of melting. I call this a primary oxide skin. The thin film on the liquid meniscus rolls up against the primary skin on the charge, creating an asymmetric thin/thick bifilm of the same area and shape of the charge material. In this way bifilm cracks, the size of newspapers can be generated to float about in the molten metal. Depending on the treatment of the melt, the bifilms based on primary oxide skins may or may not finish up in the finally prepared metal to be poured to make the casting.

Fig. 1.9 illustrates how entrainment of an oxide film can result in a variety of submerged defects. Fig. 1.9A illustrates the important feature that entrained films usually constitute a main bifilm which exhibits transverse bifilms at intervals on only one side. This structure is a

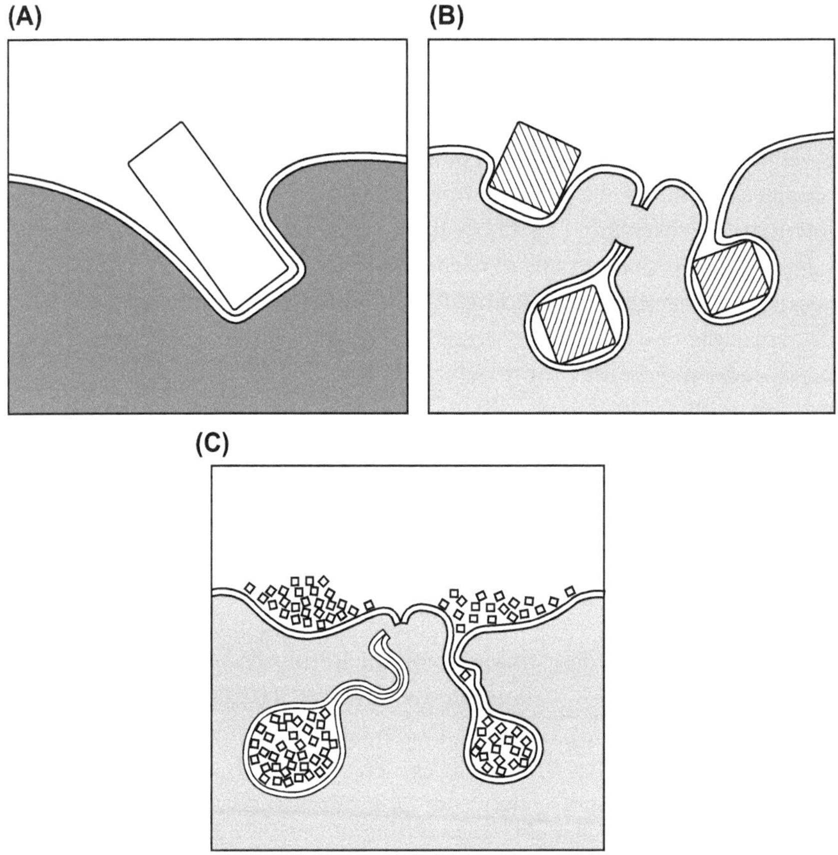

Figure 1.8 Entrainment of solids into a liquid metal: (A) charge materials; (B) optimum production of metal matrix composites (MMCs); (C) usual production of MMCs.

consequence of the two impinging films being of different areas, the side having the larger area having to contract by folding, and thus forming the transverse folds. Naturally, the transverse bifilms are mainly on the one side, the side enclosed by the film of larger area, in contrast to the smaller area film which retains its straightness. This expected natural outcome of the two impinging oxide films not being of the same size, which will, in general, always be true, is a kind of signature for the presence of a bifilm. It has interesting consequences for the structure of faults in some important second phases as we shall see.

Other consequences of the entrainment of the surface oxide film are seen in Fig. 1.9 to be (b) the entrainment of bubbles as integral features of the bifilm, being simply local regions where the two films have not come

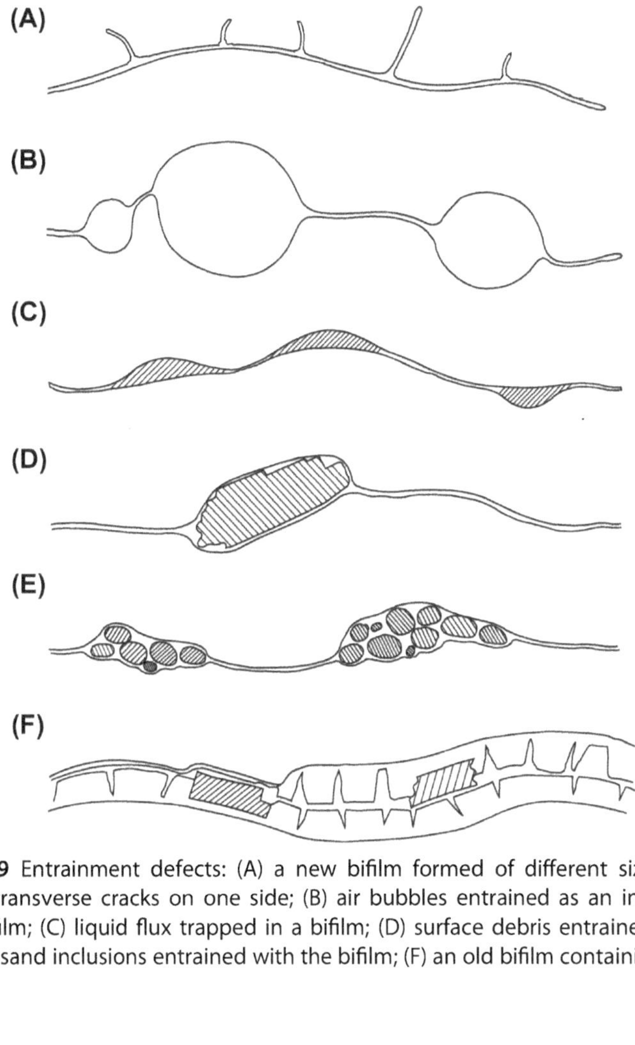

Figure 1.9 Entrainment defects: (A) a new bifilm formed of different sized halves, creating transverse cracks on one side; (B) air bubbles entrained as an integral part of the bifilm; (C) liquid flux trapped in a bifilm; (D) surface debris entrained with the bifilm; (E) sand inclusions entrained with the bifilm; (F) an old bifilm containing integral debris.

together because of the presence of entrapped gas; or (c) the entrainment of a surface liquid flux, or (d) solid debris, or (e) molding sand, or finally (f) an entrained old oxide, possibly from the surface of the ladle or furnace, in which sundry debris has accumulated by falling on to the liquid surface from time to time and has been incorporated into the oxide structure. During the tipping of a steel furnace when pouring the finished steel into a ladle, the ledges and launders of the furnace dislodge their accumulated debris which rains down into the churning mass of steel. Although during teeming there seems no shortage of candidates for a generous population of entrained foreign inclusions, most seem to vanish, almost certainly by melting and joining the layer of slag on top of the melt. Thus, despite the rain of dirt accompanying the pour, we can be grateful that entrained sundry foreign matter is rare in steels.

It is as well to keep in mind that a wide spectrum of sizes and thicknesses of oxide film probably exist. With the benefit of the hindsight provided by much intervening research, a speculative guess dating from the first edition of CASTINGS (1991) is seen to be a reasonable description of the real situation (Table 1.2). The only modification to the general picture is the realization that the very new films can be even thinner than was first thought. Thicknesses of only 20 nm seem common for films in many aluminum alloys. The names "new" and "old" (probably better stated "young" and "old") are a "rough and ready" attempt to distinguish the types of film and have proved to be a useful way to categorize the two main types of film.

The limited thickness of a newly created surface film is a consequence of its rapid formation, the film having little time to grow and thicken. Research quoted by Birch (2000) on Zn–Al pressure die casting defects using a cavity fill time from 20 to 100 ms can be back-extrapolated to indicate that in the case of this alloy the surface films formed in times of about 10 ms. Most recent work on entrained young oxides find a limiting thinness in the region of 20 nm implying that the films are only about 10 molecules thick.

To emphasize the important characteristic crack-like feature of the folded-in defect, the reader will notice that it will be often referred to as a "bifilm crack" or "oxide crack." It is hoped that the reader will understand and forgive this tendency toward tautology. A typical entrained film is seen in Fig. 1.10A, showing its convoluted nature. This irregular form, repeatedly folding back on itself, distinguishes it from a crack resulting from stress in a solid; its morphology identifies it as a defect that could only have formed in a turbulent liquid. When examining for the first time the gap between the double films at high magnification in the scanning

Table 1.2 Forms of oxide in liquid aluminum alloys.

Type	Description	Possible source	Growth time	Thickness
New	Confetti-like fragments	Pouring	0.01s	1 nm–1 μm
Old 1	Flexible extensive films	Transfer ladle	10 s–1 min	10 μm
Old 2	Thicker films, less flexible	Melting furnace	10 min–1 h	100 μm
Old 3	Rigid lumps and plates	Holding furnace	10 h–10 days	1000 μm

(A) (B)

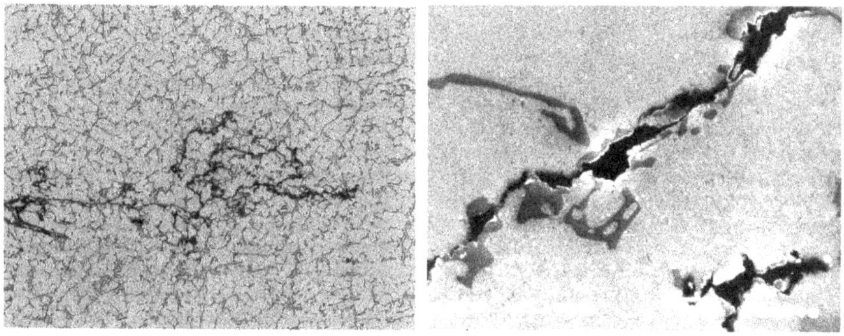

Figure 1.10 (A) Convoluted film in Al–Si alloy; (B) SEM close-up by Green (Green and Campbell, 1994) of "canyon-like" central void in Al–Si alloy oxide bifilm.

electron microscope (Fig. 1.10B), Nick Green, now Professor Green, reported it to "look like a bottomless canyon." This layer of air (or other mold gas) is nearly always present, trapped partly by the irregular folds and partly by the microscopic roughness of the impinging films.

As we have mentioned, the surface oxide can be entrained simply by a contraction of the liquid surface so as to create a problem of too much area of solid film for the surface area available. However, if more severe disturbance of the surface is experienced, as typically occurs during the pouring of liquid metals, pockets of air can be accidently trapped by chance creases and folds at random locations in the films (Fig. 1.9B) because the surface turbulence event is usually chaotic (waves in a storm rarely resemble sine waves). The resultant scattering of porosity in castings seems to originate from these pockets of entrained air, being usually the most common source of porosity in castings (so-called "shrinkage" and so-called "gas" precipitating from solution are merely additive effects that may or may not contribute additional pore growth). The creation of this source of porosity has now been regularly observed in the study of mold filling using video X-ray radiography. It explains how this rather random distribution of porosity typical in many castings has confounded the efforts of computers which have been programmed to simulate only solidification. Such porosity is commonly and understandably misidentified by the radiologist as "shrinkage porosity." In reality, it is nearly always tangled masses of oxide bifilms, together with the accompanying pockets of entrapped air. For such defects, often observed on radiographs, we need to practice saying not "it is shrinkage" but "it *appears* to be shrinkage," preferably followed by the phrase "but most likely is *not* shrinkage." Gradually, we shall appreciate that the porosity observed on radiographs is *rarely* shrinkage because in general foundry personnel know how to feed castings. More importantly, modern computer simulation packages are highly accurate at predicting "true" shrinkage, so that, in general, genuine shrinkage problems are not expected.

As bifilms constitute cracks in the liquid and are known to be of all sizes and shapes, they can become by far the largest defects in the final casting. Melting and casting conditions can produce either dense populations of submicron bifilms or, on occasions, just a few defects the size of newspapers. Bifilm cracks can therefore easily be envisaged as reaching from wall to wall of a casting, causing a leakage defect in a casting required to be leakproof or causing a major structural weakness in a casting requiring strength or fatigue resistance. Leakage across an apparently solid 200 mm thick steel wall was not easy to explain before the concept of the bifilm.

In addition to constituting defects in their own right, if bifilms are given the right conditions during the cooling of the casting, the loosely encapsulated gas film can act as an excellent initiation site for the subsequent growth of gas bubbles, shrinkage cavities, hot tears, cold cracks, etc. The nucleation and growth of such consequential damage will be considered in later sections. The important message to take on board at this stage is that without the presence of bifilms, this standard retinue of casting defects would probably be impossible. Liquid metals free from bifilms could make castings free from defects. This is a message worth keeping in the forefront of our minds.

The entrainment of surface films, leading to the creation of a double film in the liquid, can occur in very different ways and lead to quite different types of bifilm crack:

(i) Surface turbulence, one of the most common mechanisms for bifilm formation, in which the irregular or random motion of the liquid front ensures that the breaking and reforming of the surface film gives limited time for film growth, so the entrained films are thin, small, irregular, and numerous and occur widely during all types of pouring;

(ii) Bubbles of air or mold gases rising through the melt create long bubble trails as collapsed oxide tubes. As most running systems entrain at least 50% air, there is no shortage of these damaging leak paths between ingates and the top of the casting. Because bubble trails can be meters long, the entrainment of air is one of the most seriously damaging phenomena associated with castings. Steel ingots suffer mainly from bifilm mechanisms (i) and (ii).

(iii) By submergence of charge material used to make the melt (Fig. 1.8A): oxide films on charge materials are often thick, with large area, especially if returns from a sand casting operation are employed. I call these "primary oxide skins" to distinguish them from most other types of oxide film. The presence of primary skins of charge materials in melts is especially problematical in the melting of aluminum alloys; their near-neutral buoyancy means they neither sink nor float within reasonable times required during production conditions. The necessary submergence of primary skins in melts of light alloys is a fundamental disadvantage of crucible and bath-type furnaces as used by the majority of the light metal casting industry. Attempts to remove these strong, extensive, and heavy films from the melt include the use of flotation with small bubbles of inert gas as in rotary degassing. However, it is impossible to know whether the skins have all been successfully removed before using the metal for casting. I always try to avoid such inappropriate melting equipment. The dry hearth furnace is the only melting equipment capable of avoiding the incorporation of primary skins into the melt.

(iv) By laminar flow, in which the meniscus is stationary: The bulk liquid may be flowing parallel to the meniscus. The stationary meniscus then grows a thick surface film, although the underlying melt may be traveling at high speed. This curious behavior occurs when metal is poured as a smooth stream: the liquid flows rapidly, but the oxide film on the stream is stationary, acting as a tube through which the liquid flows. As the falling jet fills a pool, the rising metal rolls up against the oxide tube, creating a bizarre tubular bifilm. These serious defects are probably rather common in castings where the metal falls between levels. This is one of the key reasons to avoid any fall of the liquid when designing filling systems for castings.

(v) By laminar flow of liquid over horizontal solidified metal, as a flooding action: This seems to be the mode of advance of the vacuum arc remelted (VAR) ingot (Section 1.6). If the liquid metal surface remains at a level for sufficient time to partially freeze, despite the vacuum, it will grow a thick oxide film. The partially frozen surface has sufficient strength, therefore, to support a new layer of liquid which flows over the top, rolling its new thin oxide down over the thicker older oxide to create an extensive horizontal asymmetric bifilm crack. The VAR ingot is consequently unforgeable until the bifilm cracks are machined off. The VAR issues are dealt with in more detail in Section 1.6.3.

Entrainment creates bifilms that may never come together properly and so constitutes air bubbles immediately. Alternatively, if bifilms are originally tightly wrapped, they can gradually unwrap by a variety of drive mechanisms acting to separate the two halves of the bifilm. The unwrapping process is a major phenomenon whereby a fairly harmless compact inclusion can grow, opening fold by fold, so as to become a serious engineering crack. This phenomenon is called "unfurling." It is dealt with in detail in Section 1.4.4.

An important point to note is that a very small and insignificant bifilm may subsequently generate a large defect such as a crack or hot tear. We are, in a sense, programmed to look for such developments because of our predisposition to understand crack growth in terms of nucleation and growth. On the other hand, in many cases, the bifilm itself is large, so that any apparently consequential macro defect such as a pore or a hot tear actually *is* the bifilm. In this case, only separation of the two halves of the bifilm occurs; no growth is involved.

The situations in which a small bifilm *nucleates* the crack or in which the bifilm *is* the crack both may be possible in cast metals but the observed crack being formed *entirely* from the bifilm seems more probable. Research is needed to clarify these possibilities.

1.4.1.1 Visual evidence for bifilms

It is a new thought for most of us that the double oxide film defects are everywhere in metals. We are not describing occasional single "dross" or "slag" defects or other occasional accidently entrained inclusion. The bifilm defect occurs naturally, and in copious quantities, every time a metal is poured. Many metals are crammed with bifilm defects. The fact that they are usually so thin has allowed them to evade detection for so long. Until recently, the ubiquitous presence of these very thin double films has not been widely accepted because until recently no single metal quality test has been able to resolve such thin but extensive defects.

Even so, over recent years, there have been many significant confirmations.

The presence of bifilms is easily seen on fracture surfaces, particularly if a good scanning electron microscope (SEM) is available.

Fig. 1.11 is an unusual polished section of a cast aluminum alloy displaying a tangled bifilm imaged by SEM in the author's laboratory by Divandari (2000). The top half of the bifilm happens to have been polished away in places, revealing the inside surface of the underlying remaining half of the

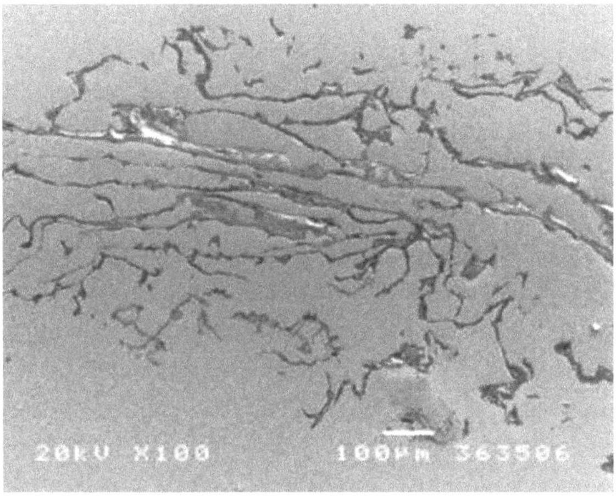

Figure 1.11 Polished surface of Al–Si alloy showing upper part of the double film removed in places, revealing the inside surface of the lower film (Divandari, 2000).

bifilm, demonstrating the oxide film to be a *double* film, with little or no bonding between the two films.

The component halves of a bifilm appear on each half of a fracture surface with nice butterfly wing symmetry (Fig. 1.12). Griffiths and Lai (2007) show such symmetry on all their fractured tensile bars of Mg alloy. Each oxide film is characterized by absence of metallic features such as ductile dimples but does show folds and creases typical of a film formed on a moving liquid, although the folds and creases may be so thin as to be difficult to see except under the best operating conditions of an SEM.

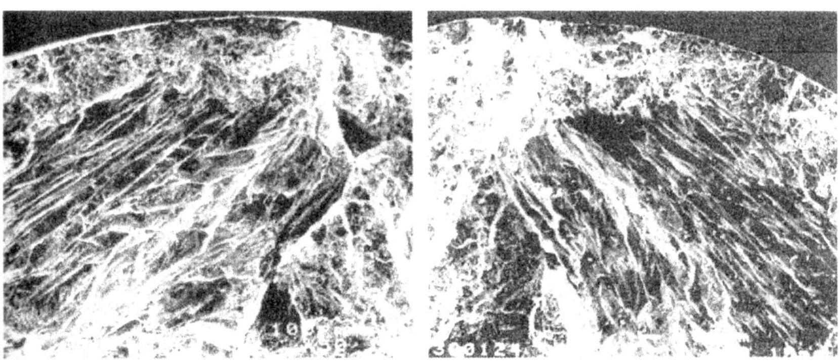

Figure 1.12 The secondary electron image of the near-mirror-images of the spinel bifilm halves on each half of the tensile failure surface of an Al–Si alloy (Cao and Campbell, 2004).

The growth of the area of oxide as the liquid surface deforms seems a general feature of entrainment. It is a one-way irreversible process. The consequent crinkling and folding of a disturbed surface is a necessary characteristic of entrained films formed on a turbulent liquid and is the common feature that assists to identify films as bifilms (rather than films formed during solid state cooling) on fracture surfaces. Fig. 1.13A and B are good examples of MgO films on AZ91 magnesium alloy (Wang, 2009). A typical thick and granular older film of spinel, Al_2O_3MgO, on an Al−11Si alloy is seen in Fig. 1.13C and contrasts with younger films on aluminum alloys which are more film-like and consist of pure alumina, Al_2O_3. The different appearance and chemistry of films is often useful to be certain that the films were formed on the liquid and not grown later during the cooling of the solidified metal.

Another clear example of bifilms is seen in the use of the reduced pressure test (RPT) for aluminum alloys. At low gas contents, many

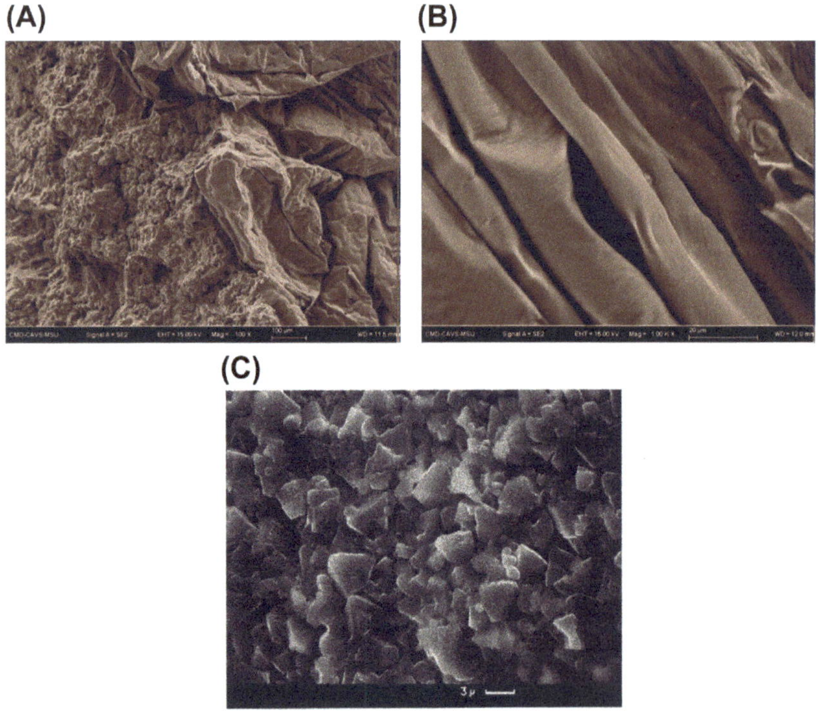

Figure 1.13 Fracture surfaces showing (A) and (B) MgO films on magnesium alloy AZ91 (Wang, 2009) (C) a thick granular spinel film on an Al−7Si-0.6 Mg alloy. *Courtesy Philip Bell 1981.*

operators have been puzzled by the appearance of hairline cracks, often extending over the whole section of the RPT casting. They had problems to understand the cracks because the test is commonly assumed to be merely an assessment of hydrogen porosity. An example of cracks is seen in Fig. 1.14.

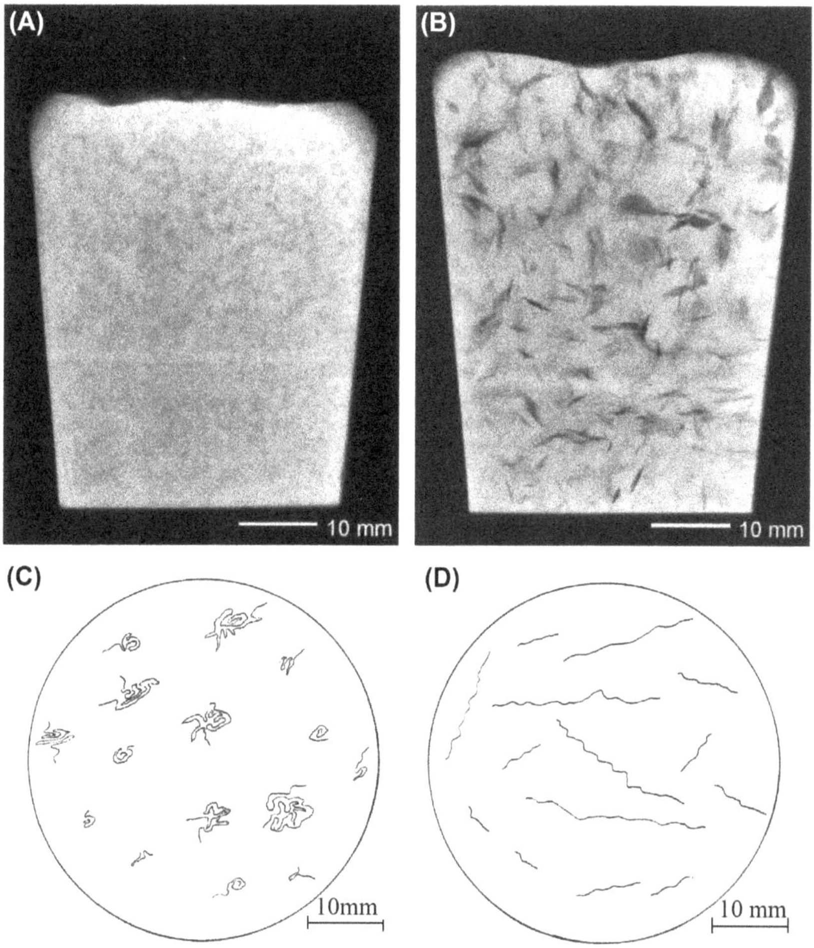

Figure 1.14 Radiographs of reduced pressure test (RPT) samples of the same as-melted Al–Si alloy solidified under pressure of (A) 1 atm; and (B) 0.01 atm (Fox and Campbell, 2000). (C) Bifilm cracks furled so as to be fairly harmless as in (A); and (D) the same bifilms unfurled to show their damaging potential, resembling engineering cracks as in (B).

Fig. 1.14 shows the results of an unusual development of the RPT using X-ray radiography. The rationale behind the use of radiography on this occasion was as follows. The bifilms are normally difficult or impossible to see by X-ray radiography when solidified under 1 atmosphere pressure. If, however, the melt is subjected to a reduced pressure of only 0.1 atm, the entrained layer of air inside the bifilm should expand by 10 times. Under 0.01 atm, the layer should expand 100 times, etc. In this way, the entrained bifilms should become visible. The result is shown in Fig. 1.14.

In this work a novel reduced pressure test machine was constructed so that tests could also be carried out using chemically bonded sand molds to make test castings as small slabs with overall dimensions approximately 50 mm high, 40 mm wide, and 15 mm thick. The parallel faces of these mini slabs allowed immediate X-ray examination without further preparation.

Fig. 1.14A shows a sample that was solidified in air indicating evidence of fine-scale low-density regions appearing as dark, faint compact images of the order of 1 millimeter in diameter. As the test pressures were lowered, the compact regions unfolded and grew into progressively longer and thicker streaks, finally reaching 10–15 mm in length at 0.01 atm (Fig. 1.14B). (This unfolding of the bifilms is an "unfurling" process to be described later.) The "streak-like" appearance of the porosity is due to an edge-on view of an essentially planar defect (although residual creases of the original folds are still clear in some bifilm images). The fact that these defects are shown in such high contrast at the lowest test pressure suggests that they almost completely penetrate the full 15 mm thickness of the casting and may only be limited in size by the 15 mm thickness of the test mold. The more extensive areas of lower density porosity are a result of defects lying at different angles to the major plane of the casting.

It is important to keep in mind that images (a) and (b) of Fig. 1.14 are the *same* metal poured at the *same* time at the *same* temperature. The only difference is the pressure they experienced while in the liquid state. You can see the faint shadows of the compact, convoluted forms, and the magical transformation within about 60 s to a structure crammed with cracks. The difference between them is perhaps the best evidence for the existence and the nature of bifilms that can be envisaged. This is hard, incontrovertible evidence for bifilms.

At 0.01 atm the thickness of bifilms as measured on the radiographs for those defects lying in the line of sight of the radiation, was in the range 0.1–0.5 mm. This indicates that the original thickness of bifilms at 1 atmosphere was approximately 1–5 μm.

Fig. 1.14 shows relatively large bifilms. They contrast with more usual samples in which the bifilms appear to be often less than 1 mm in size, and when solidified at 1 atm pressure are barely visible on radiographs. The work by Fox and the author (2000) on increasing the hydrogen content of such melts in the RPT at a constant reduced test pressure typically reveals the inflation of clouds of bifilms, first becoming unfurled and slightly expanded by the internal pressure of hydrogen gas, and finally resulting in the complete inflation of the defects into expanded spheres at high hydrogen levels, fulfilling at last their appearance as hydrogen pores as traditionally expected. Such an effect of thin cracks transforming into spherical bubbles is often observed, and can only be reasonably explained by the presence of bifilms. Fig. 1.15 illustrates examples of cracks and pores from different melts.

It is clear, therefore, from all that has been presented so far that a melt cannot be considered to be merely a liquid metal. In fact, the casting engineer must think of it as a slurry of various kinds of debris, mostly bifilms of various kinds, all containing entrained layers of air or other gases. For metals intended for applications requiring reliability, these defects need to be eliminated.

In a definitive piece of research into the fatigue of filtered and unfiltered Al—7Si-0.4 Mg alloy by Nyahumwa et al. (1998, 2000), test bars were cast by a bottom filling technique and were sectioned and examined by optical metallography. The filtered bars were relatively sound. However, for unfiltered castings, extensively tangled networks of oxide films were observed to be randomly distributed in almost all polished sections. Fig. 1.10 shows an example of such a network of oxide films in which micropores (assumed to be residual air trapped among bifilms resulting from the chaotic entrainment process) were frequently observed to be present. In the oxide film networks, it was observed that oxide film defects constitute cracks showing no bond developed across the oxide-oxide interface. In the higher magnification view of Fig. 1.10 the width between the two dry surfaces of folded oxide film is seen to vary between 1 and 10 μm, in confirmation of the previous low pressure test results. However, widths of cracks in places approached 1 mm where the bifilm had opened sufficiently to be considered as a pore.

The scanning electron microscope (SEM) has been a powerful tool that has revealed much detail of bifilms in recent years. One such example by Green (1995) is seen in Fig. 1.16, revealing a film folded many times on the fracture surface of an Al—7Si-0.4 Mg alloy casting. Its composition

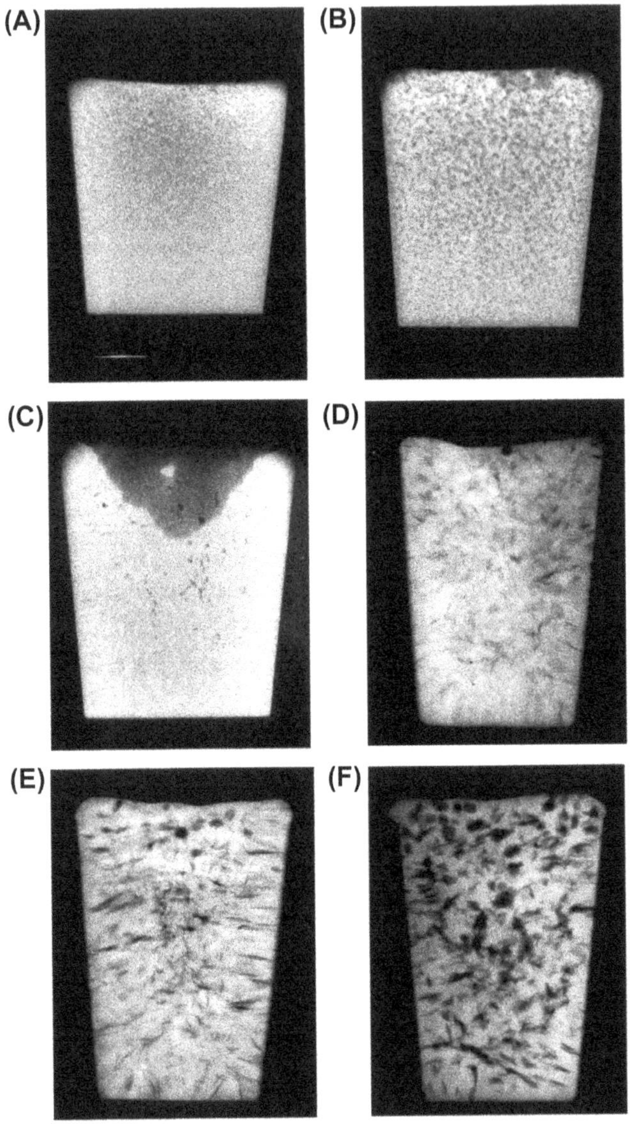

Figure 1.15 Radiographs of reduced pressure test samples illustrating (A) and (B) high defect density with moderate hydrogen in solution, spherodising pores from fine bifilms; (C) and (D) reduced hydrogen; (E) and (F) higher hydrogen with larger, columnar-aligned edge bifilms and central bifilms.

Figure 1.16 Multiply folded film on the fracture surface of an Al–Si alloy. *Courtesy N R Green.*

was confirmed by microanalysis to be alumina (Al_2O_3). The thickness of the thinnest part appeared to be close to 20 nm. It was so thin that despite its multiple folds the microstructure of the alloy remained clearly visible through the film.

1.4.1.2 Surface turbulence

As liquid metal rises up a mold cavity in the form of a vertical plate, with the meniscus constrained between the two walls of the mold, an observer looking down from above (suitably equipped with transparent protective shield of course!) will see that the liquid front advances upwards by splitting of the surface film (Fig. 1.7A). This occurs because, as the metal rises, the surface film is trapped between the wall of the mold and the liquid surface, effectively held in place by friction, and therefore cannot rise with the rising liquid. For this reason, film on the rising surface is torn apart, moving sideways to become the skin of the casting. The new oxide skin is, in turn, pinned in place against the mold walls. As the rising surface splits and opens its surface film, the film is continuously renewed so that the process is effectively continuous.

It seems that this vertical advance actually occurs by the action of one or more horizontal transverse traveling waves (I call them "unzipping waves") that can be seen to wander backwards and forwards along a meniscus rising (out of the plane of the figure) between walls, or often seen spiraling upwards in a cylindrical billet. This interesting phenomenon will be discussed in detail later.

The important feature of this continuous and progressive upward (counter-gravity) mode of advance of the liquid front is that the surface film is always expanding but never becomes entrained in the bulk of the liquid. Therefore, the liquid metal remains undamaged. Clearly, castings require to be made by only uphill movement of the metal if the entrainment of bifilm cracks is to be avoided.

A problem arises if the velocity of flow is sufficiently high for the melt to rise high above the general level of the liquid surface, so that it subsequently falls back under gravity. Any downward fall introduces the danger of entrapment of its own surface. This is the action of a fountain or breaking wave. It is represented in Fig. 1.17. This process of entrainment of the liquid surface, and in particular, its surface film, has been modeled by computer simulation by a number of workers. The various approaches have been reviewed by the author (JC, 2006).

To gain some insight into the physics involved we can carry out the following analysis. The pressure tending to perturb the surface is the inertial pressure $\rho V^2/2$ where ρ is the density of the melt, and V is the local flow velocity. The action of the surface tension γ tends to counter any perturbation by tensioning the surface, pulling the surface as flat as possible. The maximum pressure that surface tension can hold back is achieved when the surface is deformed into a hemisphere of radius r, giving a pressure $2\gamma/r$ resisting the inertial pressure (Fig. 1.17). Thus in the limit that the velocity is just sufficiently high to push the surface beyond the hemisphere, and so just exceeding the maximum restraint that the surface tension can provide, we have $\rho V^2/2 = 2\gamma/r$, so that the critical velocity V_{crit} at which

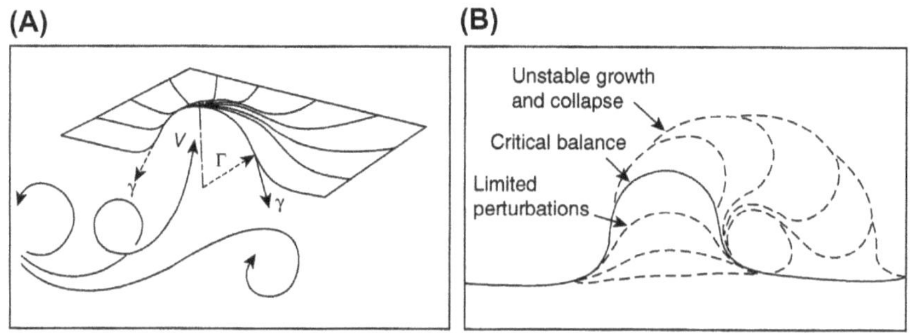

Figure 1.17 (A) The balance of forces in the liquid surface; (B) perturbations of the surface by inertial forces in the melt remain reversible up to the critical hemispherical limit, after which the perturbation becomes unstable, irreversibly continuing into a breaking wave which may entrain the surface during its collapse.

the surface will become unstable, and start to suffer *surface turbulence* (my name for the phenomenon of breaking waves and entrainment of the surface into the bulk) is

$$V_{crit} = 2\,(\gamma/r\rho)^{1/2} \tag{1.4}$$

Subsequent experiments have shown that in spite of its rough derivation, the above relation is a surprisingly accurate result. However, the approach can be considered unsatisfactory in the sense that the radius r has to be assumed, but there is little wrong with the physics. The derivation is reproduced here because of its directness and simplicity.

As mentioned above, Eq. (1.4) is not an especially convenient relation since the critical radius of curvature r of the front at which instability occurs is not known independently. However, we can guess that it will not be far from a natural radius of the front as defined, for instance, by a sessile drop. (The word "sessile" means "sitting" and is used in contrast to the "glissile" or "gliding" drop.) The sessile drop sitting on a nonwetted surface exhibits a specific height h defined by the balance between the forces of gravity, tending to flatten it, and surface tension, tending to pull the drop together to make it as spherical as possible. We can derive an approximate expression for h considering a drop on a nonwetted substrate as illustrated in Fig. 1.18.

The pressure to expand the drop is the average pressure $\rho gh/2$ acting over the central area hL, giving the net force $\rho gh^2L/2$. For a drop sitting quietly, with its forces in equilibrium, this force is equal to the net force due to surface tension acting over the length L on the top and bottom surfaces of the drop, $2\gamma L$. Hence

$$\rho g h^2 L/2 = 2\gamma L$$

$$\text{giving } h = 2(\gamma/\rho g)^{1/2} \tag{1.5}$$

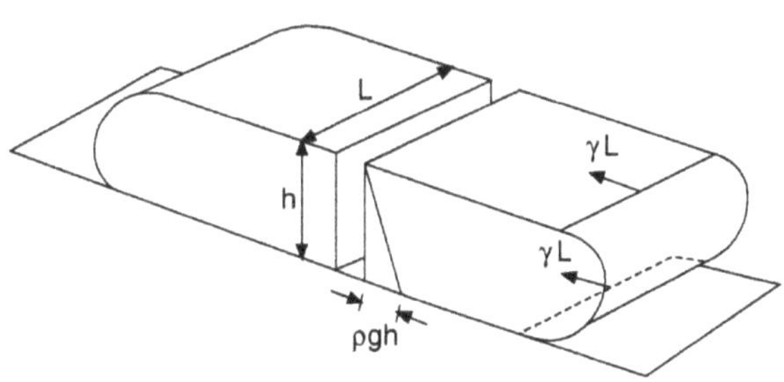

Figure 1.18 Balance of forces in a sessile drop sitting on a nonwetted substrate.

Assuming that $h = 2r$ approximately, and eliminating r from Eq. (1.4) we have

$$V_{crit} = 2\,(\gamma g/\rho)^{1/4} \tag{1.6}$$

This is the more elegant way to derive the critical velocity involving no assumptions and no adjustable parameters. It is analogous to the concept illustrated in Fig. 1.19 in which liquid metal rises through a vertical ingate to enter a mold.

Clearly, for case (a) in Fig. 1.19A, when the velocity at the entrance to the gate is zero the liquid is perfectly safe; there is no danger of enfolding the liquid surface. However, of course, the zero velocity filling condition, although ultra safe, is unfortunately not particularly productive for the manufacture of castings.

In contrast, case (c) can be envisaged in which the velocity is so high that the liquid behaves as a jet. The liquid is then in great danger of enfolding its surface during its subsequent fall. Regrettably, this condition is common in casting metals. It is not recommended for making good castings.

The interesting intermediate case (b) illustrates the situation where the melt travels at sufficient velocity to rise to a level that is just supported by its surface tension. This is the height of a sessile drop. Any higher, and the melt would subsequently have to fall back, and would be in danger of enfolding its surface. Thus, the critical velocity required is that which is only just sufficient to project the liquid to the height h of the sessile drop. This height is simply that derived from conservation of energy, equating kinetic energy gain to potential energy loss, $mV^2/2 = mgh$, we obtain $h = V^2/2g$. Substituting this value in the value for the height of the sessile drop above (Eq. 1.5) gives the same result for the critical velocity (Eq. 1.6).

From Eq. (1.5) we can find the height from which the metal can fall, accelerated by gravity, before it reaches the critical velocity given by Eq. (1.6). When falling from greater heights, the surface of the liquid will be in danger of being entrained in the bulk liquid. The critical fall heights for entrainment of the surface are listed in Table 1.2.

At first sight it seems surprising that nearly all liquids, including water, have a critical velocity in the region of 0.3–0.5/ms. This is a consequence of the fact that the velocity is a function of the *ratio* of surface tension and density, and, in general, both these physical constants tend to change in the same direction from one liquid to another. Furthermore, the effect of the one quarter power in Eq. (1.6) further suppresses differences. Most of the important engineering metals have critical velocities close to

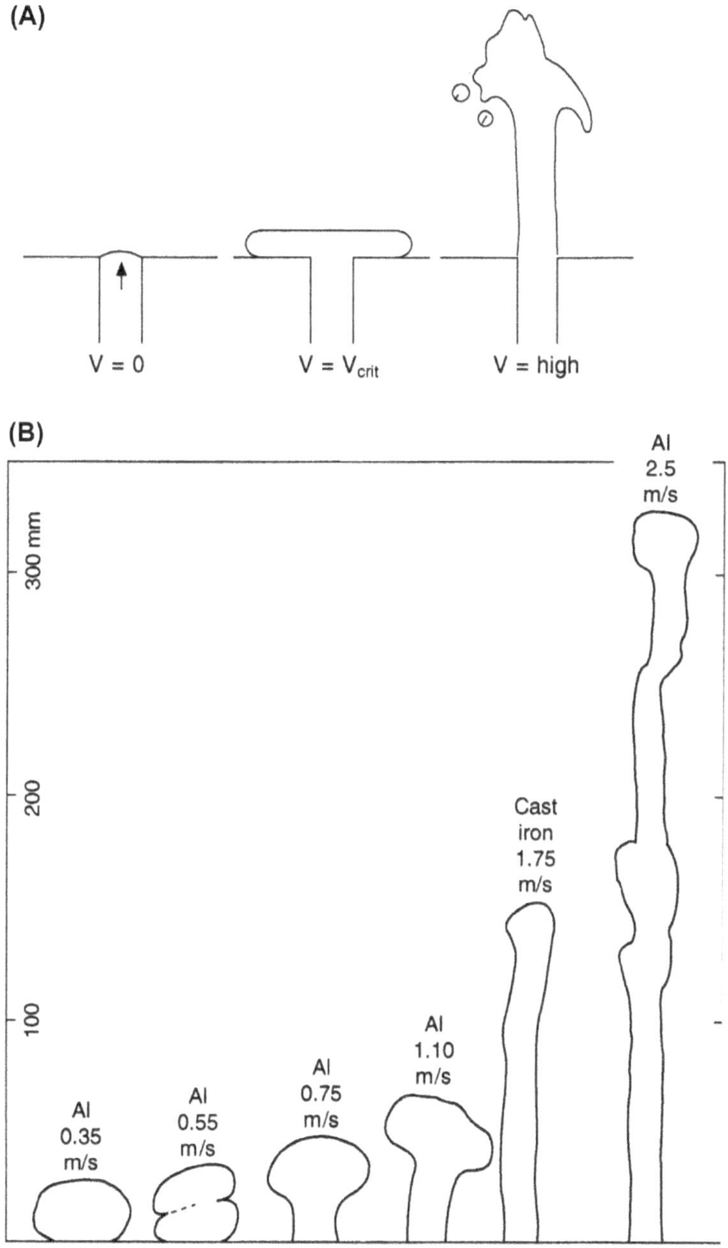

Figure 1.19 (A) Metal emerging into a mold at zero, critical and high velocities; (B) experimentally observed profiles (Runyoro et al., 1992).

0.5 m/s so that this value I now generally assume in the calculation of filling systems for castings of all metals and alloys.

The value of 0.5 m/s for liquid aluminum has been confirmed experimentally by Runyoro et al. (1992). He observed by video the emergence of liquid aluminum at increasing speeds from a vertically oriented gate in a sand mold. An outline of some of his observed melt profiles is given in Fig. 1.19. At just over the critical velocity, at 0.55 m/s (Fig. 1.19B), the metal falls back only slightly, forming its first fold around its middle, taking on the appearance of a traditional English bread called a "cottage loaf." At progressively higher velocities the potential for damage during the subsequent fall of the jets is evident.

In further work shown in Fig. 1.20, Runyoro (1992) observed a dramatic rise in surface cracks identified by dye penetrant testing, and the precipitous fall in bend strength of castings filled above 0.5 m/s. I was so shaken by the alarming nature and clarity of this result that I completely changed the direction of research work at the University of Birmingham for the next 10 years, focusing on the role of surface turbulence generated by filling system design on the mechanical performance of castings.

For instance, the critical velocity for Cu−10Al alloy at 0.5 m/s was subsequently confirmed by Helvaee and Campbell (1996), and a similar value for ductile cast irons shown in Fig. 1.21 was reported by the author (2000). The critical velocity for aluminum has been explored by computer simulation (Lai et al., 2003) in which area of the melt surface was computed as a function of increasing ingate velocity. Assuming the same values for physical constants as used here, a value of precisely 0.5/ms was found (Fig. 1.22), providing reassuring confirmation of the concept of critical velocity by a totally independent technique.

Since those early days further confirmations of the critical entry velocity into molds has been found for magnesium to be in the range 0.25−0.50 m/s (Bahreinian et al., 2005). Mirak et al. (2011) located a value for the Mg−9Al−1Zn alloy 0.30 ± 0.05 m/s. Furthermore, in much subsequent industrial work in foundries, in which filling systems have been redesigned to reduce ingate velocities to near 0.5 m/s, the improvements in quality of castings has been demonstrated in metals and alloys of all types, from light alloys to high temperature alloys, steels and titanium-based alloys.

1.4.2 Weber Number We

The concept of the critical velocity for the entrainment of the surface by what the author has called *surface turbulence* is enshrined in the *Weber*

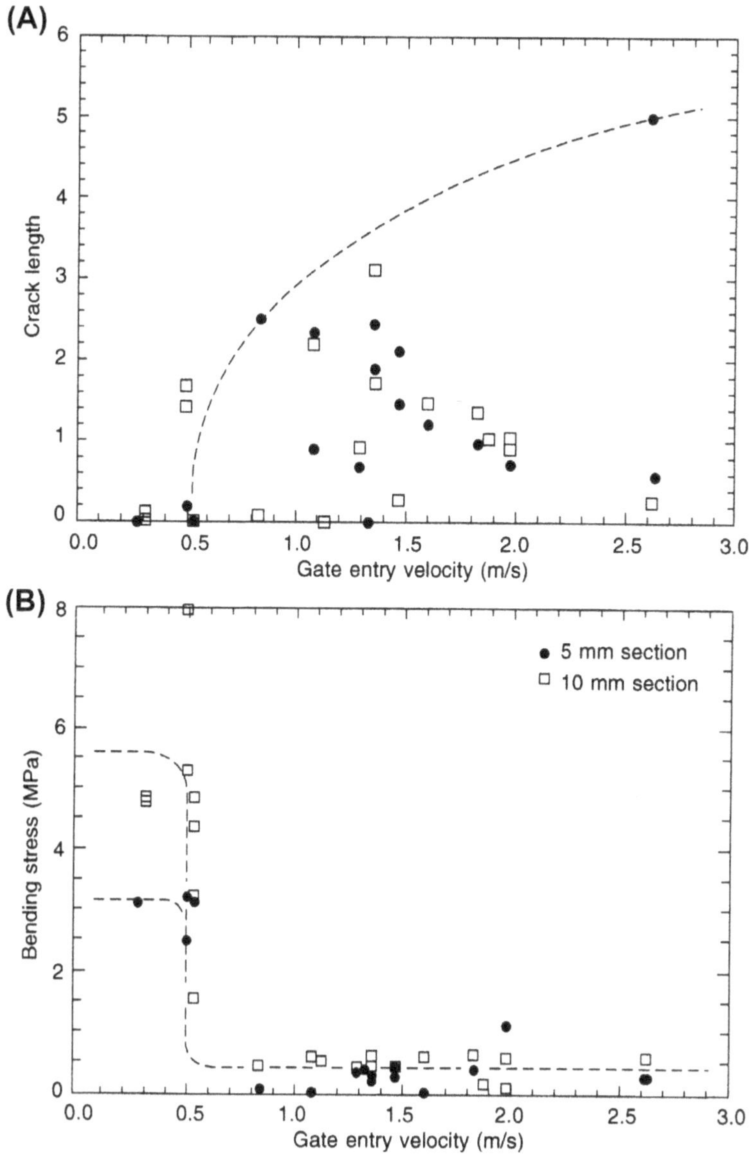

Figure 1.20 (A) Crack lengths by dye penetrant testing in Al alloys cast at different ingate velocities; (B) the bend strengths of the same plate castings (Runyoro, 1992).

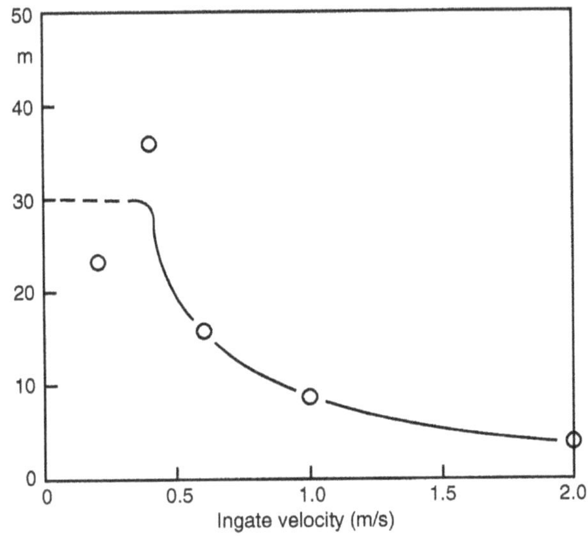

Figure 1.21 Reliability of ductile cast iron cast at different ingate velocities in term of the two-parameter Weibull modulus (m) (Campbell, 2000a).

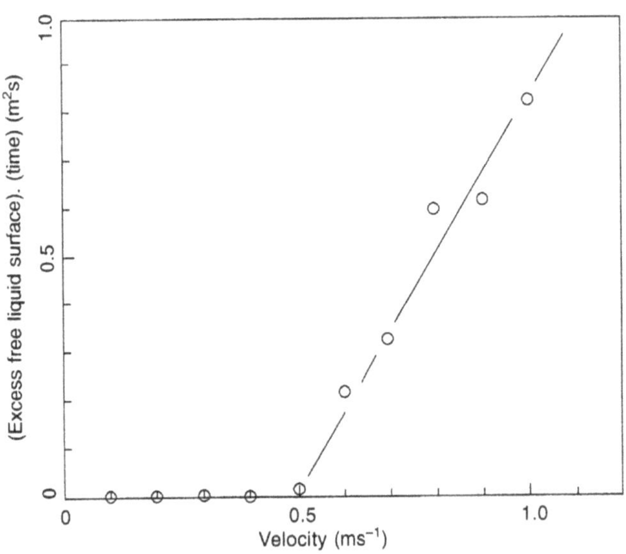

Figure 1.22 Computer simulation of the filling of a mold with liquid aluminum, showing the excess free area of the liquid as a function of ingate velocity (Lai et al., 2003).

Number, We. This elegant dimensionless quantity is defined as the ratio between (i) the inertial pressure in the melt, assessed as $\rho V^2/2$, and (ii) the pressure due to surface tension $\gamma(1/r_1 + 1/r_2)$ where r_1 and r_2 are the radii of curvature of the surface in two perpendicular directions. If, as a cylinder curved only in one plane, the pressure becomes γ/r, or for a hemisphere where the orthogonal radii are equal, $2\gamma/r$. Thus, for a spherical liquid meniscus

$$\mathrm{We} = \rho \mathrm{L} \mathrm{V}^2/\gamma \tag{1.7}$$

Only the dimensioned quantities are included in the ratio, the scalar numbers (the factors of 2) are neglected, as is usual for a dimensionless number. Dimensionless numbers tend to be used over many orders of magnitude which is why scalar numbers become insignificant. The radius term becomes the so-called "characteristic length" parameter, *L*. For flows in channels, *L* is typically chosen to be the hydraulic radius, defined as the ratio of the occupied area of the channel to its wetted perimeter. For a filled cylindrical channel this is half the radius of the channel (diameter/4). For a filled thin slot, it is approximately half the width of the slot. (As will be appreciated by casting engineers, the hydraulic radius is a concept closely allied to that of "geometrical modulus" as commonly used by casting engineers to gauge the heat content of a section of a cast part.)

When $We = 1$ the inertial and surface forces are roughly balanced. From the definition of *We* in Eq. (1.7), and assuming for liquid aluminum a height of 12.5 mm for a sessile drop, giving a radius $L = 12.5/2$ mm we find $V = 0.25$ m/s, a value only a factor of 2 different from the critical velocity found earlier. When it is recalled that the real value of dimensionless numbers lies in their use to define numbers to the nearest order of magnitude (i.e., the nearest factor of 10), this is impressively close agreement. The agreement is only to be expected, of course, since the condition that $We < 1$ defines those conditions where the liquid surface is tranquil, and $We > 1$ defines the conditions in which surface turbulence is expected.

The *We* number has been little used in casting research. This is a major omission on the part of researchers. It is instructive to look at the values that *We* can adopt in some casting processes.

Work on aluminum bronze casting has shown empirically that these alloys retain their quality when the metal enters the mold cavity through vertical gates at speeds of up to 0.06 m/s. However, this extremely cautious velocity is almost certainly erroneously low as a result of confusion introduced from other faulty parts of the filling system in the original experiment (notably problems with the pouring basin, and the well at the sprue/runner junction). Halvae and Campbell (1997) using a better technique, later confirmed the theoretical predicted critical velocity for aluminum bronze as 0.4/ms.) However, if for the moment we accept the low velocity 0.060/ms into the mold cavity, we can assume that the radius of the free liquid surface is close to 5 mm (corresponding to half of the height of the sessile drop given in Table 1.3), the density approximately 8000 kg/m^3, and surface tension is approximately 2.3/Nm, then *We* in this case is seen to be approximately 0.1 representing a condition dominated by surface tension, and thus ultra safe from any danger of surface turbulence.

In a mold of section thickness of 5 mm fitted with a glass window, Nieswaag (1990) observed that an advancing front of liquid aluminum just starts to break up into jets above a speed of approximately 0.37 m/s. This is reasonably close to our theoretical prediction 0.5 m/s (Table 1.3). Assuming that the radius of the liquid front is 2.5 mm, and taking 2400 kg/m^3 for the density of liquid aluminum, the corresponding critical value of *We* at which the break-up of the surface first occurs is therefore seen to be approximately 0.9. The users of vertically injected squeeze casting machines now commonly use 0.4 m/s as a safe filling speed (for instance Suzuki, 1989), corresponding to We close to 2.0.

UK work in the 1950s (IBF Subcommittee, 1960) showed that cast iron issues from a vertically oriented gate as a fountain when the sprue is only about 125 mm high and the gate about 12 mm in diameter (deduced

Table 1.3 The critical heights and velocities of some liquids.

Liquid	Density kgm^{-3}	Surface tension Nm^{-1}	Critical height mm	Critical velocity ms^{-1}
Ti	4100	1.65	12.8	0.50
Al	2385	0.914	12.5	0.50
Mg	1590	0.559	16.0	0.42
Fe	7015	1.872	10.4	0.45
Ni	7905	1.778	9.6	0.43
Cu	8000	1.285	8.1	0.40
Zn	6575	0.782	7.0	0.37
Pt	19,000	1.8	6.2	0.35
Au	17,360	1.14	5.2	0.32
Pb	10,678	0.468	4.2	0.29
Hg	13,691	0.498	3.9	0.27
Water	1000	0.072	5.4	0.33

from careful inspection of the photograph of the work). We can calculate, therefore, that *We* is approximately 80 for this well-advanced condition of surface turbulence.

Going on to consider the case of a pressure die casting of an aluminum alloy in a die of section thickness 5 mm and for which the velocity *V* might be 40 m/s, *We* becomes approximately 100,000. It is known that the filling conditions in a pressure die casting are, in fact, characterized by extreme surface fragmentation, and possibly represent an upper limit which might be expected while making a cast product. Owen (1966) noted from examination of the surfaces of zinc alloy pressure die castings that the metal advanced into the mold in a series of filaments or jets. These advanced splashes were subsequently overtaken by the bulk of the metal, although they were never fully assimilated. These observations are typical of much work carried out on the filling of the die during the injection of pressure die castings.

In summary, it seems that We numbers in the range 0−2.0 define the range of flow conditions that are free from surface turbulence. Although 1.0 is theoretically perfectly safe, it seems that a value nearer 2.0 appears to be transitional, indicating that a safe working critical velocity might be as high as 1.0 m/s. By the time the Weber number reaches 100, surface turbulence is certainly a problem, with jumping and splashing to a height of 100 mm or so. However, at and above values of 100,000 the concept of surface turbulence has probably given way to modes of flow characterized by jets and sprays, and eventually, atomization.

1.4.3 Froude number Fr

The Froude number is another useful dimensionless quantity. It assesses the propensity of conditions in the liquid for making gravity waves. Thus, it compares the inertial pressures $\rho V^2/2$ that are perturbing the surface with the restraining action of gravity, ρgh, where g is the acceleration due to gravity, and h is the height of the wave. The Froude number is therefore defined as

$$\mathrm{Fr} = \mathrm{V}^2/\mathrm{gh} \tag{1.8}$$

It is important to emphasize the regimes of application of the *We* and *Fr* numbers. The Weber number deals with small scale surface effects controlled by surface tension such as ripples, droplets and jets. These surface tension waves are in contrast to the Froude number, which deals with larger scale gravity waves, including such effects as slopping and surging in more open vessels, and in the "see-saw" oscillations of flow observed in U-tube conditions of some filling systems. We shall see later how *Fr* is useful to assess conditions for the formation of the hydraulic jump, a troublesome feature of the flow in large runners.

In the meantime, however, Froude number is useful to assess conditions where the undulation of the surface may cause its surface area to contract, forcing an entrainment condition, as in Fig. 1.7B. This damaging behavior has been so far overlooked as having a major potential role in entrainment. The continued positive expansion of the advancing front is a necessary and sufficient condition for the avoidance of entrainment. Conversely, the contraction of the front in the sense of a loss of area of the "free surface" of the oxide-covered liquid is a necessary and sufficient condition for the entrainment of bifilms.

Interestingly, the Weber number condition, indicating the onset of surface turbulence in the form of a potentially breaking wave, is a necessary but not sufficient condition for entrainment of bifilms. Similarly, it is most probably a necessary but not sufficient condition for the entrainment of bubbles. It is also likely that the surface turbulence condition aids the permanent entrainment of material entrained by surface contraction that might otherwise detrain during a subsequent expansion. Thus the surface turbulence condition in the form of the critical velocity criterion remains a major influence on cast metal quality.

1.4.4 Reynolds number Re

Many texts deal with turbulence, especially in running and gating systems, citing Reynold's number:

$$\mathrm{Re} = \rho \mathrm{VL}/\mu \tag{1.9}$$

where V is the velocity of the liquid, ρ is the density of the liquid, L is a characteristic linear dimension of the geometry of the flow path, and μ is the viscosity. The definition of Reynold's number follows from a comparison of inertial pressures $C_d \, \rho \, V^2/2$ (where C_d is the drag coefficient) and viscous pressures $\mu V/L$. The reader can confirm that the ratio of these two forces does give the dimensionless number *Re* (neglecting, as always for dimensionless numbers, the presence of the numerical factors $2C_d$). The characteristic length L is, as before, usually defined as the hydraulic radius.

Clearly, the inertial forces are a measure of those effects that would cause the liquid to flow in directions dictated by its momentum. These effects are resisted by the effect of viscous drag from the walls of the channel containing the flow. The walls, if sufficiently near, act to *constrain* the flow in a linear fashion. At values of *Re* below about 2000, viscous forces prevail, causing the flow to be smooth and laminar, i.e., approximately parallel to the walls. At values of *Re* higher than this, flow tends to become turbulent; the walls are too far away to provide effective constraint, so that momentum perturbations overcome viscous restraint. The flow subsequently degenerates into a chaos of unpredictable swirling patterns.

It is vital to understand, however, that all of this flow behavior takes place in the bulk of the liquid, underneath the surface. During such turbulence in the bulk liquid, therefore, the surface of the melt may remain relatively tranquil. Turbulence as predicted and measured by *Re* is therefore strictly *bulk turbulence*. It does not apply to the problem of assessing whether the surface film will be incorporated into the melt. This can only be assessed by the concept of *surface turbulence* associated with *We* and *Fr*.

Flow in running systems is typically turbulent as defined by Reynold's number. Even for very narrow running systems designed to keep surface turbulence to a minimum, bulk turbulence would still be expected. In this case for a small aluminum alloy casting where $V = 2$ m/s, in a runner only 4 mm high, hence $L = 2$ mm, *Re* is still close to 7000. For a fairly small steel casting weighing a few kg where the melt is flowing at 2 m/s in a runner 25 mm high, giving approximately $L = 10$ mm, and assuming $\mu = 5.5 \times 10^{-3}$ Nsm^{-2} and $\rho = 7000$ kgm^{-3}, *Re* is approximately 20,000. For a large casting a meter or more high and weighing several tonnes it is quickly shown that *Re* is in the region of 2,000,000. Thus, bulk turbulence exists in running systems of all types, and can be severe.

Only in the very narrow pores of ceramic foam filters is flow expected to be laminar as indicated in Fig. 1.23. For velocities of 2 m/s for liquid aluminum in a filter of pore diameter 1 mm $Re = 2000$. For liquid steel, which is somewhat more viscous, $Re = 500$. These values show that the flow is not far from an unstable turbulent condition even in the narrowest of channels found in most running systems. Only moderate increases in speed of flow would promote turbulence even in these very small channels.

The map of flow regimes presented in Fig. 1.23 is a guide to our thinking and our research about the very different flow behavior that liquids can adopt when subjected to different conditions of speed and geometry. Regions below the threshold $We = 1$ will, in general, be good gravity-poured castings; those above will be expected to contain entrainment defects.

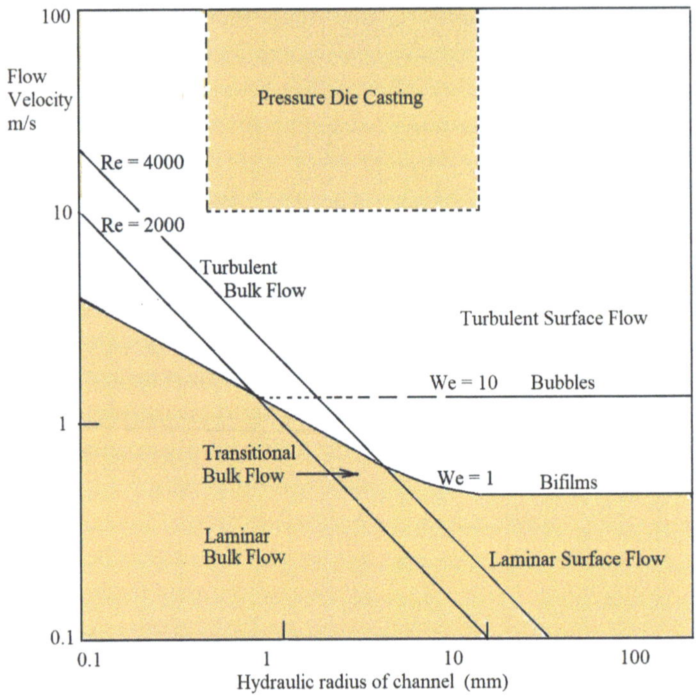

Figure 1.23 Map of flow regimes in the casting of metals.

1.4.5 Oxide skins from melt charge materials

Prior to melting, the charge materials need to be selected and weighed. Normally, charge materials are heavily oxidized with an oxidized coat of extreme thickness and toughness. As the charge sinks beneath the melt surface, such "primary oxide skins" take down with them the thinner oxide film on the melt surface to create a highly asymmetrical bifilm. When the charge is melted, the bifilms are released to float clear, resembling leather skins in suspension. These heavy bifilms are of the same size as the original pieces of charge, and so are often of huge size (fractions of meters across).

For dense metals such as irons and steels the oxide skins submerged in this way often separate quickly to form a dross or slag that can be removed from the top of the melt.

For Al alloys for which the submerged oxide necessarily entrains with it a layer of air as a bifilm, the oxide skins are nearly neutrally buoyant, so the melt clears only with great reluctance. Problems of the introduction of skins from the surfaces of charge materials are most commonly seen in the melting of aluminum and its alloys. Whether or not the skins on the charge materials find their way into the cast products depends on the type of furnace used for melting and to some extent on the melt treatment.

Furnaces in which the solid charge materials are melted in a crucible or bath produce a poor quality of melt. For such melting practice, all the oxide skins on the charge necessarily finish floating about in the melt. Once these primary oxide skins are in the melt it is not easy to get them out. This is especially problematic for all light metal melting, but less serious for dense metals such as copper and steel alloys in view of their higher density compared to their oxides. The dense metals have the advantage of reasonably rapid detrainment of defects by flotation.

In the case of charge materials such as ingots that have been chill cast into metal molds the surface oxide skin introduced in this way is relatively thin. However, charges that are made from sand castings that are to be recycled (usually called "foundry returns") represent a worst case. The oxide skins on sand castings have grown thick during the extended cooling period of the casting in the aggressively moist and oxidizing environment of the sand mold. The author has seen complete skins of cylinder head castings dragged out of the liquid metal, complete with clearly recognizable combustion chambers, ports and bosses. The melt can become so bad as to resemble a slurry of old sacks. Unfortunately, this is not unusual.

In a less severe case where normal melting was carried out repeatedly on 99.5% pure aluminum Panchanathan (1965) found that progressively poorer mechanical properties were obtained. By the time the melt had been recycled eight times, the high elongation of relatively pure Al had fallen from approximately 30% to 20%. This is easily understood if the oxide skin content of the metal is progressively increased by repeated casting.

Dry hearth furnaces can, in principle, give a significantly improved quality, and are especially recommended for Al alloys. Such units are often based on a tower or shaft design to provide a classical counter-flow heat exchange process; the charge is preheated during its course down the tower, while furnace waste gases are cooled as they flow upwards. The base of such a tower is usually a gentle slope of refractory on which the charge in the tower is supported. The slope is known as the dry hearth. Such a design of melting unit is not only thermally efficient, but has the advantage that the oxide skins on the surface of the metallic charge materials are left in place on the dry hearth. The liquid metal flows, practically at its melting point, out of the skin of the melting materials, down the hearth, traveling through its own extended oxide tube, and into the bulk of the melt. Here it needs to be brought up to a useful casting temperature. The oxide tube forms as the liquid flows, and finally collapses, more or less empty, when the flow is finished. The oxide content of melts from this variety of furnace is low. (The melting losses and gas content might be expected to be higher in this design of melter, but because of the protective nature of the oxide that forms in the water vapor of the waste gases, the reaction between the melt and the waste gases is surprisingly low).

The additional advantage of a dry hearth furnace for aluminum alloys is that foundry returns containing iron or steel cast-in inserts (such as the iron liners of cylinder blocks or valve seats in cylinder heads) can be recycled. The insets remain on the hearth and can, from time to time, be raked clear, together with all the dross of oxide skins from the charge materials, via a side door into a skip. (A dross consists of oxides with entrapped liquid metal. Thus, most dross contains between 50% and 80% metal, making the recovery of aluminum from dross economically viable.)

The benefits of melting in a dry hearth furnace are, of course, eliminated at a stroke by the misguided enthusiasm of the operator, who, thinking he is keeping the furnace clean and tidy, and that the heap of remaining oxide debris sitting on the hearth will all make good castings, shoves the heap into the melt. Unfortunately, it is probably slightly less effort to push the dross downhill, rather than rake it out of the furnace through the dross door. The message is clear, but requires restating frequently: good technology alone will not produce good cast metal. Good training, and vigilant management remain essential. Producing good castings was never straightforward!

1.4.6 Pouring

During the pouring of some alloys in a continuous stream, as a falling jet, the surface film on the liquid grows so quickly that it forms a tube around the moving liquid. The author calls this an oxide flow tube. The stationary "flow tube" around the fast-moving jet is often clearly seen during the casting of metals.

A patent dating from 1928 by Beck describes how liquid magnesium can be transferred from a ladle into a mold by arranging for the pouring lip of the ladle to be as close as possible to the pouring cup of the mold, and to be in a relatively fixed position so that the semi-rigid oxide pipe which forms automatically around the jet is maintained unbroken, and thus protects the metal from contact with the air.

A similar phenomenon is seen in the pouring of aluminum alloys and other metals such as aluminum bronze, ductile iron and stainless steels.

However, if the length of the falling stream is increased, then the shear forces of the falling liquid against the inner surface of the oxide tube increase. This drag may become so great that after a second or so the oxide tears, allowing the tube to detach from the lip of the ladle. The tube then accompanies the metal into the mold, only to be immediately replaced by a second tube, and so on. A typical 10 kg aluminum alloy casting poured in about 10 s can be observed to carry a total area of between 0.1 and 2.0 m^2 of oxide into the melt from 10 or more repeated flow tube additions. This is an impressive area of oxide to be dispersed in a casting of average dimensions only 100 × 200 × 500 mm, especially when this is *only one* source of oxide. The oxide in the original metal, together with the oxides entrained by the surface turbulence of the pour, will be expected to augment the total significantly.

During the pouring of steel from a melting furnace, a process known as "tapping the furnace," the metal can fall several meters into the ladle. The turbulence generated in this heavy liquid is of monumental and awesome power. After the pour, a layer of slag to a depth of 100 mm or more is seen to build up quickly on the surface of the steel. We can assume this slag is the accumulation of bifilms which float out quickly. If the bifilm is of the order of 10 nm thick, the 100 mm depth of slag clearly consists of the depth of 10^7 (10 million) bifilms. The reader can guess that such figures have never been checked.

1.4.7 The critical fall height

The critical fall height is that height from which the melt falls to gain just sufficient energy to enfold its surface, creating an entrainment defect. This height happens to be identical to the height of the sessile drop of the melt.

Why is the critical fall height the same as the height of a sessile drop? The critical fall height can be seen to be a kind of re-statement of the critical velocity condition. It is because the critical velocity V is that required to propel the metal vertically upwards to the height at which it is still just supported by surface tension (Figs. 1.18 and 1.19). This is the same velocity V that the melt would have acquired by falling from that height. (A freely traveling particle of melt starting from a vertical ingate would execute a parabola, with its upward starting and downward finishing velocities identical).

When melts are transferred by pouring from heights less than the critical heights predicted in Table 1.3 (the heights of the sessile drop) there is no danger of the formation of entrainment defects. Surface tension is dominant in such circumstances, and can prevent the folding inwards of the surface, and thus prevent entrainment defects. It is unfortunate that the critical fall height is such a minute distance, being close to 10 mm for most metals. Most falls that an engineer might wish to design into a liquid metal handling system are nearly always greater, if not vastly greater. However, the critical fall height is one of those extremely inconvenient facts, a ridiculously small fall distance, threatening to create a penalty that casting engineers must learn to live with.

At greater falls, the surface oxide is carried under the surface of the liquid, together with entrained air. Turner (1965) has reported that, above a pouring height of 90 mm, air begins to be taken into an Al alloy melt with the stream, to reappear as bubbles which rise to sit under the surface oxide film. The We number in this case seems to be of the order of 10 (not easily calculated precisely for this early work) and this bubble threshold is shown schematically in Fig. 1.23.

More recent experiments by Din and the author (2003) on Al–Si alloy have demonstrated that in practice the damage to the tensile properties of castings caused by falls up to 100 mm appears controlled and reproducible provided the melt has first gone through a ceramic foam filter. However, above 200 mm, random damage was certain. These heights are well above the critical fall heights predicted above, and almost certainly are a consequence of the stabilization of the surface of the falling jet by the presence of a film, plus the benefit of laminar flow of the stream after passing through the filter. The mechanical rigidity of the tubular film holds the jet in place, and the smoothness of the falling jet provides minimal disturbance on entry, effectively delaying the onset of entrainment by the plunging action. Clearly, more work would be helpful to clarify the allowable fall heights in different conditions.

In a study of water models, Goklu and Lange (1986) found that the quality of the pouring nozzle affects the surface smoothness of the plunging jet, which in turn influences the amount of air entrainment. They found that the disturbance to the surface of the falling jet is mainly controlled by the turbulence ahead of and inside the nozzle that forms the jet. These findings were powerfully confirmed in an excellent study by Majidi et al. (2018).

During the pouring of a casting from the lip of a ladle via an offset stepped basin kept properly full of metal, then the oxide will probably not be entrained in the pouring basin if the pouring head is sufficiently low, as is achievable during lip pouring. However, in practice, it seems that for fall distances of more than perhaps 50 or 100 mm as are commonly seen in foundries, freedom from damage cannot be relied upon. The pouring basin will usually represent a threat to a filling system from this source.

Clearly, the benefits of defect-free pouring are easily lost if the pouring speed into the entry point of the filling system is too high. This is often observed when pouring castings from unnecessary height. A succession of increasingly bad pouring conditions can be listed as current examples: In aluminum foundries, pouring is often by robot. In iron foundries it is commonly via automatic pouring systems from fixed launders sited over the line of molds. In steel foundries it is common to pour from bottom poured ladles that contain over a meter depth of steel above the exit nozzle (the situation for steel from bottom-teemed ladles is further complicated by the depth of metal in the ladle decreasing progressively as the ladle empties). In a practical instance of a jet plunging at 10/ms into steel held in a 4 m diameter ladle, Guthrie (1989) found that the Weber Number was 2.7×10^6 whereas the Froude number was only 2.5. Thus despite very little slopping and surging, the surface forces were being overwhelmed by inertial forces nearly two million times more effectively, which would certainly have created a very dirty reoxidized steel.

In all types of foundries, if a simple conical pouring bush is used to funnel the liquid stream into the sprue, the surface oxide is automatically entrained and carried into the casting. In this case, of course, all of the oxide formed on the stream will enter the casting. It has to be admitted that pouring basins of sophisticated design, such as the usually recommended offset step basin, are not without their problems and their limitations, but the current widespread use of conical pouring basins undermines all other attempts to improve casting quality. This is especially true in steel casting where the conical basin is filled directly from a bottom-pour ladle. In this condition it is now known that at least 50% of the material taken into the cone is air. The proportion of air seems capable of rising to 80% on occasions. On sunny days in the steel foundry, shafts of sunlight penetrating the foundry reveal the motes of dust in the atmosphere, being sucked in toward the pouring cup from over a meter away, and zooming down into the conical basin during pouring. There can be no doubt about the huge volumes of air sucked into the stream. At night when the foundry is quiet, the gurgling sounds from deep down in the filling system confirm the churning of air into the metal.

Water models clearly reveal that the steel and air mix disastrously, resembling an emulsion. It is little wonder that steels are full of oxide inclusions. The continued use of conical basins is a disgrace, undermining much metallurgy; they have to be eliminated if casting quality can have any chance to improve.

The recent development to avoid such massively unhelpful technology is the use of contact pouring, in which the conical basin is eliminated by direct contact between the ladle nozzle and the entrance to the sprue, the channel which conveys the liquid steel into the mold. Further refinements of this technology include the use of carefully sized channels to avoid the danger of additional air entrainment en-route to the mold (Campbell, 2018).

In summary, large pour heights are extremely damaging, but are widespread in steel casting. It is clear that the lower the pour heights the less damage is suffered by a melt. In addition, of course, less metal is oxidized, thus directly saving the costs of unnecessary melt losses. Ultimately, however, it is, of course, best to avoid pouring altogether. In this way losses are reduced to a minimum and the melt is maintained free from damage.

The adoption of counter-gravity designs for foundries, in which metal is never poured at any point of the process, is no longer a pipe dream. This practical and economical technique for casting production is already in use in a number of foundries worldwide. It is an ultimate solution that cannot be recommended too strongly. Every foundry should have the installation of counter-gravity in their next 5-Year Plan.

1.4.8 The oxide lap from surface flooding

The steady, progressive rise of the liquid metal in a mold may be interrupted for a number of reasons. There could be (i) an inadvertent break during pouring, or (ii) an overflow of the melt (called elsewhere in this work a "waterfall effect") into a deep cavity at some other location in a shaped mold, or (iii) the arrival of the front at a very much enlarged area, thus slowing the rate of rise nearly to a stop.

If the melt stops its advance for any of these reasons, the thickness of the oxide on the melt surface is no longer controlled by the constant splitting and re-forming action. It now simply thickens. If the delay to its advance is prolonged, the surface oxide may become a rigid crust.

When filling restarts (for instance when pouring resumes, or the overflow cavity is filled) the thickened film may now to too rigid to move. It will therefore fracture at some weak point, and fresh melt then roll over the thickened surface film, laying down its own new, thin film. In this way an extensive double film defect, a bifilm, will be created. A massive horizontal crack has now been formed across the casting. The bifilm will be highly asymmetrical, consisting of a lower thick film and an upper thin film. When this rolling wave mode of entrainment was explored by Evans et al. (1997) the rolling wave could not be found on an advancing liquid aluminum front. It seems that some cooling is required to promote a little solidification, so the weight of the freshly arriving liquid can be supported by the stationary metal. For this reason, the defect can be called a "cold lap" (Fig. 1.24).

Asymmetric films are interesting, in that precipitates are usually observed to prefer one film as a substrate for formation and growth, but not the other. Examples are briefly described later in the sections relating to confluence welds and oxide flow tubes. The asymmetric bifilm is a common defect in cast metals.

A key aspect of the stopping of the front, is that the double film defect that is thereby created is a single, huge planar defect, extending completely

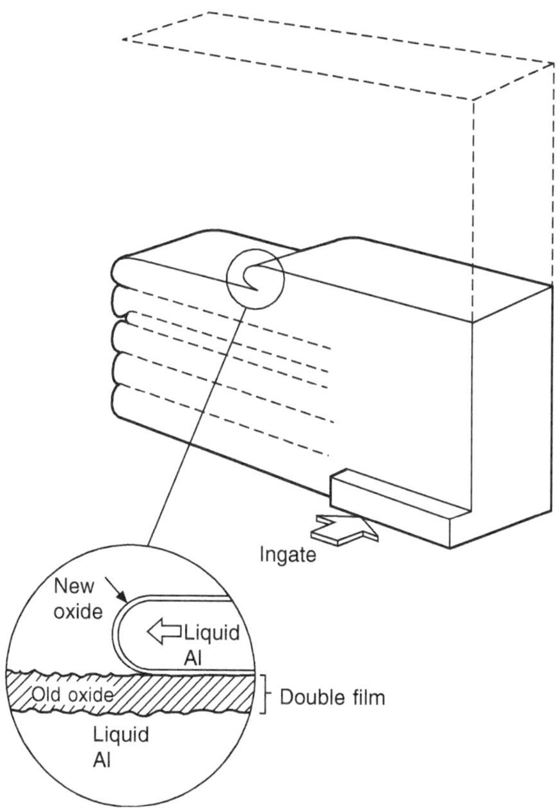

Figure 1.24 Filling of a mold showing the formation of laps containing an asymmetric bifilm crack.

through the product. Also, its orientation is perfectly horizontal. (Notice it is quite different from the creation of double film defects by surface turbulence. In this chaotic process the defects are random in shape, size, orientation and location in the casting.) Flooding over the surface in this way is relatively common during the filling of molds, but is extremely harmful in creating massive horizontal cracks. The vacuum arc remelting process, unfortunately, automatically encourages such conditions, fatally undermining its use for safety-critical applications.

From an exterior view of the casting, the newly arriving melt flooding over the thick stationary oxide will only have the pressure of its own sessile drop height as it attempts to run into the tapering gap left between the old meniscus and the mold wall. Thus, this gap is imperfectly filled, leaving a horizontal lap defect clearly visible around the perimeter of the casting. This surface lap defect corresponding to the arrest of the progress of the melt is the only external sign of a major defect that might extend completely through the casting wall. This is the geometry of lap defect which is usually seen on continuously cast steel. Although, in general, it seems that the bifilm crack extends only around 5−10 mm into the metal, this cast surface is often removed prior to subsequent working to avoid surface defects in the finished rolled product.

Whether the resulting bifilm extends completely across the casting probably depends on whether the arrested front has fully solidified or not. If no solidification has occurred, the bifilm defect may be confined to the lap on the exterior of the casting as in continuously cast steel described above. Notice that in this way both sides of the lap are liquid, separated by an oxide double film. We have created an *oxide lap*.

If the arrest of the advance of the melt had been further delayed the meniscus could have lost so much heat that it had become partially or completely solid. In this case the lap would take on the form of a *cold lap* (the name "cold shut" is a piece of unhelpful jargon recommended to be consigned to the scrap heap. I hate jargon.).

The distinction between *oxide laps* and *cold laps* is sometimes useful. Whereas both may be eliminated by speeding up filling, and/or avoiding any arrest of progress of the liquid front, only the cold lap may be cured by increasing the casting temperature, whereas the oxide lap may become worse. Even so, it will be clear to the reader that the distinction between the two can be blurred, so that some clarifying research into laps in different metals would be helpful.

1.4.9 Oxide lap as a confluence weld

Even in those castings where the metal is melted and handled perfectly, so that no surface turbulence creates entrainment defects, the geometry of the casting may mean that the metal stream has to separate and subsequently join together again at some distant location. This separation and re-joining necessarily involves the collision of the oxide films on the advancing fronts of both streams, with the consequent danger of the streams having difficulty in joining successfully. I called this junction of streams a *confluence weld* (Campbell, 1988). Most castings necessarily contain dozens of confluence welds, and it is important to recognize that the confluence weld does not necessarily, or even commonly, lead to any significant defect. However, occasionally, and *only* occasionally, it does lead to a severe defect. The conditions in which these issues arise are outlined below.

As an instance, the author recalls that in the early days of the development of the Cosworth Process, a small aluminum alloy pipe casting was made for very high pressure service conditions. At that time, it was assumed that the mold should be filled uphill from the electromagnetic pump as slowly as possible, the metal entering via a slot gate at the bottom of the horizontal pipe, and arriving at the top of the pipe just as the melt was about to freeze. In this way directional feeding back toward the ingate would be encouraged. When the pipe was finally cast it looked perfect. It passed radiographic and dye penetrant tests. However, it failed catastrophically under a simulated service test by splitting longitudinally, exactly along its top, where the metal streams were assumed to join. The problem defeated our team of casting experts, but was solved instantly by our Foundry Manager, George Wright, the company's very own dyed-in-the-wool traditional foundryman. He simply turned up the voltage on the pump, greatly increasing filling rate (neglecting the niceties of setting up favorable temperature gradients to assist feeding). The casting was perfect. The problem never occurred again. Readers will note a moral (or two) in this story.

As with many phenomena relating to the mechanical effects of double oxide films, the conditions for avoidance of confluence welds are rather straightforward. The concept is illustrated in Fig. 1.25.

In the case of two liquid fronts that progress toward each other by the splitting and re-forming of their surface films, the situation just after the instant of contact is fascinating and key to the understanding of this problem. At this moment the splitting will occur at the point of contact because the film is necessarily thinnest at this point but, of course, no oxygen can access

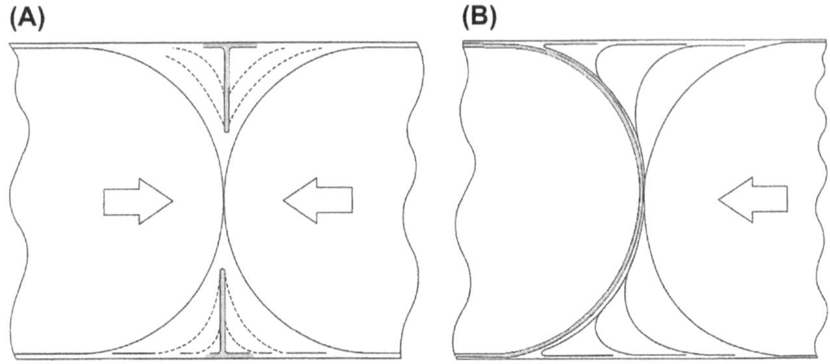

Figure 1.25 Mechanism of the confluence weld (A) in the case of the meeting of two moving fronts which fuse to create a perfect weld (the residual entrained bifilms will tend to flatten against the walls of the mold); (B) a through-thickness crack because one front had stopped.

the microscopic area where the metal is in contact. As the streams continue to engage, the oxide on the surfaces of the two menisci continue to slide back from the point of contact, but because of the exclusion of oxygen from the contact region, no new film can form. Remnants of the double film which has been created during the merging of the two streams occupy a quarter to a third of the outer part of the casting section. Although these exist as a possible crack extending inward from each surface, this is most unlikely in practice. These remnant films will be thin because of their short growth period. Having little rigidity, being more akin to tissue paper of gossamer lightness, they will be folded against the oxide skin of the casting by the random gusts and gales of internal bulk turbulence. There they will attach, adhering as a result of little-understood atomic forces; probably van der Waal forces. Any such forces, if they exist, are likely to be only weak but the vanishingly thin and weak films will not need strong forces to ensure their permanent capture in the stagnant boundary layer against the wall of the mold. Thus, finally, the weld is seen to be perfect. This situation is expected to be common in cast products.

The case contrasts importantly with the approach of two liquid fronts, in which one front comes to a stop, while the other continues to advance toward it. In this case the stationary front builds up the thickness of its oxide layer to become strong and rigid. When the "live" front meets it, the thin film on the newly arriving front is now pinned against the rigid, thick film and is held in place by friction. Thus, the continuously advancing stream expands, wrapping around the rigidized meniscus, forming a layer of thin new film to be laid down on the old thick substrate. Clearly, a double film defect constituting a crack has been created completely across the wall of the casting. Again, the double film is asymmetrical.

Note once again that for the conditions in which one of the fronts is stationary, the final defect is a lap defect in which the crack is usually in a vertical plane (although, of course, other geometries can be envisaged). This contrasts with the surface flooding defect where the orientation of the crack is usually horizontal.

Fig. 1.26 shows a location in an Al alloy casting where a confluence weld was known to have occurred and formed a crack. When observed under the scanning electron microscope the original thick oxide could be seen trapped against the tops of dendrites that had originally flattened themselves against the double film. The poor feeding in that locality had drained residual liquid away from the defect, sucking large areas of the film deeply into the dendrite mesh. One of the remaining islands of film pinned in its original place by the dendrites can still be seen; the draped appearance suggests the dragging action of the surrounding film as it was pulled and torn away as the surrounding liquid and its film disappeared into the depths.

Garcia et al. (2007) studied the problem of severe leakage defects in an Al alloy cylinder head casting, discovering the problem to be a confluence weld formed as a result of the back pressure of gas in the mold causing the

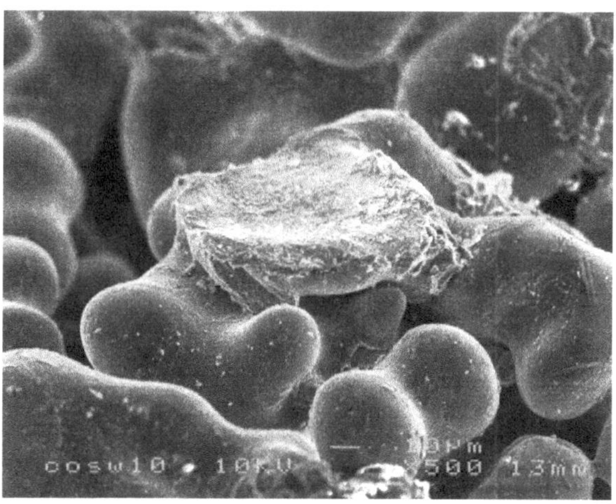

Figure 1.26 SEM fractograph of a confluence weld at a stopped front. An island of thick oxide remains, the residue of the film on the liquid which had been sucked into the depths of the dendrite mesh because of a local feeding problem. Courtesy of Professor N R Green.

advancing front to stop momentarily. They noted alpha-Fe intermetallics had nucleated on the stopped front, but the "live" front was clear of precipitates. Their successful identification of a confluence weld defect contrasts with the study by Lett (2009) who attempted to form defective confluence welds, but in which their casting conditions did not cause either impinging liquid front to stop. The result was the absence of confluence welds.

It is worth observing that the confluence weld is usually a true "oxide lap" (not a "cold lap"). No freezing has necessarily taken place, and the liquid fronts do not need the underlying support of some solidification since the occur in places usually supported by the mold. As a true "oxide lap" it cannot be cured by increasing temperature but only by speeding up the mold filling processes.

In summary, if the two fronts can be kept "alive" the confluence is expected to be a perfect weld. If one fronts stops, the result is a crack. Thus, the problem can be eliminated by keeping the liquid fronts moving. This is simply ensured by casting at a rate which is sufficiently high. Care is needed of course to avoid casting at too high a rate at which surface turbulence may become an issue. Providentially, there is usually a comfortably wide operational window in which the fill rate can meet the requirements to avoid both defects.

1.4.10 The oxide flow tube

Prior to dealing with the formation of oxide flow tubes in cast metals, it is worth diverting briefly to illustrate the principle involved by an elegant example. A piece of foundry equipment is available in which liquid aluminum alloy is caused to flow uphill in a launder trough. This impressive feat is achieved by a traveling magnetic field from coils under the trough, which act to drag the liquid up the slope. The stream of upward-flowing alloy develops an oxide film, as a stationary tube around itself. Interestingly, when the magnetic field is switched off, the metal reverses down its oxide tube under gravity, emptying the tube, which collapses onto the floor of the trough. When the field is switched on once again, the same tube re-fills and re-inflates, continuing to convey the stream uphill. The filling and emptying sequence seems repeatable indefinitely. Analogous oxide tubes can occur when attempting to fill molds to make a shaped casting.

Turning now to consider oxide flow tubes in real casting situations, if, during the filling, a "waterfall" condition occurs at some point in the mold, the falling stream becomes surrounded by its oxide flow tube. As metal rises around the tube, the oxide on the rising metal rolls out against the oxide of the tube, creating an asymmetric bifilm. The bifilm naturally takes the curious form of a vertical cylindrical crack, constituting an almost unbelievably unexpected morphology for a crack (Fig. 1.27). Any metallurgical engineer familiar with normal fairly planar cracks generated by stress would be hard pressed to believe that a long, vertical, cylindrical crack can exist in metals, and even more hard pressed to explain it.

The falling stream does not need to fall vertically. Streams can be seen that have slid down gradients in such processes as tilt casting, or have merely progressed along a horizontal surface of the mold, are common occurrences and, if on the surface of a casting, can be clearly visible.

The jets of flow in high pressure die castings can be seen to leave permanent legacies as oxide tubes as seen in section in Fig. 1.28. Many authors have noted such defects: Lavelle (1962) found such a high density of oxide

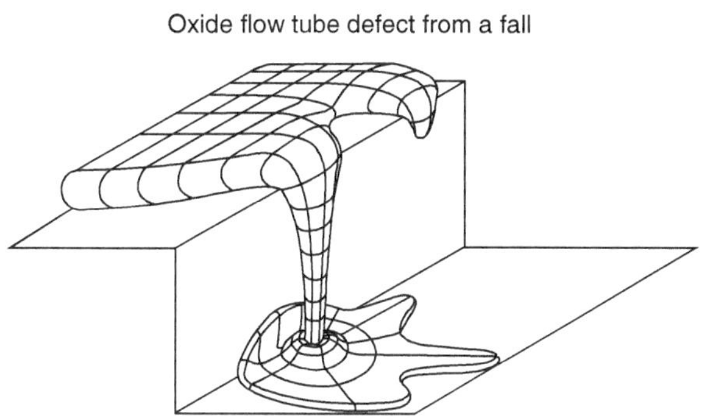

Figure 1.27 Waterfall effect in a mold causing (i) stationary top surface forming a horizontal lap and crack defect; (ii) a falling jet creating a cylindrical oxide flow tube leading to a cylindrical crack; and (iii) random surface turbulence damage in the lower levels of the casting.

Figure 1.28 Oxide flow tubes separating regions of the solidifying structure of a high pressure Al alloy die casting. *Courtesy Ghomashchi and Chadwick (1986).*

laps in a high pressure die casting that the casting fractured with a fibrous appearance. More recently, Ahmed and Kato (2008) using a scanning acoustic microscope noted the damage to fatigue properties by "cold flakes" that could detach from the surface of the casting to reveal a crack.

The surfaces of both falling and horizontal streams can be effectively stationary, and thus develop a characteristic asymmetrical double film. On such a double film in a vacuum cast Ni-based superalloy, the author has seen sulphide precipitates formed only on one side of the defect, indicating that only one side of the double film was favorable to nucleation and growth of the sulphides (microphotographs were not released for security reasons). In this case, no work was carried out to clarify whether the active substrate was the thick or thin side of the bifilm.

In all cases it will be noticed that such interruptions to flow, where, for any reason, the surface of the liquid locally stops its advance, a large, asymmetric double film defect is created. These defects are always large, and always have a recognizable, predictable geometrical form (i.e., they are cylinders, planes, meandering streams etc). They are quite different to the random size and shape of double films formed in the chaos of surface turbulence.

1.4.11 Microjetting

When studying the behavior of Al—7Si-0.4 Mg alloy cast into plaster molds, Evans and Campbell (1997) progressively reduced the wall thickness of the castings to increase the effect of surface tension to constrain the flow, reducing surface turbulence, and thus increasing reliability of the mechanical properties of the cast material. As predicted, properties were clearly improved as the section was reduced from 6 to 3 mm. However, as the section was reduced further, instead of the reliability increasing further as expected, it fell dramatically.

In an effort to explain this surprising behavior a video camera was inserted into the mold to observe the liquid meniscus as it traveled toward the camera. In the very small sections the smooth profile of the meniscus was observed to be punctured by cracks, through which tiny jets of metal spurted ahead, only to be quickly engulfed by the following flow. The image resembled advancing spaghetti.

The flow had become chaotic on a microscale. Jets of liquid issued from cracks in the oxide on the front, only to be caught up within a fraction of a second by the general advance of the front, and so become incorporated back into the bulk liquid. The jets, of course, became oxidized, so that the advancing liquid was naturally expected to become contaminated with a random assortment of tangled double films.

It seems likely that the effect is the result of the strength of the oxide film on the advancing front in thin section castings. In thin sections, the limited area of the front limits the number of defects present in the film. (The effect seems analogous to the behavior of metal whiskers, whose remarkable strength derives from the fact that they are too small to contain any significant defects.) Following this logic, a small area of film may contain no significant defect, and so may resist failure, and will therefore hold back the flowing liquid. Pressure therefore builds up behind the film, until finally it ruptures, the split releasing a jet of liquid. (To explain further, the effect is not observed in thick sections because the greater area of film assures the presence of plenty of defects, so the film splits easily under only gentle pressure so that the advance of the melt is therefore smooth.)

Similar microjets have been observed to occur during the filling of Al alloy castings via 2 mm thick ingates. Single or multiple narrow jets, a millimeter or so in diameter, have been seen to shoot 100 mm or more across the mold cavity from the narrow slot ingates prior to the arrival of the main flow of metal (Cunliffe, 1996).

Microjetting has also been observed in detail under the scanning electron microscope (Storaska and Howe, 2004). When small droplets of Al—Si alloy have been melted in the electron beam the expansion of the liquid causes the oxide skin of the droplet to be subject to a massive stress in the region of 15 GPa. This must be close to the theoretical maximum tensile strength of the oxide. Some droplets are sufficiently small to have a good chance of not containing any defect in their skin, with the result that the skin slowly creeps, gradually expanding to reduce the internal pressure. Others, however, are seen to fracture suddenly, jetting out the compressed liquid.

The microjetting mechanism of the advance of liquid metals in castings has so far only been observed in aluminum alloys, and the precise conditions for its occurrence are not yet known. It does not seem to occur in all narrow channels. The gaseous environment surrounding the flow may be critical to the behavior of the oxide film and its failure mechanism.

Where it does occur, however, the mechanical properties of the casting are seriously impaired. The reliability falls by up to a factor of 3 as casting sections reduce from 3 to 1 mm. Unless microjetting can be understood and controlled, the effect might impose an ultimate limit on the reliability of very thin section castings. This would be a bitter disappointment, and hard to accept. To avoid the risk of this outcome more research is needed.

Apart from surface turbulence, there are other important ways in which bifilms can be entrained in melts. The different mechanisms are described in the following sections.

1.4.12 Entrainment of bubbles

As we have seen, entrainment defects can be caused by the folding action of the (oxidized) liquid surface. Sometimes only oxides are entrained, as double film defects we call bifilms when the pouring height exceeds about 10 mm, corresponding to an impact velocity 0.5 m/s (but in practice up to 1.0 m/s generally appears to be sufficiently low to avoid bifilm entrainment). Thus the bifilm is formed rather easily, under only gentle condition of the disturbance of the liquid surface.

Sometimes the bifilms themselves contain small pockets of accidently enfolded air, so that the bifilm is decorated by arrays of small trapped bubbles (Fig. 1.9B). Only the small spherical bubbles will usually be seen on a polished section or by X-ray radiograph; the bifilm itself being not visible. The bubbles decorating bifilms are typically in the region of 0.1 mm in size. These are integral with, and extensions of, the central interface of the bifilm which consists of a layer of air entrapped in the microscopic roughness of the oxide surfaces.

When the liquid surface suffers breaking waves and splashes which are sufficiently energetic, not only bifilms may be entrained, but larger pockets of air or other environment gases may also be entrained. These are mesoscale defects of the order of 10 mm diameter. The entrainment of these substantial bubbles requires a higher energy than that for the entrainment of bifilms. The different conditions for the entrainment of bubbles can be inferred from the observations of Turner in 1965. He reported that, above a fall height of 90 mm, air begins to be taken into an Al alloy melt with the pouring stream, to reappear as bubbles on the surface. The impact velocity of the melt to create the bubble entrainment condition is 1.34 m/s. Majidi et al., (2018) has confirmed this value by demonstrations on both bubble entrainment in water and liquid metals.

Whereas the large surface area and small volume of the bifilm causes it to float or sink reluctantly, the relatively large volume of gas in the bubble gives the bubble a life of its own. It can have sufficient buoyancy to force its way upwards through the tangle of other bifilms, and possibly other features such as dendrites or grains.

Samuel (1993) reports reduced pressure test samples of aluminum alloy in which bubbles in the middle of the reduced pressure test casting are clearly seen to be prevented from floating up by the presence of oxide films. This lack of buoyancy of small bubbles, generally smaller than about 5 mm in diameter for most metals, results in them being unable to force their way through oxide film barriers. For this reason, they are commonly found entrapped just under the oxide skin of the surface of the casting. In Al alloys they are known therefore as *subsurface pores*. During the heat treatment of steels, the surface oxidizes and flakes away to a depth of a millimeter or more, so that the pores are broken into. Whereas, therefore, they may have originally been *subsurface defects*, they now become *surface defects*.

Larger entrained bubbles are, of course, originally spherical or cap-shaped, but if trapped in the microstructure nearly always resemble crumpled prunes. The reason is almost certainly the result of the deformation of the bubble during the period of intense turbulence whilst the mold is filling. A spherical bubble would have a minimum surface area. However, when deformed its area necessarily increases, increasing the area of oxide film on its surface. On attempting to regain its original spherical shape the additional irreversible area of film is now too large for the bubble, so that the skin becomes wrinkled. As the forces of turbulence hammer the bubble, each deformation of its shape would be expected to add additional area. A further factor adding nonsymmetry, possibly even more important, may be the reduction in volume of the bubble with time, as air is consumed by on-going oxidation and nitridation. In this case the analogy with a large shiny round plum shrinking to become a smaller wrinkled prune, may be rather accurate.

The irregular shape of bubbles has led to them being confused with shrinkage pores. Furthermore, perfectly round gas bubbles have been observed by video X-ray radiography of solidifying castings to form features which have been observed to grow into a defect indistinguishable from a shrinkage pore. This process has been observed in real time of a solidifying casting, in which some bubbles find themselves in a region of casting suffering a fall in pressure late during solidification. The perfectly spherical bubbles expand, growing out among the dendrites, thereby accruing a "furry" growth of interdendritic porosity as residual liquid is drawn away from their surface in this poorly fed region. Such developments obscure the key role of the entrained bubble as the originating source of the apparent shrinkage problem.

Most bad filling system designs are capable of generating not just one bubble, but tens, hundreds or thousands of large bubbles. Unfortunately, this is all too common in castings of all types, from small shaped Al alloy products to tonnage-sized steel ingots.

Furthermore, clusters of entrained *air bubbles* in cast alloys are routinely labeled as "*gas*," so that, in error, degassing treatments to reduce hydrogen or other gases are called for and carried out. I always stress the need for accurate language in the foundry; it is a great help to aid understanding and guide appropriate action. So-called "gas" is nearly always "air bubbles" and should be called "air bubbles," not "gas."

1.4.12.1 Bubble trails

When simply floating up through the liquid metal, large bubbles have enough buoyancy to power an ascent, continuously breaking the oxide skin on their crowns, so that the oxide film on the bubble slides around the bubble surface, coming together under the bubble, dry side to dry side as a long bifilm, the *bubble trail* created in its wake. The bubble trail was predicted, searched for, and found, much to everyone's surprise (Campbell, 2011).

Since air, water vapor and other core gases are normally all highly oxidizing to the liquid metal, a bubble of any of these gases will react aggressively, oxidizing the metal as it progresses, and leaving in its wake the collapsed tube of oxide like an old sack.

This long, tubular defect is usually anchored well back in the mold, at the point at which the bubble was first entrained. The trail therefore acts as a tether, restraining the movement of the bubble. If the bubble has enough buoyancy, its progress causes the oxide skin on its crown to split. The upward motion of the bubble now causes its oxide skin to slide around the bubble, coming together under the bubble in a mass of longitudinal pleats (Fig. 1.29). Any spiraling motion of the bubble will twist and additionally tighten the rope-like tether that continues to lengthen as it rises.

The coming together of the opposite skins of the bubble during the formation of the trail ensure that the films make contact dry-side to dry-side, and so constitute our familiar classical bifilm crack, but cutting through the casting as a sword through butter. Bubble trails are an especially damaging consequence of the entrainment process. Effectively, they are a huge, long bifilm crack. A tangled mess of trails, which has tumbled into a corner of a casting, is not uncommon. On an X-ray radiograph the region

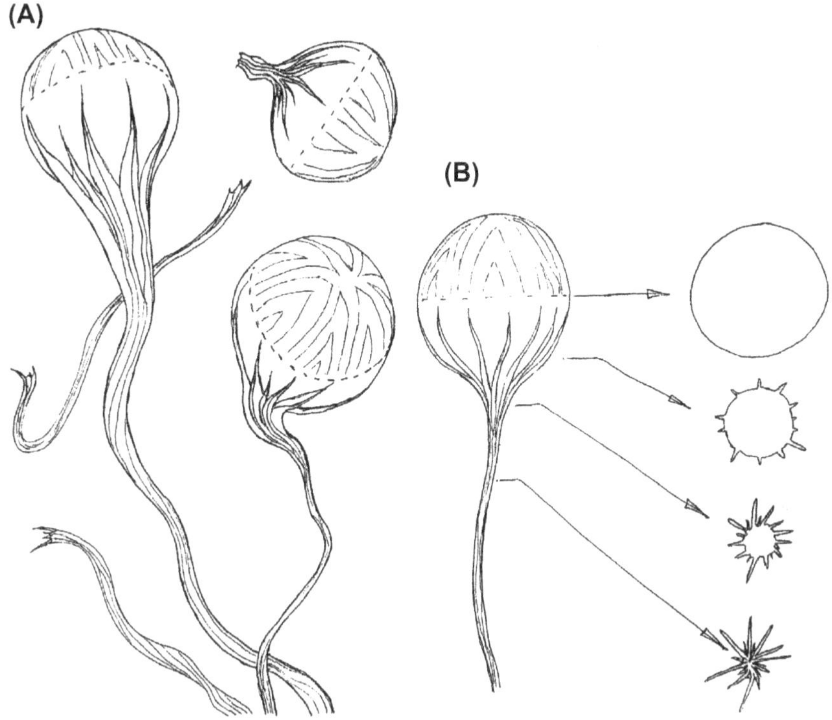

Figure 1.29 Schematic illustration of rising bubbles and their associated trails and fractured trails; plus cross-sections illustrating the progressive collapse of the trails.

is nearly always wrongly identified as a region of "shrinkage" porosity despite the absence of corroboration from the computer simulation.

Large bubbles that are entrained during the pouring of the casting are rarely retained in the casting. This is because they float rapidly, arriving quickly at the top surface of the casting before any freezing has had time to occur. Because their buoyancy is sufficient to split the oxide at its crown as an integral part of its trail-creating mechanism, it is similarly sufficient to burst the oxide skin of the casting that constitutes the last barrier between the bubble and the outside world, and so escape. This detrainment of the bubble unfortunately leaves the legacy of the bubble trail.

During his studies of bubble trails in plate castings of aluminum alloys, Divandari (1998) created a series of artificial air bubbles, and observed their passage though the liquid metal by video radiography. Naturally, the bubble was highly visible, but its trail could not be seen. However, the gradual solidification of the outer edges of the plate casting imposed a tensile stress on the interior of the plate that increased with time. After a few minutes the level of stress in the central region of the plate had risen so high that the trail suddenly flashed into view in a fraction of a second, expanding as a long crack, following the path taken by the bubbles, carving a path through what had appeared previously to be featureless solidifying metal.

Although bubble trails are a kind of elongated bifilm, their curious form as a long, collapsed tube tends not to collapse completely, but retains a small central hole, because of the nonzero residual rigidity of the oxide film. For this reason, it can form a highly effective and nontrivial leak path from the bottom to the top of the cast metal. These lengthy defects can appear elegant under the scanning electron microscope as seen in Fig. 1.30. However, when dozens of trails are scrambled and raveled together they are practically indistinguishable from ordinary bifilms.

1.4.12.2 Bubble damage

Masses of bubbles can be introduced to the casting by a poor filling system design, so that bubbles arriving later are trapped in the tangled mesh of trails left by earlier bubbles. Thus, a mess of oxide trails and bubbles is the result. I have called this messy mixture of bubbles and oxide bubble trails "*bubble damage*" (Fig. 1.31). In the author's experience, bubble damage is the most serious and most common defect in castings, accounting for perhaps 80% of all casting defects. When it is recollected that each bubble rising through the casting will donate an oxide crack of the order of 1 meter

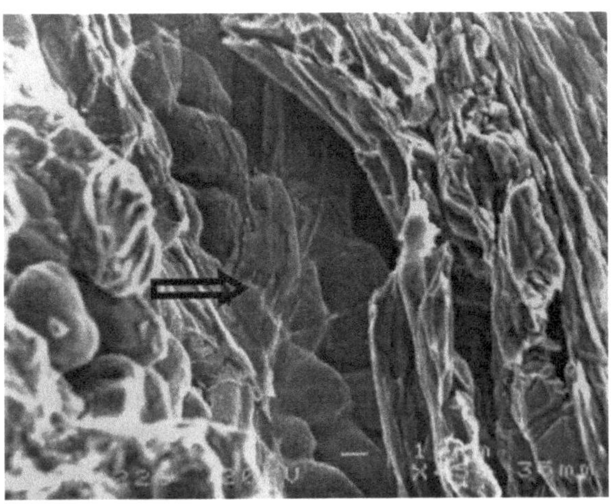

Figure 1.30 SEM fractograph of a detail of a bubble trail in Al—Si alloy with its central noncollapsed cavity arrowed. *Courtesy Divandari 2000.*

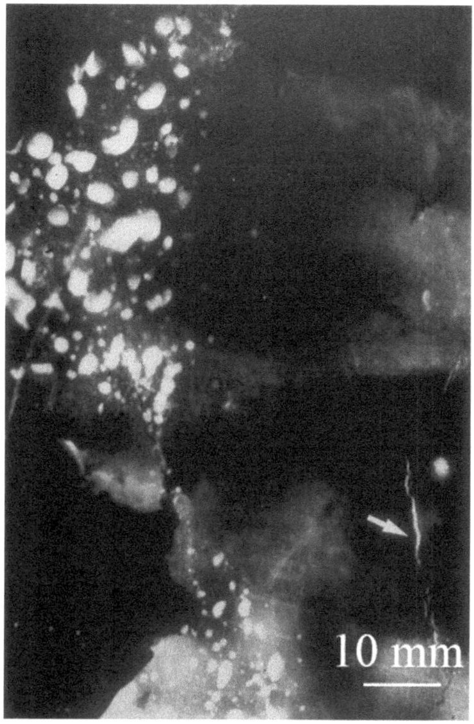

Figure 1.31 Radiograph of a thin-wall copper casting showing extensive bubble damage.

long and 10 mm wide, the scale of the damage is seen to be awesome. It is no wonder that the current computer simulations cannot predict the problems in many castings. In fact, it seems that relatively few of our important defects are attributable to the commonly blamed "gas" or "shrinkage." The bubble trail has long been our unrecognized foe.

Occasionally the bubble trail may break. At the top of Fig. 1.29 is shown a bubble that has been torn free from its trail. Such bubbles, with the stump of their trail showing the bubble to be tumbling irregularly as it rises, have been directly observed by X-ray video.

The author can recall the deep disappointment of otherwise excellent and valuable castings being scrapped by bubble damage. The X-ray radiograph of a Cosworth cast aluminum air intake for a helicopter turbine exhibited a single, perfectly vertical trace from a bottom ingate to the top of the casting. It could only have been the result of a bubble which had been trapped for some time in the tubing between the electromagnetic pump and the mold, and which for some reason had freed itself during the filling of the casting. On a later occasion, the dye penetrant test of an expensive nickel base casting exhibited several bubble trains which meandered up the sides of the casting, emanating from ingates in the base. The event dates from the early days of gravity filling systems equipped with a vortex gate. It was not known at the time that these devices effectively guaranteed the concentration of entrained air into large bubbles; these would accumulate and finally penetrate the ceramic foam filter at the top of the vortex cylinder. In contrast to these scrapped castings suffering only very few bubbles, during the filling of a large casting or ingot, it would be usual for hundreds or thousands of bubbles to be entrained as is easily and convincingly demonstrated by water models. The extent of the damage is difficult to envisage. It is no wonder that our steels are full of oxide inclusions.

In cast irons containing resin-bonded cores, core gases initially contain significant quantities of water vapor. If the pressure in the core rises high enough to blow a bubble into the liquid, the ensuing bubble trail is an oxide trail. These oxides are mainly silicates, and have rather low melting points. They tend therefore to eventually to ravel and be battered into compact formless heaps of slag in the tornado of escaping gases that blasts its way through the wall of the casting to form a permanent open link: a leakage defect. However, after a short time, the core has become sufficiently hot to drive off all its moisture, the supply of oxidizing gases is exhausted. The evolving gases now become carbon-rich volatiles. At that stage of gas evolution, graphite bubble trails start to form (Campbell, 2008). Cast irons seem capable of developing complicated patterns of bubble damage.

For steel ingots, bubble damage is probably the most serious problem when top pouring or conventional bottom gating via a conical trumpet intake. The bubbles arrive as a welter of defects, boiling up with the liquid as it fills the mold. Many of the early bubbles will have enough buoyancy to escape. However, the long trails have a high drag coefficient, and so will not be expected to float out quickly. When the casting is finally solidified, porosity and cracks may or may not be detected by X-ray, and the remaining vast number of trails and bifilms is certain to be invisible. Clean ingot steels will always be an elusive goal if air bubbles cannot be eliminated from the incoming metal.

In shaped castings, bubble trails seen on horizontal cut sections of the casting are nearly always mistaken for shrinkage porosity as a result of their characteristic cusp-like or star-like morphology when the trail is practically completely collapsed (Fig. 1.29). When seen on polished sections, the cusp forms that characterize bubble damage are often confused with cusps that are associated with the shapes of interdendritic shrinkage porosity. However, they can nearly always be distinguished with complete certainty by their difference in size. Careful inspection of the dendrite arm spacing will usually reveal the interdendritic cusps are more than 10 times smaller than the cusps that are caused by the folds of oxide in bubble trails. Clearly, the two are totally unrelated and exhibit quite distinct forms.

If the bubbles completely escape the cast metal, as will be common, the remaining trails will float around, finally settling sometimes at great distances from their source. Irregular masses of oxides in odd corners of castings have been positively identified as groups of tangled bubble trails. The work by Divandari (Fig. 1.32) confirms that bubble trails, in filtered melts known to be free from oxide tangles, can detach and float freely, finally appearing at distant locations. The bubbles have moved on and escaped from the casting, but their trails have remained in suspension. They have broken free from their moorings (the point at which the bubble was first entrained) and have traveled, tumbling and raveling as they go, carried by the sweeping and circulating flow of the liquid during the filling process. Texan founders will recognize an analogy with tumbleweed.

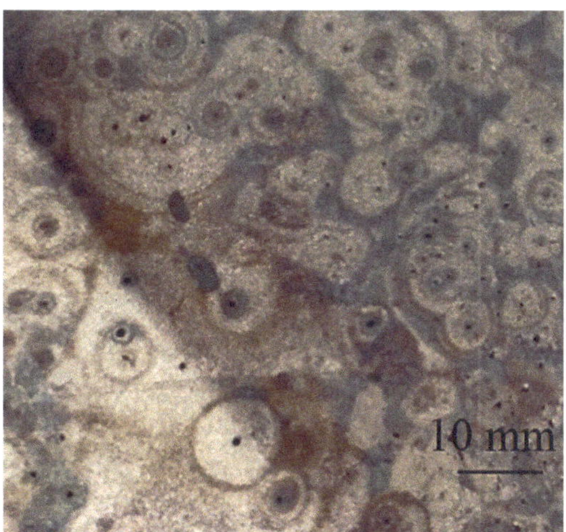

Figure 1.32 Emerging bubble trails revealed by lightly abrading away the oxidized surface of a tin-bronze casting showing the central core and burst halo of oxide. *Courtesy of the artist 2019.*

Another common feature of bubble damage is that large bubbles escape because of their powerful buoyancy, whereas small bubbles become trapped just under the top skin of the casting. They are prevented from escaping only by the thickness of the oxide skin on the casting and their own oxide skin (a double oxide film problem as usual!). Both these films must be broken before the bubble can escape to the outside world. The dividing line between large and small bubbles seems to be in the region of 5 mm diameter for most metals. Such small bubbles are commonly observed in video radiographic studies of mold filling. They float up and sit under the top surface, resting only the distance of the double thickness of oxide depth under the skin of the casting.

Close optical examination of the interiors of bubbles and bifilms in an aluminum alloy casting often reveals some shiny dendrite tips characteristic of shrinkage porosity. This adds to confusion of identification, because the action of a fall in pressure in a casting due to solidification shrinkage will often expand an existing air bubble or bifilm, unfurling and opening it, and if shrinkage conditions persist, will finally suck its oxide films deep into the dendrite mesh. The result is a dendrite-lined cavity. Whether the origin of the cavity was a bubble or a bifilm will usually no longer be discernible. Subsequent examination will usually reveal only fragments of the originating oxides among the dendrites.

This process has been observed in video radiographic studies of castings. An unfed casting has been seen to draw in air bubbles at a hot spot on its surface. The bubbles floated up in succession within the wall of the casting, initially bursting and disappearing at the top of the casting. However, later bubbles became trapped during their rise by dendrites growing in from the walls. As solidification progressed, the onset of shrinkage conditions caused the originally spherical air bubbles to gradually convert to shrinkage cavities. The perfectly round and sharp radiographic images were seen to become 'furry' and indistinct as the liquid meniscus was sucked into the surrounding mesh of dendrites. Finally, the defect resembled extensive shrinkage sponge; its origin as a gas bubble long since disappeared.

Other real-time radiography has shown bubbles entrained in the runner, and swept into the casting. Bubbles of various sizes up to about 5 mm diameter have been observed to float to the top of the casting, coming to rest under the oxide skin of the cast surface. These bubbles had clearly broken free from their trails, probably as a result of the extreme turbulence during the early part of the filling process. However, the upward progress of one bubble in the region of 5 mm diameter appeared to be suddenly arrested, the bubble circulating in the center of the casting, behaving like a balloon on a string. The string, of course, was assumed to be its bubble trail acting as a tether, even though this was invisible in the radiographic image. The bubble was marginally just too small to tear free from its trail. In addition, it may have lost some buoyancy as a result of loss of oxygen during its rise, or perhaps more likely, it ascended as far as it did because of assistance from the force of the flow of the melt. When the force of the flow abated higher in the mold cavity, the bubble's buoyancy alone was insufficient to split its crown of oxide skin, so that its upward progress was halted.

Where many bubbles have passed through the runner into the mold, a cross section of the runner will reveal some central porosity. The porosity would normally be expected to be shrinkage porosity. However, close examination will confirm that this porosity does not result from shrinkage, but is a mass of double oxide films, the bubble trails. The trails congregate in the center of the channel because they are pushed there by the dendrites growing inwards from the walls. In Al alloys they appear as a series of dark, nonreflective oxidized surfaces interleaved like the flaky, crumpled pages of an old newspaper, sepia-colored as a result of stains from mold gases.

In his pioneering systematic study of the formation of bubble trails in aluminum castings by X-ray video technique, Divandari (1999) fractured castings through bubble trails to reveal their interior form. Fig. 1.30 shows the inside of a trail in Al—7Si-0.4 Mg alloy. The longitudinally pleated film is clear, as is the presence of shrinkage cavities that have expanded away from the defect because the casting was not provided with a feeder. Newer, thinner film has sucked back the residual liquid among the dendrites seen in Fig. 1.30B, and in places stretching the film over the dendrites as seen in Fig. 1.30C. The thicker, older part of the film is seen pulled in tight creases, and where torn is seen to be replaced by newer, thinner film.

During the video radiography of some stainless steels I have witnessed an area initially identified as shrinkage porosity, but under the microscope was seen to be a mixture of bubbles and cracks. Such classic bubble damage is a remarkable combination that, without the concept of the bifilm and the bubble trail, would be impossible to explain; tensile stress cannot have been present in the liquid phase because the bubbles would simply have expanded to relieve the stress, thus how can stress cracks have arisen? The answer, of course, is that the cracks are not due to stress but are bifilms in the form of bubble trails, and have no problem to sit side-by-side with bubbles. In these strong steels the high cooling strain leads to high stresses that open up the bifilms to reveal their true nature as cracks.

Finally, it is worth considering what length a bubble trail might reach. If we assume a bubble of 10 mm diameter rising through liquid aluminum, and if aluminum oxide grows to a thickness of 20 nm, it is not difficult to estimate that the 20% oxygen in the air is used up after creating a trail of about 0.5 m. This is quite sufficient to cause a major problem in most castings. However, this estimate is uncertain because Raiszadeh and Griffiths (2008) show that after the oxygen is consumed in a bubble, the nitrogen then starts to react to form aluminum nitride. The maximum trail length might therefore be nearer 2.5 m. These distances might be subject to some reduction because the reaction with air will continue, reacting to deplete the oxygen and nitrogen in the lower sections of the bifilm while the bubble advances at its tip.

For castings poured in vacuum vacuum, bubbles are still to be expected to form, although the situation is a little more complicated. The entrained atmosphere in this case will be at a pressure somewhere between 10^{-3} and 10^{-6} atm (1 torr-10^{-3} torr) indicating that the "vacuum" corresponds more accurately only to "dilute air." In addition, the local vacuum inside the mold at the instant of pouring is likely to be significantly higher than the pressure indicated on the vacuum gauge of the furnace, as a result of mold out-gassing. Thus, considering the case of the vacuum casting of Ni-base superalloys, at a nominal depth of 100 mm in the liquid of density close to 8000 kg/m^3, a bubble of 20 mm diameter will collapse down to somewhere in the region of 5 − 0.5 mm diameter depending on the vacuum pressure before its internal pressure rises to equal that of the surrounding melt. (We are neglecting the somewhat balancing effects of (i) a small reduction in size because of surface tension, and (ii) a small increase in size from the expansion of the mold gases, since molds for many investment casting of Ni-based alloys are already at 1000 C or more compared to the metal nearer 1600 C) The bubbles will be smaller at greater depths, and larger toward the top of the casting of course.

As the bubbles use up their oxygen and nitrogen, the bubbles will shrink further. When they contain only the remaining 1% argon from the air, the final size of the bubble will be between 1 and 0.1 mm depending on the quality of the vacuum. The phenomenon is illustrated in Fig. 1.33.

The considerably reduced concentration of gas in the vacuum bubble will result in rapid collapse of its first stage of reduction, reducing any potential for bubble trails, or perhaps generating more nitride film based trails. (It is

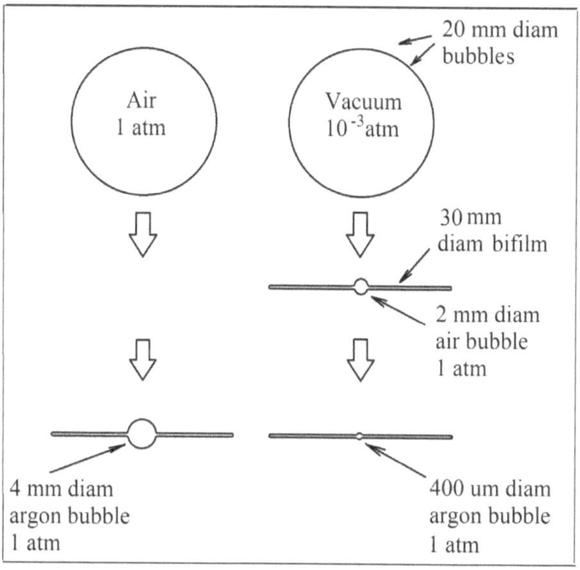

Figure 1.33 Relative defect sizes of a collapsed air and vacuum bubbles.

known that Ni-based superalloys sometimes suffer from nitride problems.) Most probably, the collapse of the bubble will not result in the elegant disc seen in Fig. 1.24, but would be an irregular jumble of oxides. The collapsed vacuum bubble is therefore not expected to form an elegant "fish eye" on the fracture surface of a vacuum casting as suggested by Fig. 1.33, but will finish as a somewhat enlarged bifilm, of irregular shape, and prevented from total collapse by its content of argon (A quite separate mechanism is suggested in Section 3.4 for the formation of a fish eye, but for fracture surfaces of castings poured in air.). The interesting feature of Fig. 1.33 is the illustration of the size of the residual argon bubbles. Clearly, in nearly all entrainment conditions, the residual argon volume is surprisingly large. Also, it will be a permanent feature of the microstructure.

1.4.13 Other entrainment defects

1.4.13.1 Extrinsic inclusions

In addition to bifilms (cracks) and bubbles (porosity), there are a number of other, related defects that can be similarly entrained (i.e., introduced into the matrix from the outside) such as illustrated in Fig. 1.9. These externally introduced features are variously but aptly called *entrained inclusions, extrinsic inclusions or external inclusions* (they are also occasionally known by the equally accurate but awkward term *exogenous inclusions* which we shall avoid).

Such features are, of course, completely different in character to second phases that precipitate from the melt as part of the cooling and solidification processes. The differences warrant careful examination, and are discussed below.

Figs. 1.8 and 1.9D represent one of many different kinds of extrinsic or external inclusion, such as a piece of refractory from the wall of a furnace or ladle or a piece of old surface oxide. External inclusions can only be entrained together with a wrapping of young surface film.

Attempts to introduce particles of various kinds into aluminum alloys have been made from time to time. All of these attempts are seen to be undermined by the problem of the enveloping young oxide with its associated layer of air, preventing the introduced particle from ever making contact with the melt as in Fig. 1.8. This behavior necessarily occurs despite attempts by such authors as Mohanty and Gruzleski (1995) to add TiB_2 particles to liquid aluminum beneath the surface of the melt. They found that such particles did not nucleate grains, but were pushed by dendrites to the grain boundaries confirming the presence of a nonwetting protective film together with its layer of air separating the TiB_2 particle from the matrix metal. A particle of TiB_2 in contact with the matrix would be expected to behave quite differently; it would be incorporated into the growing solid, and actually nucleate new solid grains.

Similarly, the vortex technique for the addition of particles to make a particulate metal matrix composite (MMC) is similarly troubled. It is to be expected that the method would drag the oxide film down into the vortex, together with the heaps of particles supported on the film, so that the clumping of particles, now nicely and collectively packaged, wrapped in film, is probably unavoidable. The "clumping" of particles observed in such MMCs survives even long periods of vigorous stirring (Fig. 1.8C). Such features are seen in the work by Nai and Gupta (2002) in which the central particles in clumps of stirred-in SiC particles are observed to have fallen out during the preparation of their polished metallographic sample, leaving small cave-like structures lined with SiC particles that are somehow still held in place. In addition, whether in clumps or not, during the stirring action the particles are almost certainly held in suspension by the network of bifilms that will have been entrained, and now dispersed throughout the entire volume of the MMC, in the same way that networks of films have been observed to hold gas bubbles in suspension (Samuel, 1993). It seems also likely that the oxide network throughout the MMC will reduce the fluidity of the material, and reduce the achievable mechanical properties, as discovered by Emamy et al. (2009).

The investigation by Miranda and collaborators (2013) is another example of efforts to introduce strengthening features to alloys such as carbon nanotubes (CNTs). The immense strength of the CNTs results in approximately 20% benefit to strength, but a 20% reduction in elongation. The relatively poor strength improvements and loss of ductility are the predictable consequences of the additions entering via the surface of the liquid metal in their oxide "paper bag." The CNTs protruding from the surfaces of tensile fractures, assumed by the authors to be evidence of their presence in the matrix, appear to be emerging from the interiors of bifilm cracks.

In a further example, the manufacture of aluminum alloy MMCs based on aluminum alloy containing SiC particles introduced under a so-called "high vacuum" were investigated by Emamy and Campbell (1997). These authors found that under the hydrostatic tension produced by poor feeding conditions the SiC/matrix interface frequently decohered, suggesting the presence of a nonwetted oxide film. This contrasted with the behavior of an Al$-$$TiB_2$ MMC in which the TiB_2 particles were created by a flux reaction and added to the matrix alloy without ever having been in contact with the air (this is the technique used to make Al + B grain refiners for aluminum alloys, ensuring that the "paper bag" nonwetting phenomenon is avoided, and the added particles can be wetted by the melt to perform their function as nucleants). These particles exhibited much better adhesion at the particle/matrix interface when subjected to tension; relatively few pores were initiated when subjected to the hydrostatic tensile stress generated by the poor feeding.

Fig. 1.9D illustrates the entrainment of a piece of *old oxide*, probably from an earlier part of the melting or holding process. Such oxide, if very old, will have grown thick, and no longer justifies description as a film. It can be crusty or even plate-like. It is simply just another external inclusion. If entrained during pouring, its passage through the liquid surface will ensure that it becomes wrapped in a new thin film, the film of air entrapped by the new thin oxide film separates the old inclusion from contact with the melt. This wrapping will allow the matrix to "decohere" easily from the inclusion, leading to the initiation of porosity or cracks.

However, if the old oxide was entrained some time previously during the melting or holding process, its entrained wrapping will itself become old, having lost the majority of its air layer by continued reaction. In this case it will have become fairly "inert" in the sense that it cannot easily initiate porosity; all its surfaces will have reacted with the melt and thus have come into good atomic contact; the oxide wrapping will have contracted hard onto its entrapped inclusion, and its air layer will have become largely lost and welded closed.

When considering the wetting and submergence of inclusions into liquid metals, a precisely analogous and well-known macroscopic process already exists which is quite well understood. It is the casting-in of inserts. Cast iron or steel inserts are often cast into aluminum alloy castings as bearings for shafts or surfaces to withstand wear or spread high localized loads. The insert is, of course, merely a rather large inclusion. The studies by Aguardo et al. (2013) confirm that no metallurgical bond is formed between a steel and an Al—7Si-0.3 Mg alloy. Effectively, the two metals never come into contact. The authors assume that bonding is prevented by the Fe_3O_4 oxide layer formed on the insert during the filling of the mold. They overlook the vanishingly thin alumina layer on the aluminum alloy which would also roll into place during the filling process, further contributing to the separation of the liquid metal and the insert.

As an aside, the lack of metallurgical bond between steel inserts in aluminum alloy castings has been accepted as inevitable by Ucsnik et al. (2013). As a result, they report the development of strength between the insert and the casting by the welding-on to the insert of an array of steel pins - an arrangement on the insert not unlike a porcupine. Arrays of different geometries and spacings are optimized to create strong joints. It is likely that this curious geometry has advantages over the innumerable alternative geometries which could be selected, such as the simple grooves on a cylindrical insert, but the authors do not favor us with an explanation.

The numerous examples which can be cited of the lack of metallurgical bonding as a result of the interposition of oxide films contrast with the very few examples known to the author in which bonding is successful. One such process is used for the insertion of a cast iron wear ring in an aluminum piston for an internal combustion engine. A cast iron ring was held in a liquid aluminum alloy for sufficient time for the aluminum to diffuse into the surface of the ring. At a critical time, the ring was pulled out of the melt and positioned in the liquid aluminum alloy freshly poured into the steel piston die. It might be expected that the alumina film formed on the ring during its transfer through the air (during the short time of its extraction from the first liquid and immersion in the second) and its enfolding around itself of the film on the second liquid, forming a bifilm, would prevent any bonding. However, if the procedure was carried out correctly, the alumina bifilm was seen, magically, to detach and float off. In the past, when the process was carried out by an operator, it required extreme skill and perfect timing. However, despite all, the bond between the ring and the aluminum alloy piston was successful.

The secret of the detachment of the bifilm relies, of course, on the fact that the film has an underlayer of aluminum as the remnant of the wetted coat from the first liquid immersion. This underlayer may be still molten, or in any case quickly re-melts when dipped into the liquid piston alloy. This fluidized foundation releases the film, which, being extremely thin and weak, and now unsupported, easily tears in the turbulence of the second dip. Its bifilm nature, with its entrained air, causes it to float to the surface. The appearance of this wispy gray veil confirms success to the operator.

1.4.13.2 Flux and slag inclusions

Fluxes are commonly added to light alloy melts. For simple chloride fluxes, the oxide is largely insoluble and inert toward the flux. The flux therefore remains as a liquid layer on top of the oxide film on the liquid alloy. The many common fluxes are based on NaCl and KCl. They are assumed to reduce oxidation and adhere to and absorb inclusions. These efforts to reduce oxidation and inclusions in the melt are not entirely without problems.

Unfortunately, the fluxes are often found in the castings. The fluxes are deliquescent, so that when opened to the air by machining they absorb moisture, leading to localized pockets of corrosion and the unsightly exudation of corrosion products on machined surfaces. Also, during routine examination of fracture surfaces, the elements chlorine and fluorine are quite often found as chlorides or fluorides on aluminum and magnesium alloys. However, because chlorides and fluorides are insoluble in aluminum, it presents the question of how such elements came to be located in the matrix.

If they were present as entrained liquid droplets in the melt, they would float out quickly as a result of their large density difference and their spherical, low drag shapes (Fig. 1.34). However, this does not seem to be true. Rapid detrainment does not occur. The flux inclusions are long-lived in the liquid metal. How is this possible?

It seems that such liquids, originally on the top of the surface oxide film, have been mechanically entrained during pouring, wrapped in the surface oxide film as it is plunged beneath the surface, forming a flux sandwich in an entrained oxide bifilm. Spherodisation of the liquid flux is now

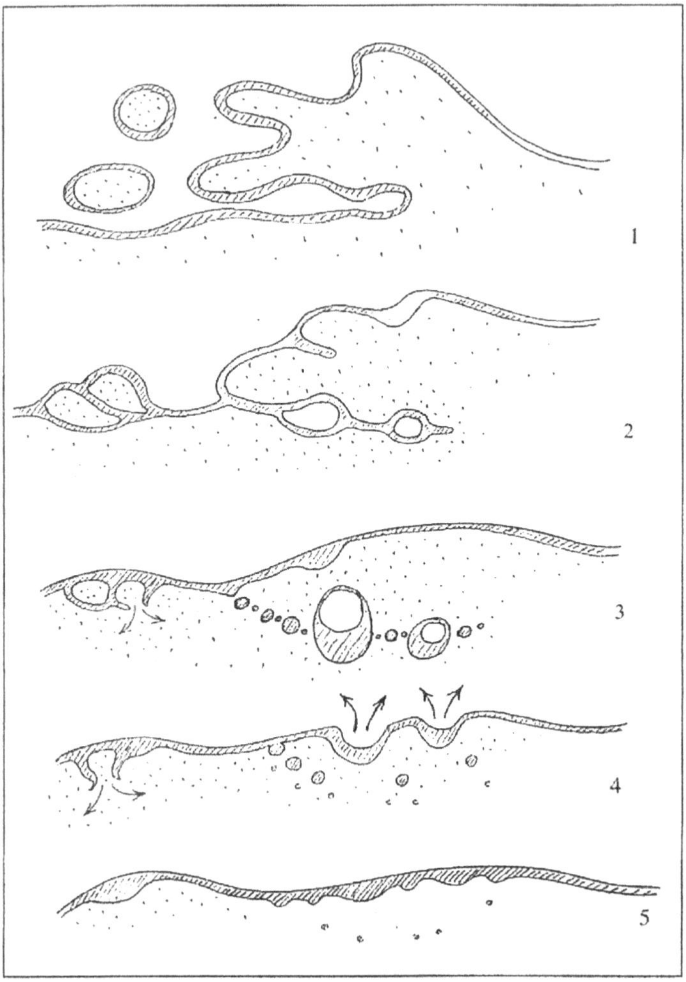

Figure 1.34 Entrainment of a liquid surface film. No bifilm is formed (contrast Fig. 1.9). Entrained metal, gases and liquid droplets are all fluid phases and therefore quickly detrained, leaving the metal mainly free of defects.

impossible and the package is slow to float or settle because of the extended area provided by the bifilm.

The consequential long residence times allow transport of these contaminants over long distances in melt transfer and launder systems. Sokolowski et al. (2000) find that chloride flux inclusions travel distances of tens of meters in launders, whereas if they were simply spherical droplets they would be expected to separate completely from the melt within a meter or so of travel along a launder. In the Windsor Al alloy foundry, they found that the electromagnetic pumps at the end of the launder system were eventually blocked by a mix of oxides and fluxes deposited in the working interiors of the pumps. It is likely that the inclusions are forced out of suspension by the combined centrifugal and electromagnetic body forces in the pump. It is hardly conceivable that fluxes themselves, as relatively low viscosity liquids, could form a blockage. However, the accretion of a mixture of solid oxide films bonded with a sticky liquid flux would be expected to be highly effective in choking the system.

When bifilms are created by folding into the melt, the presence of a surface liquid flux would be expected to have a powerful effect by effectively causing the two halves of the film to adhere together by viscous adhesion. This may be one of the key mechanisms explaining why fluxes are so effective in reducing the porosity in aluminum alloy castings. The bifilms may be glued shut. At room temperature the bonding by the solidified flux may aid the strength and ductility of the metal to some extent. On the other hand, the observed benefits to strength and ductility in flux treated alloys may be the result of a genuine reduction in bifilms by absorption into the flux layer, or by some intermediate reaction such as agglomeration (because of their sticky nature in the presence of a liquid flux) prior to flotation. These factors will require much research to disentangle.

Whether fluxes are completely successful to prevent oxidation of the surface of light alloys does not seem to be clear. It may be that a solid oxide film always underlies the simple chloride fluxes and possibly some of the fluoride fluxes. Even if an underlying oxide film were unstable, and gradually dissolving in highly reactive fluoride flux, the early entrainment of the melt surface prior to the complete solution of the oxide would explain the oxide presence despite the thermodynamic instability.

Even now, some foundries successfully melt light alloys without fluxes. Whether such practice is really more beneficial deserves to be thoroughly investigated. What is certain is that the environment would benefit from the reduced dumping of flux residues from so-called cleaning treatments.

There are circumstances when the flux inclusions may not be associated with a solid oxide film simply because the oxide is rapidly soluble in the flux. Such fluxes include cryolite $AlF_3.3NaF$ as used to dissolve alumina during the electrolytic production of aluminum, and the family of other fluorides K_2TiF_6, KBF_4, K_3TaF_7 and K_2ZrF_6. In these special cases, the surface layer may be a uniform liquid phase, provided the temperature is sufficiently high for the flux itself to be above its melting point.

For irons and steels when a liquid slag layer is present it is normally expected to be liquid, or mainly liquid, throughout. Where a liquid slag surface is entrained Fig. 1.34 shows the expected detrainment of the slag and the accidently entrained gases and entrained liquid metal. Such spherodisation of fluid phases is expected to occur in seconds as a result of the high surface energy of liquid metals and their rather low viscosity.

1.4.14 Furling and unfurling

The verb "*to furl*" is used in the same sense that a sailing boat will *furl* its sails, gathering them in; or it may *unfurl* them, letting them out to adopt their full smooth area as they fill with wind.

If the surface of the liquid suffers turbulence (i.e., the Weber number significantly in excess of unity) the liquid is likely to enfold our familiar bifilm defects. Their size will probably depend on the strength (i.e., the tear resistance) of the film and the power of the turbulent flow. For some high duty stainless steels and Ni alloys the size might be 100 mm across. For Al—Si alloys containing low levels of Mg the bifilms seem to be more usually in the range 0.1—1 mm and occasionally up to 10 or 15 mm diameter, although bifilms derived from the oxide skins of charge materials for melts can, as we have seen, be the size of newspapers.

Once submerged, the bifilms will be subjected to the conditions of bulk turbulence beneath the surface of the liquid. This is because Reynolds number, Re, is nearly always over the critical value of 2000 in liquid metal filling systems. (Exceptions can be counter gravity and tilt pouring systems.) The launching of our delicate, gossamer-thin double film into a maelstrom of vortices in heavy metallic liquids will ensure that it is pummeled and raveled into a compact, convoluted form almost immediately. Furled in this way it will be effectively reduced in size from a planar crack of diameter perhaps 1—10 mm to a small ragged ball of perhaps a 10th of this size, i.e., a diameter in the range 0.1—1 mm. These are the crumpled dross-like defects, as "furled" cracks, commonly seen on polished metallographic sections of aluminum alloy castings (Figs. 1.10A and 1.14).

It is worth emphasizing that although the furled bifilm now has a highly contorted form; crumpled, convoluted and raveled in an untidy random manner, it has not lost its crack-like character. The form of the crack contrasts sharply with the types of cracks familiar to the engineer. Clearly, the furled bifilm bears no resemblance to a crack formed in solid metals by the application of stress. It is a crack morphology unique to liquid metals and castings.

In its furled form, bifilms may be able to pass through filters whose normal pore size in the region of 1 mm diameter. Some will be arrested as a result of untidy trails of film that may become caught by the filter, since the raveled forms will be winding and unwinding continuously. This is particularly expected inside ceramic foam filters, as the bifilm negotiates the branching passages inside the filter. Here, the constraint of the narrow channels probably promotes laminar flow (Fig. 1.23). It is possible that many compact bifilms will be unraveled and flattened against the internal surfaces of the filter if they become hooked up somewhere in a laminar flow stream. This seems likely to be a potentially important filtration mechanism whereby compacted bifilms become caught at one point and straightened downstream. When completely unfurled against the internal surfaces of the filter, they are likely to be invisible if the filter is subsequently sectioned and inspected. A few hundred flattened bifilms, each 20 nm thick, will tend to be invisible against the mm thickness walls of the filter.

At practically all other locations, during the whole of the voyage into the mold, the bulk turbulence will ensure that the bifilms are continuously tumbled, and therefore enter the mold in a fully furled form.

The freshly filled mold therefore initially finds itself with a distribution of compact, convoluted bifilms. Their tensile strength will, on average, be about half that of a pore of the same average diameter as the bifilm ball because of the effect of the mechanical interlinking of the crumpled crack as seen in Fig. 1.35. If the furled crack is subject to a tensile stress it cannot

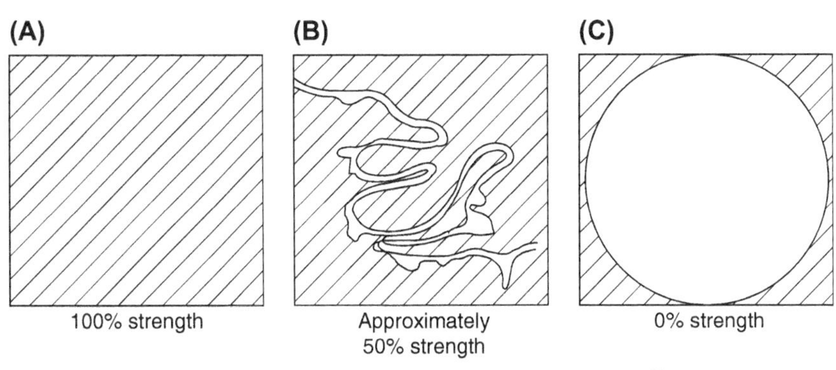

Figure 1.35 A rather obvious illustration (for which no apology is offered), showing the relative strengths of (A) matrix; (B) furled bifilm; and (C) inflated bifilm or pore.

unfurl easily. There is much interlocking of sound metal, so that much plastic deformation and shearing will be involved in the failure.

Because of the combination of a certain amount of strength together with their compact size, the convoluted bifilms are rendered as harmless as possible at this stage. As a result, the casting will exhibit tolerable strength.

However, once through the energetic turbulence of the filling system pipework, when the bifilm enters the much larger mold cavity, it finds itself in a new, quiescent environment. If there is any movement of liquid because of solute or thermally driven convection, such movements are smooth and gentle, occurring at values of Re now well below 2000.

Conditions are now right for the bifilm to unfurl. Starting from its crumpled form composed largely of flat, closed cracks, the stages of the unfurling process may include its gradual inflation. The flat cracks inflate to become lenticular, forcing the opening of neighboring regions. The opening, unfolding progressively, fold after fold, and effectively growing outwards in planar extent, returning to its original unfurled size. By this action the defect now re-establishes itself as a planar crack, akin to an "engineering" crack, and so can impair the strength and ductility of the metal to the maximum extent. The unfurling process can, of course, progress further if conditions are right, and can inflate to become a spherical pore (Fig. 1.36). This astonishing ability of the bifilm to open like a flower explains many of the problems of castings, as will become clear throughout the book.

The arrival of the bifilm in a convoluted condition explains how castings which are quickly solidified have high properties, compared to those cooled slowly, often in heavy sections, in which bifilms have had chance to unfurl, which therefore display relatively poor properties.

The short account above describes unfurling as a result of increasing internal pressure as would be commonly expected from the inward diffusion of a gas such as hydrogen. However, it is fascinating that as many as seven or eight potential mechanisms can be envisaged which separate the component halves of the bifilm and thereby drive the unfurling process. They are listed below. The evidence mainly comes from recent studies of aluminum alloys where the effects have been most thoroughly studied to date. However, of course, the general principles are expected to apply to all metals to some degree. They include two inflation mechanisms by gas or shrinkage; two straightening mechanisms by grains and dendrites; and two straightening mechanisms by intermetallics or second phases, and two total failure modes of hot tearing and in-service cracking as a result of applied strains.

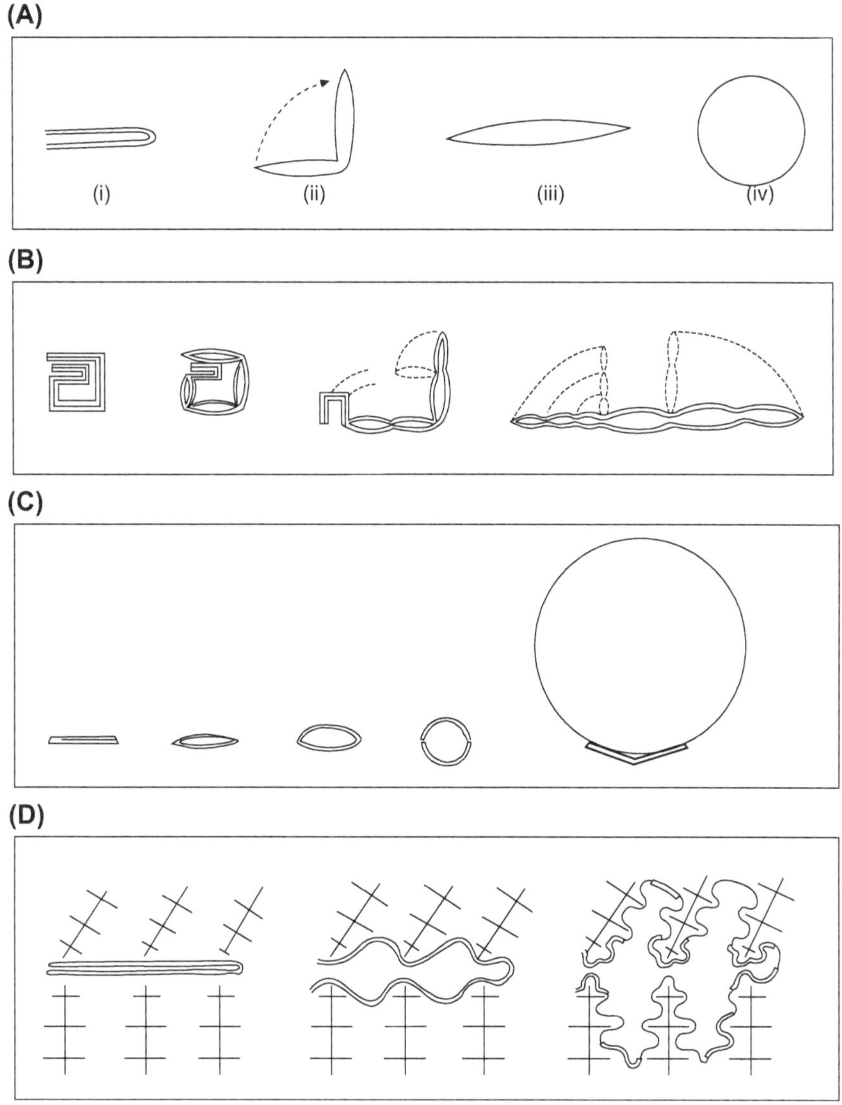

Figure 1.36 (A—D) stages in the unfurling and inflation of bifilms.

1.4.14.1 Inflation by gas

The precipitation of hydrogen in the gas film between the oxide interfaces, can inflate the defect. The inflation is likely to take place in two main stages. The first we may call micro-inflation, with short lengths inflating individually. These short lengths are defined and isolated from adjacent lengths by the crease lines of nearby folds. When sufficiently inflated, these short lengths will start to exert an unfolding pressure on adjacent lengths. As gas continues to inflate the bifilm, the bifilm will unfold progressively from one fold to the next, and the next, and so on. The unfurling action will therefore most probably be rather jerky and irregular, until the whole bifilm is unfolded. At this stage, of course, if enough hydrogen continues to precipitate, the planar bifilm crack will eventually inflate into a spherical pore. The stages of growth are illustrated in Fig. 1.36A for a simply folded bifilm. Fig. 1.36B illustrates the stages for a more convoluted bifilm.

If there is much gas in solution, or if there is plenty of time for diffusion, gathering gas from a wide surrounding region, the gas pore may eventually outgrow its original bifilm. Fig. 1.36C shows how it can unfurl to become a large spherical pore if growing freely in the liquid, or an interdendritic pore as in Fig. 1.36D if growing later during freezing. Note that in this situation the two very different morphologies (c) and (d) are both hydrogen gas pores. On sectioning either pore it is probable that the originating fragment of bifilm will never be found. It will probably be an insignificant oxide fragment lying out of the plane of section.

Interestingly, at the stage at which the pore starts to outgrow the boundaries of its originating bifilm, it will extend its own oxide film using its own residual oxygen gas. The new area of the defect will then consist of new, very thin film. When all the oxygen is consumed the films will grow by consumption of the residual nitrogen in the air as discovered by Raiszadeh and Griffiths (2008). When this is also entirely consumed, growth might occur by adding clean, virginal metal surface. The many round pores in cast metals indicate that the linear dimensions of many bifilms are probably smaller than the diameters of such pores. It is a reminder of the wide spread of bifilm sizes to be expected.

The formation of gas porosity in cast metals occurs on bifilms, and therefore avoids any nucleation problem: in most metals there seems no shortage of supplies of bifilms, some of which will open extremely easily and grow into gas pores; others will open less easily. The ease of formation of gas pores, but remaining subject to some variability, has long been a source of mystery to researchers. This mystery is taken up in detail later.

1.4.14.2 Expansion by solidification shrinkage (3-D strain)

Shrinkage acts to unfurl bifilms because it can reduce the pressure in a local region of the casting. It acts as a hydrostatic strain, which in a confined unfed region can lead to a hydrostatic stress. Thus, in this region it may suck the two halves of the bifilm apart. If this happened at an early stage of freezing the defect would grow no differently to a hydrogen pore, the inflation process proceeding from one fold to the next. Exactly this effect is seen in Fig. 1.14 in which small furled bifilms are seen to unfurl freely in the liquid, unhindered by dendrites, and therefore becoming significant cracks up to 15 mm long.

Returning to study Fig. 1.14 in more detail: the figure shows two reduced pressure test (RPT) samples of the same Al alloy cast at the same time from a single melt. The left hand sample solidified under atmospheric pressure, so its bifilms, furled by the sampling and pouring, remained furled. They are seen as faint mottled shadows in the radiograph. The right hand liquid sample is the same metal, subjected to a reduced pressure, unfurling the compacted bifilms into the clear, engineering cracks. The close density of the cracks should not be overlooked. They are so densely packed that they probably cannot float out, they mutually contact each other as a kind of lattice of cracks. It is not surprising that cracks can travel easily through a metal in which the bifilms have been allowed to fully unfurl. This condition simulates that which occurs in many castings, where feeding of the residual liquid through the solidifying mesh of dendrites is not easy, resulting in a reduced pressure; a suction effect to inflate bifilms.

If the shrinkage effect is severe, the final stage of unfurling would be followed by the bifilm eventually growing to be spherical, as seen in Fig. 1.36A and C. If the opening took place at a late stage of freezing the defect would attempt to open when surrounded by dendrites as shown in Fig. 1.36D. Both of these modes of opening could be driven by either gas or shrinkage, or both working in cooperation. Thus the rounded and interdendritic forms cannot be concluded to be either "gas" or "shrinkage." In reality, both forms can have both origins. The round form is a reliable indicator that the pore grew prior to the arrival of dendrites, while the dendritic pore is a reliable indicator that this pore was a late arrival. No more can be concluded.

In the case of the dendritic morphology, withdrawal of the residual liquid into the depths of the dendrite mesh (being either pushed by gas from inside the pore, or pulled by reduced pressure outside the pore) would stretch the film over tips of surrounding dendrites (Fig. 1.30C), or may be sucked completely out of sight into interdendritic depths (Fig. 1.36D).

1.4.14.3 Transgranular straightening (flattening) by dendrite growth

Oxides are pushed ahead of growing dendrites essentially because the dendrites are unable to grow through the "air layer" between the films. The result is the bifilms are automatically unraveled and flattened, effectively teased out and organized into planar areas among the dendrite arrays. The teasing action is all the more effective because the bifilms are usually attached to the skin of the casting, under the roots of the dendrites. Thus, as the dendrites grow, pushing ahead the heap of films, those that have fixed attachments near their roots are organized into flatness between the rows of dendrites (Figs. 1.37 and 1.38). In addition, oxides are pushed into

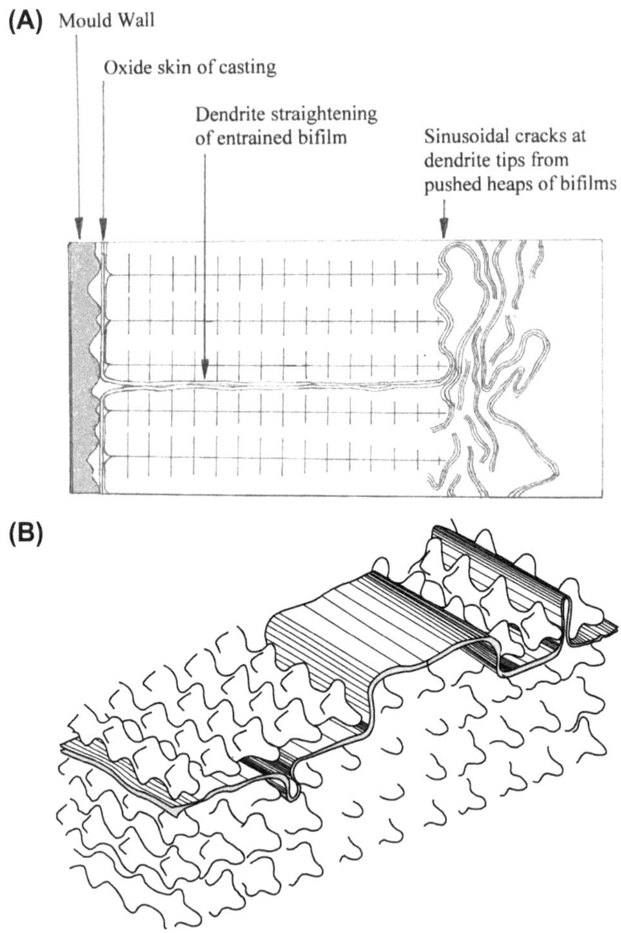

Figure 1.37 Dendrite flattening of bifilms (A) the pushing of bifilm heaps; (B) the fairly flattened bifilm.

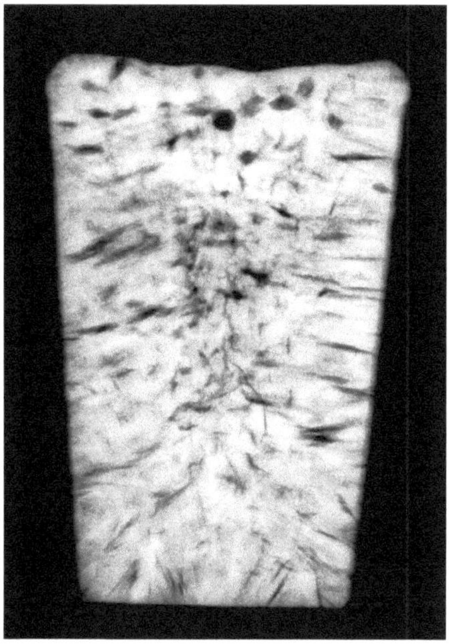

Figure 1.38 Radiograph of a reduced pressure test casting showing bifilm cracks flattened by columnar growth of dendrites from the side walls (Fox and Campbell, 2000).

interdendritic and grain boundary regions. This is an important mechanism since it can occur in most cast alloys, since most have surface oxide films which are easily entrained. The pushing action, sideways or ahead, results in large flat facets on fracture surfaces, or, occasionally, regularly undulating hilly regions of a fracture surface, resembling sine waves, each "hill" marking a locations of a dendrite tip (Fig. 1.37).

A room-temperature fracture surface is seen in Fig. 1.39 for an Al-4.5Cu alloy (Mi et al., 2002) in which the fracture surface is completely covered in thin alumina films, as "half" of the original double film, or bifilm.

The films have been pushed from "East" and "West" by the advance of dendrites, forming a central "North-South" ridge as a heap of excess film pushed into the middle of the fracture surface. Fig. 1.40 shows fracture surfaces of test bars taken from the same casting. The test bar failing by brittle fracture 1.40A was observed by video radiography to have suffered a major turbulent breaking wave during mold filling, whereas 1.41A was observed to fill smoothly, its fracture surface exhibiting only rather small bifilms from the population that existed in the original melt and which caused the scattered microporosity and ductile dimples. The ductile dimples indicated that there

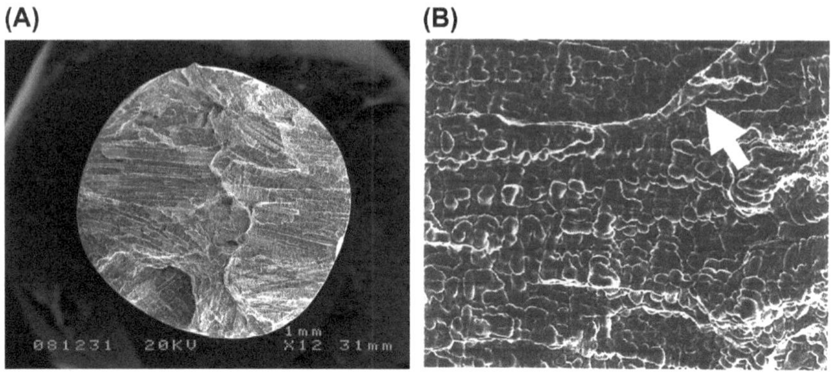

Figure 1.39 SEM of an Al alloy fracture surface whose apparently "brittle" failure was the result of bifilms from a turbulent wave extending over 100% of the surface. (A) Large bifilms pushed and straightened by dendrites growing from both right and left, resulting in excess heaps of bifilm as a central ridge; (B) A close-up revealing dendrites with no obvious oxide film overlay, but confirmed by local regions where the film is doubly folded (indicated by arrow) (Mi et al., 2002).

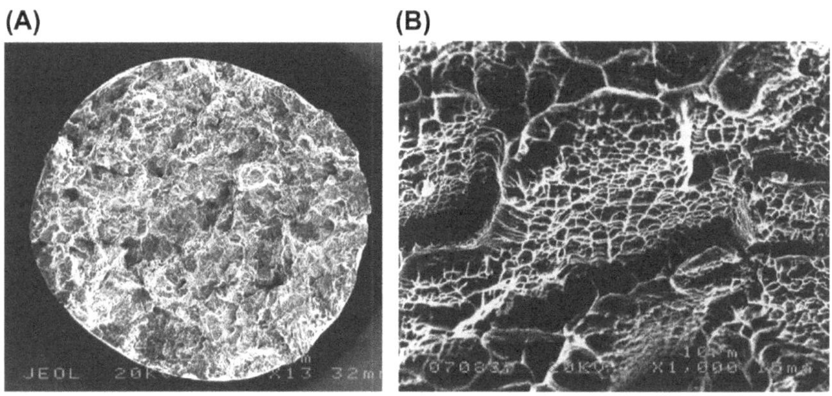

Figure 1.40 The same casting as shown in Fig. 1.39, but a region which had filled tranquilly; (A) A second fracture surface from a portion of the same casting which had suffered less turbulent filling, exhibiting ductile dimples, evidence of the population of small bifilms in the original poured metal. (B) Close up of dimples. (Mi et al., 2002).

had been real metal-to-metal contact between the two halves of the fracture surface, in complete contrast to surface 1.40B where not a single ductile dimple could be found anywhere on the fracture surface, indicating no metal-to-metal contact over its whole area. In fact, its surface had been entirely covered with bifilms and was therefore effectively completely pre-cracked. This explained its elongation to failure of only 0.3%, compared to 3% of specimen 1.41.

Close examination of 1.40B was required to reveal tiny clues that the dendritic surface was actually covered with an oxide film derived from the turbulent liquid surface, and which had constituted half of the original bifilm. It was not immediately evident that a film was present because, if it was present, it was so thin that the dendrite structure could be clearly seen through it. However, the arrow in Fig. 1.40B indicates a small area where the film had been folded over twice; this barely visible clue confirmed that an oxide film was present. In addition, this observation confirmed that the oxide film was not merely an oxide grown on the dendrites in the solid state (for instance, after fracture). Other samples from this research exhibited rather thicker oxide bifilms that were much easier to discern (Fig. 1.41). In fact, it seems in this image that a second oxide lies beneath the top fractured oxide, thus possibly originally constituting a *double* bifilm (i.e., originally four layers of oxide).

In all cases where large bifilms are found to have experienced dendrite straightening, the grain size always appears to be large. It seems reasonable to suppose that in a solidifying casting in which large bifilms are present, the natural action of convection would be to some extent suppressed. Thus, the normal eddies of high and low temperature metal destabilizing the growth of dendrites causing their arms to melt off and thereby promoting grain multiplication, would be prevented. The bifilm barriers to flow would be present across the complete casting section, in contrast to other regions of the casting which contained only confetti-like bifilms in suspension. The confetti populations would convect freely, and thus

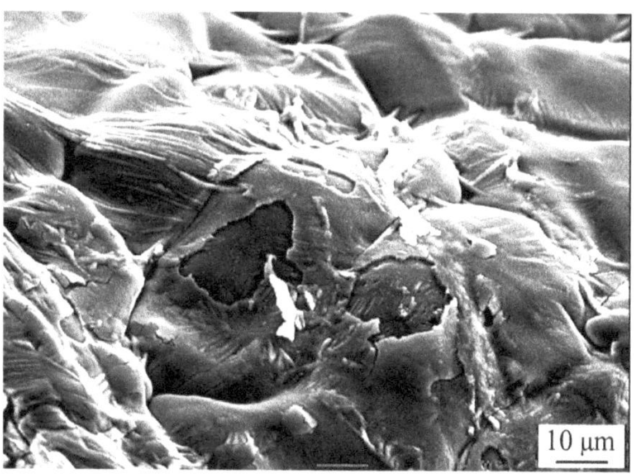

Figure 1.41 An area of the fracture seen in Fig. 1.39 in which the film is thicker, revealing a second film underneath (Mi et al., 2002).

naturally generate a refined grain size. For this reason the grain size in Fig. 1.39, with large bifilms, is large, whereas the grain size in Fig. 1.40, with small bifilms, is small.

Some spectacularly flattened transgranular areas in Al—7Si-0.3 Mg alloy are shown in Fig. 1.42. The first image (a) shows a facet so crystallographically smooth that it seems likely to have been formed by some variety of cleavage; the rugged steps across the facet might be the intersection of bifilms. The origin of this rather perfect facet remain unclear; cleavage would hardly be expected in an aluminum alloy. Its crystalline perfection might have been generated by a slip plane mechanism. Image (b) is different. There seems no doubt that this is a transgranularly flattened bifilm formed by the mechanism illustrated in Fig. 1.37. This conclusion is confirmed by the density of minute imperfections in the surface on the scale of the interdendritic spacing. In addition, careful examination in the region of the sand inclusion also confirmed that its entrained oxide film is contiguous with (i.e. effectively continously 'flows from') the film on the surface of the facet.

In Part 3 we shall consider further examples of planar defects, facets, created by dendritic unfurling, and revealed as integral features of fatigue initiation sites in metals.

1.4.14.4 Intergranular straightening by grains

During the growth of grains, bifilms are pushed ahead of the closely packed arrays of dendrites. Once again, of course, bifilms are pushed for the reason that the solidifying metal cannot grow through the "air gap" between the films of the bifilm. For the purist, even this explanation is a convenient

(A) (B)

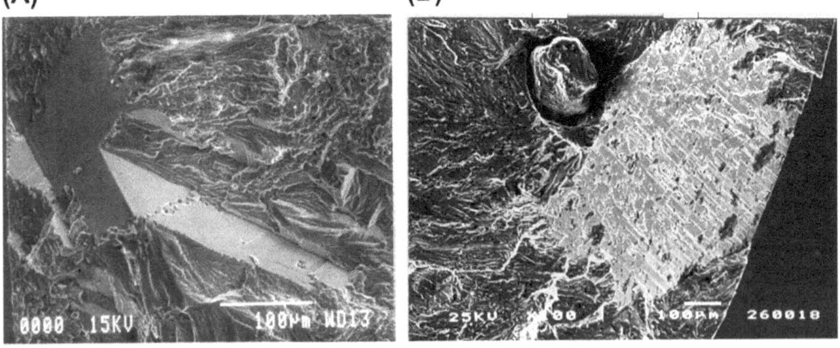

Figure 1.42 Facets on an Al alloy fatigue fracture surface generated by (A) slip planes or (B) a flattened bifilm, apparently using the same bifilm as was entrained by the sand inclusion.

shorthand for what appears to be happening. On a microscale, the advancing solidification front cannot "push"; it merely advances by the addition atoms to its tip. Because the presence of the oxide bifilm on the front, any advance now occurs by liquid finding its way by minute movements of the bifilm, allowing new liquid to arrive, seeping in the narrow gap between the dendrite and the film. Thus, minor mechanical movements of the bifilm are necessary to allow the front to advance, giving the impression that the bifilm is being "pushed." At times, however, it seems probable that the bifilm succeeds in wrapping itself so closely and completely around the growing dendrites that they are brought to a stop. The nucleation of new grains ahead of the bifilm now occurs in the liquid whose temperature will continue to fall. The mechanism is probably common in the growth of second grains during the growth of single crystal turbine blades.

Returning to more normal solidification conditions in which bifilms appear to be pushed ahead of growing grains, when a pushed bifilm is finally pushed up against a second grain, the bifilm finds itself trapped in a grain boundary. Grain boundaries therefore seem to be a rather common location for bifilms. These unbonded interfaces in the grain boundary naturally give the impression that grain boundaries are weak (whereas from atomic modeling it is known that grain boundaries have enormous strength, almost equal to that of the matrix).

In addition to this mechanical weakness, the other major feature of bifilms at grain boundaries is the tendency for the bifilm to be flat (which probably indicates a degree of unfurling based on shrinkage or gas). Planar grain boundaries on fracture surfaces are the result of the combination of planarity and the apparent weakness favoring the crack path.

In Fig. 1.43 an AlN or Al_2O_3 film is probably the cause of the planar boundaries seen in the fracture surface of a vacuum cast and HIPped Ni-base alloy IN939 (Cox et al., 2000). The castings which had suffered surface turbulence during filling had poor tensile ductility and were observed to exhibit planar boundary features on the fracture surfaces. Interestingly, Cox reported to me personally that the planar boundaries were *not* observed in control castings made without turbulence. Exasperatingly, this important observation was not reported in the published paper.

The unfurling of bifilms is also seen in so-called ductile cast irons, which causes this material to behave as brittlely as a gray cast iron (Fig. 1.44). The microstructure reveals the amazing alignment of small irregular graphite spheroids, and ferrite, both of which form most favorably on bifilms, highlighting their presence.

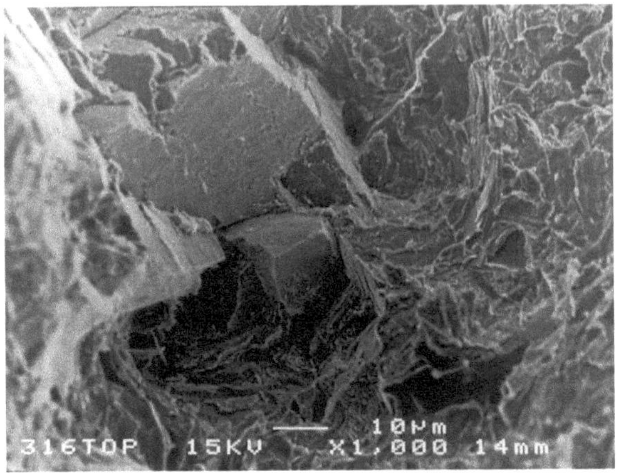

Figure 1.43 A fracture surface of a vacuum cast Ni alloy, poured turbulently, in which planar grain boundaries are caused to appear brittle.

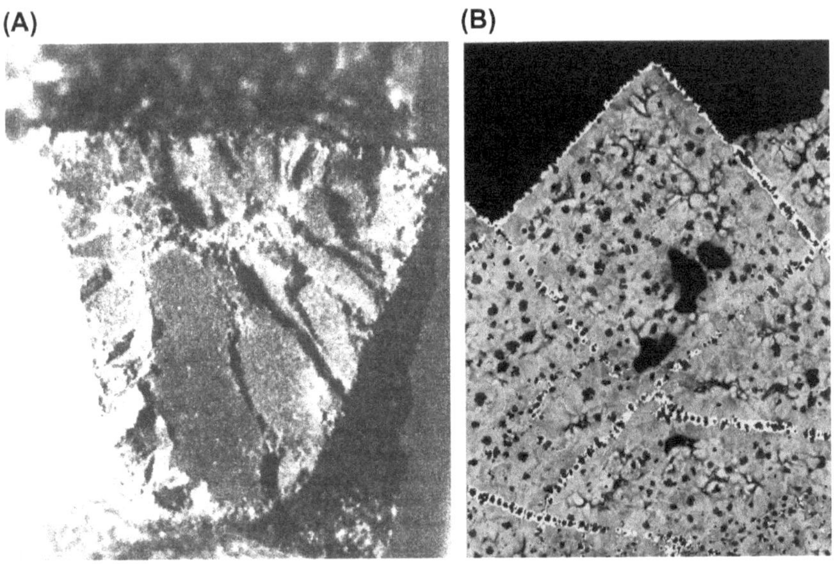

Figure 1.44 "Plate fracture" in the feeder neck of a ductile iron castings (Karsay, 1980; and (B) Polished microsection through the fracture (Barton, 1985).

The phenomenon of bifilms becoming coincident with grain boundaries is repeated in the solid state: during recrystallization and grain growth, if the advancing boundary meets a bifilm, it cannot cross the "air gap." Thus the advancing boundary is pinned in position on arrival at a bifilm.

In many if not most studies of metallurgical microstructures, some boundaries are seen to be decorated with inclusions or second phases whereas others are clear. We can assume with some confidence that the presence of bifilms is highlighted by the presence of inclusions or second phases. In the solid state the precipitation of new phases is almost always accompanied by a volume and shape change, which, of course, can most easily occur at a bifilm because its central void greatly reduces the strain energy for formation. Thus, bifilms are highly favored substrates for nearly all new phases. The obverse of this conclusion is that, perhaps surprisingly, grain boundaries are not especially favored sites for the formation of new phases (only boundaries containing bifilms are favored). It would be valuable to test this prediction.

In addition, the inclusions or second phases at grain boundaries often appear to be cracked, giving the impression that they are brittle. The cracks are, of course, merely the presence of bifilms upon which the phases grew — further evidence for the presence of the bifilms. It is known from independent evidence that many such phases are strong and crack resistant, such as carbides which are used for machining hard steels. Even more than this, most phases such as carbides and borides are actually ductile, deforming significantly under stress without the formation of cracks. Other supporting evidence for this perhaps surprising behavior will be described in due course.

Clearly, the driving forces for the unfurling of bifilms are all those factors that are already known to the casting metallurgist as precisely those factors that impair ductility; including hydrogen in solution, shrinkage, iron contamination in Al alloys (because the beta-Fe particle unfurls bifilms), and large grain size. Significantly, the only common factor implicated in the loss of ductility is the presence of bifilms.

However, in practice, we need to bear in mind that it will take time for the unfurling processes to occur. In a casting that is frozen quickly, the bifilms are frozen into the casting in their compact form, with the result that most casting alloys that are chilled rapidly are strong and ductile, as all foundry engineers know.

Conversely, castings that have suffered lengthy solidification times exhibit "shrinkage" porosity problems, show higher levels of gas porosity, or large grain size. In addition, all exhibit reduced properties, particularly reduced ductility. The action of slow freezing impairing ductility is, once again, well known to the casting metallurgist. All this is the result of the time-dependent unfurling of bifilms to convert them from harmless defects when fast frozen into major cracks in slowly solidified metals. The rate of unfurling is dealt with in more detail later.

In summary, it is clear that all the factors that refine dendrite arm spacing (DAS), especially the secondary arm spacing SDAS, improve the mechanical properties of cast aluminum alloys (Fig. 1.45). This is almost certainly not primarily the result of action of the SDAS itself, as has been commonly supposed. The SDAS is merely the measure of the local freezing time, which is the time available for the inflation and unfurling of defects. If the growth of area of the defect was not bad enough, its inflation, causing its surfaces to separate, transforming it from a crack that can enjoy some shear strength into a pore which has none, makes the defect even more damaging. It is both (i) the growth of area and (ii) the growth of volume of the defects that combine to inflict so much damage to the mechanical properties as solidification time increases. Increases in SDAS are not the cause of the loss of ductility; the SDAS is merely an indicator of the time involved. It is the built-in, independent clock (Campbell, 2015).

1.4.14.5 Straightening (flattening) by intermetallics and second phases

Intermetallics and second phases tend to favor nucleation and growth on the outer, wetted surfaces of the bifilm. This appears to happen in metals and alloys of all types.

In Al–Si alloys, both Si platelets and Fe-rich, Ti-rich and Sr-rich intermetallics are all known to grow on bifilms. The βFe particles, Al_5FeSi, are particularly good examples. Initially, when the βFe crystals are no more than a few nanometers thick, the crystals can follow the curvature and creases of the crumpled double film as seen in Fig. 1.46. However, as the crystals grow in thickness, they quickly develop rigidity (the rigidity of a beam in bending mode is proportional to its thickness to the third power). Thus, as the precipitates thicken and rigidize, they *force* the straightening of the bifilm to which they are attached.

The result is the familiar βFe plate seen on polished sections, straight as a needle (Fig. 1.47). (Occasional curved βFe plates are probably the result of restraint at the two ends of the plate, so that the plate has been unable to straighten fully, remaining stressed like an archer's bow.) The βFe plates sometimes exhibit a crack along their centerline (if βFe has precipitated on both sides of the bifilm) or between the βFe and the matrix, showing apparent decohesion from the matrix (if the βFe has precipitated only on

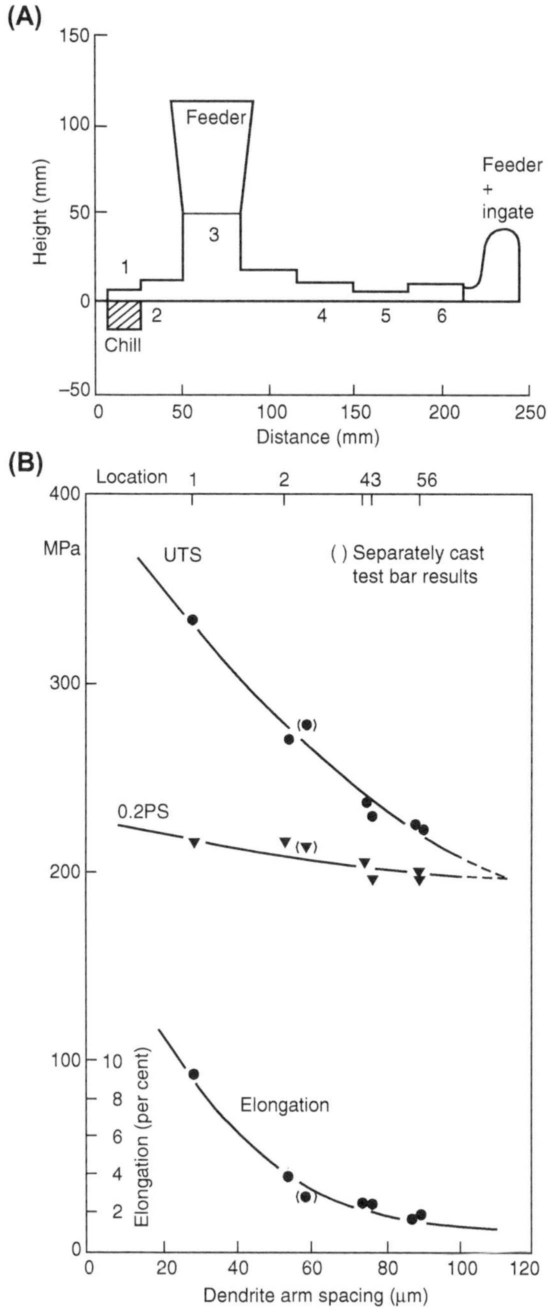

Figure 1.45 (A) Al–7Si-0.4 Mg alloy test casting and (B) its mechanical properties (Miguelucci, 1985).

(A) (B)

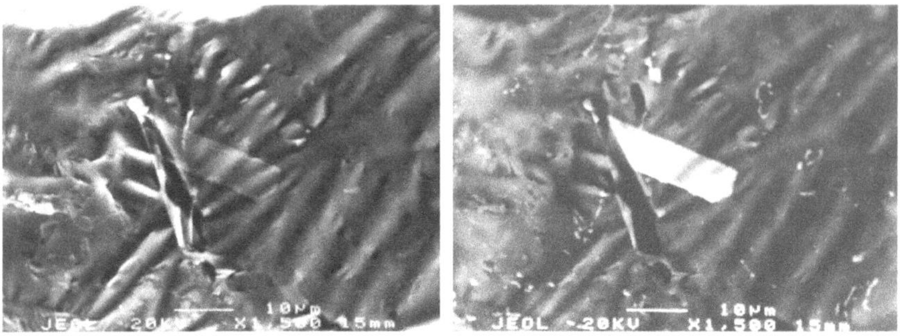

Figure 1.46 SEM image of a fracture surface showing an Fe-rich particle nucleated on the underside, and seen through the thin bifilm half-film. (A) secondary electron image; (B) back-scattered electron image.

(A) (B)

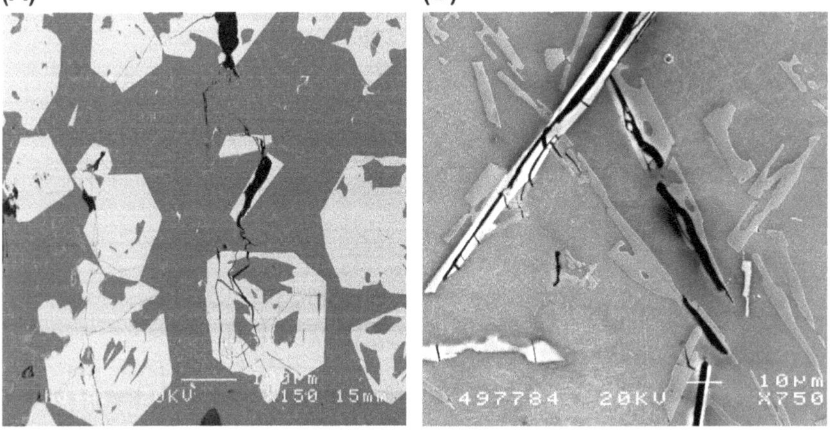

Figure 1.47 In an Al–Si alloy, (A) alpha-Fe particles with cubic symmetry growing on and wrapping around bifilms in their furled condition. (B) A monoclinic beta-Fe particle having rigidly straightened its bifilm substrate resulting in characteristic center crack and transverse cracks only on one side. Si particles are primitive cubic, so less good at straightening, but also show their bifilm substrates as central and transverse cracks.

one side of the bifilm). It seems likely that all βFe particles are cracked or decohered in this way because of the presence of the originating bifilm.

The cracks in second phase particles may not always be visible on a polished section, especially if the casting is well fed, pressurizing the two halves of the crack to keep it closed. Keeping up the pressure on the bifilms during freezing will ensure the bifilm remains closed and thus becoming an effectively invisible crack. Moderately poor feeding, or some hydrogen in solution, will part the films, allowing them to be seen on a polished metallurgical specimen. Research on this topic requires, paradoxically, to be carried out on poorly fed sample castings to reveal the bifilm crack, in contrast to foundry personnel who require their castings to appear as sound as possible.

The minor transverse cracks only on one side of the beta-Fe particles seen in Fig. 1.47 and the TCP phase in Fig. 1.48 are easily understood in terms of the formation mechanism for the bifilm. As the two oxide films impinge to create a bifilm one film is always larger than the other. The larger film necessarily therefore develop folds and pleats as it closes on and accommodates itself to its new neighbor, as seen in Fig. 1.9A. On the back of the smaller film the new phase grows without hindrance, forming a long, uninterrupted straight precipitate. The other side grows, but its progress is interrupted by the transverse folds and pleats which form short transverse cracks on one side of the main crack only. This interesting asymmetry in the structure of second phases is a powerful corroboration of the accuracy of the bifilm concept.

A further corroboration lies with the action of the alpha-Fe phase shown in Fig. 1.47A. Alpha-Fe has a cubic lattice, and so has sufficiently high symmetry that it can grow in practically any direction. This contrasts with the beta-Fe whose monoclinic lattice forces its growth as thin planar crystals. The alpha-Fe still nucleates on oxide bifilms as favored substrates, but then simply grows all over it, paying no attention to any favored growth direction: all growth directions are favored. In this way the alpha-Fe particle wraps around the bifilm while it is still in its furled state, and seals-in the furled, compact form, as seen on Fig. 1.47A.

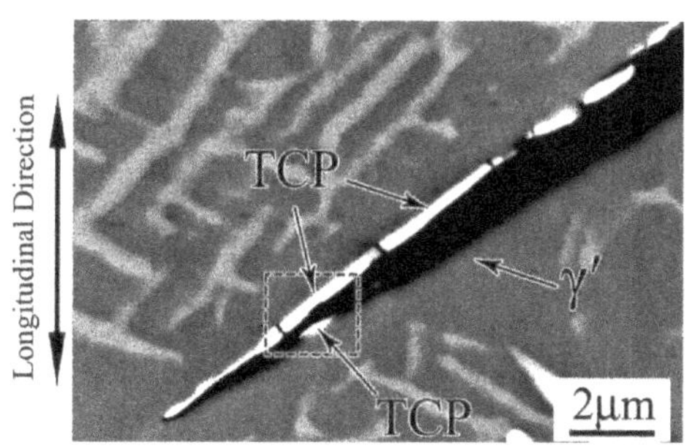

Figure 1.48 A topologically close packed (TCP) phase in a Ni superalloy formed largely on one side of a bifilm, as though decohered, the other side showing transverse cracks. *Courtesy Professor Qiang Feng (Shi et al., 2014).*

Metallurgists know that if the Fe content of Al–Si alloys is high, the formation of beta-Fe particles can be catastrophic to the ductility of the alloy. It seems now clear that the embrittlement of the alloy is the result of the bifilms being straightened to act as "engineering" cracks. However, the addition of Mn to promote the formation of alpha-Fe significantly helps to restore the ductility because the bifilms are prevented from straightening; they are sealed in their compact, furled condition.

Turning now to evidence of bifilm straightening in vacuum-cast Ni superalloys, Fig. 1.49 shows a small, gently curved, nearly planar oxide on a fracture surface. The oxide bifilm has, in turn, formed the substrate for the precipitation of a carbide.

Finally, in this section, it is worth noting one of the major examples of the straightening of bifilms is proposed to be flake cast iron (Campbell, 2009). The bifilms are straightened by being the substrates for the growth of graphite crystals which have a planar morphology. The graphite appears to form on silicate (silica oxide, SiO_2) bifilms, and during its growth,

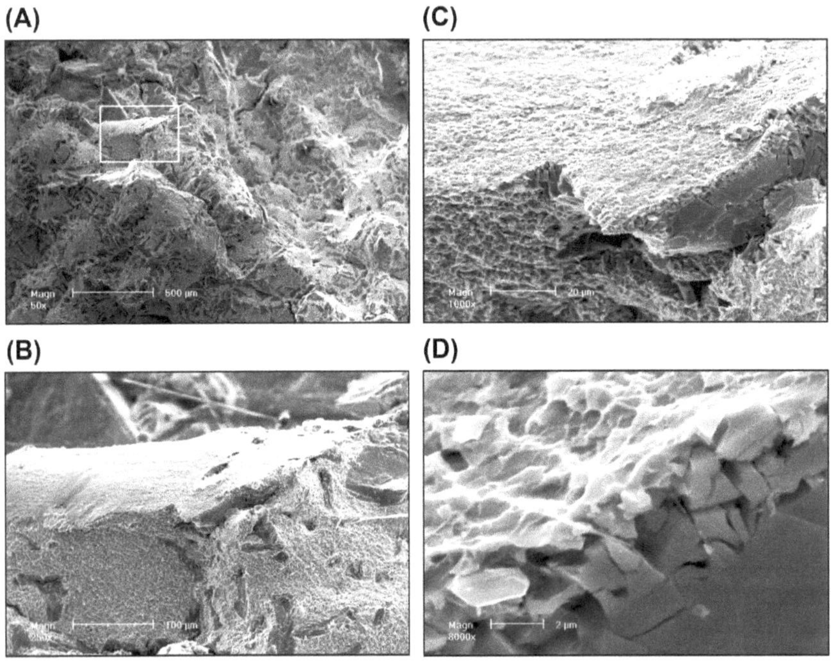

Figure 1.49 SEM images (A) to (D) show successively closer views of the fracture surface of a Ni superalloy melted and cast in vacuum. (A) The light gray oxide bifilm occupies most of the image. The selected area of oxide (B) resembles a glacier, is only about 1 μm thick but has a 20 μm deep underlay of carbides. At its high temperature of formation, the oxide film has recrystallized.

effectively straightens out the bifilms, forming the familiar graphite flakes. The confirmation of this proposal had to wait until the high resolution electron microscopy by Boisvert et al. in 2016.

1.4.14.6 Flattening by rigid flat intermetallic or second phase

Although flattening of the bifilm could, in principle, be encouraged if it were attracted to a pre-existing, flat, rigid substrate, it is not clear whether this ever happens. If such a rigid favored substrate were available in the melt, and if it were encountered by a bifilm, it is likely that the bifilm would attach itself, plastering itself against the surface. But it would be expected to do so in a messy "heap," effectively without unfurling. The so-called "heap" would still be microscopically thin, in places having multiple thicknesses of bifilms.

Pre-existing rigid substrates can be envisaged at temperatures well above the solidification temperature; they could consist of old, thick, oxide films. Such films, on being originally submerged, would have gain a wrapping of new, thin oxide film which might represent a favored substrate for adhering bifilms. In a way, such agglomeration of bifilms would represent a kind of deactivation, gradually reducing the number of bifilms in suspension.

1.4.14.7 Opening by 1-D cooling strain

As lengths of a cast product cool, parts of the casting can experience thermal contraction strains which can encourage the opening of bifilms, particularly those parts of the bifilm at right angles to the direction of strain. In this way hot tears and cold cracks ("cold" being used in the sense that the metal is below its freezing temperature, but still probably rather hot!) can form.

Hot tears are some of the most serious defects in metals. Fortunately, in the experience of the author, all hot tears can be instantly eliminated in all metals and alloys by the reduction of bifilm content. This is usually easy and quick to arrange, simply by improving the filling system design for the casting process. The huge literature on hot tearing and hot tearing tests has yet to come to terms with this fundamental cause. For instance, all current hot tearing tests would never display a hot tear if their filling system designs were improved in line with current best practice. It is hoped that good casting practice will eventually infiltrate research laboratories.

More will be discussed on hot tearing in Part II.

1.4.14.8 Opening under service stress

The opening of the bifilm in the solid state during the in-service life of the component is not, of course, an unfurling phenomenon, and does not really belong in this list. Regrettably, however, the bifilm which survives to undergo service under conditions of varying stress will usually extend as a fatigue crack until failure. It is the reality which we must not forget. Failure initiated by bifilms is all too frequent.

1.4.14.8.1 Rate of unfurling

We can gain some idea of the rate of unfurling of bifilms by gas or shrinkage from a simple mechanical model as illustrated in Fig. 1.50.

We shall assume that the unfolding of the bifilm is resisted by a force F of the same type as that resisting the motion of a sphere in a viscous liquid (as in the derivation of Stoke's law). Thus, the force would be $3\pi\eta RV$ if it were evenly distributed over the large square area $R \times R$. Since the velocity V is that at the tip of the bifilm, and the pivot of the bifilm is fixed, on average the resisting force is halved to $3\pi\eta RV/2$. The opening force is that due to the pressure P in the gas phase of the bifilm, acting over the very small area hR. Equating the small opening moment to the large viscous restraining moment (which is a function of opening speed) we have

$$2\mathrm{PRh}.(\mathrm{h}/2) = 3\pi\eta\mathrm{RV}.(\mathrm{R}/2)$$

so that we can find the opening time t from the speed V and the distance traveled πR:

$$\begin{aligned} t &= \left(3\pi^2/2\right)(\eta/\mathrm{P})(R/\mathrm{h})^2 \\ &= 15(\eta/\mathrm{P})(\mathrm{R}/\mathrm{h})^2 \end{aligned} \tag{1.10}$$

For viscosity $\eta = 1.4 \times 10^{-3}$ Nsm^{-2}, and reasonable figures for P of about 0.2 atm (0.2×10^5 Pa) above ambient (many studies of pore formation in metals have found pores appear to initiate when the internal gas pressure becomes approximately 1.2 atm), and for $R = 5$ mm and

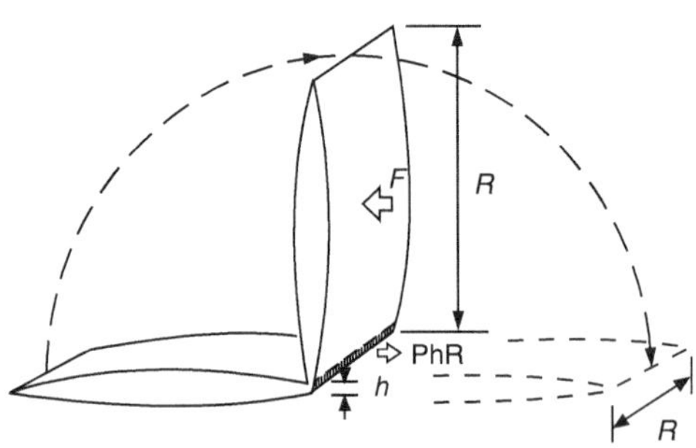

Figure 1.50 Model of the unfurling of a bifilm.

$h = 5\ \mu m$, t is only approximately 1 s. Whereas if $R = 10$ mm and $h = 1\ \mu m$ t is approximately 2 min. The rate of opening is therefore seen to be highly dependent on the geometry of the bifilm, as might be expected. Nevertheless, the rates are of the correct order of magnitude to explain the rate of loss of properties in most ordinary castings as the freezing time increases.

This wide scatter in the performance of bifilms, some unfurling easily and quickly, and others being resistant to unfurling, supports the interpretation of variable nucleation conditions surmised as a result of the sub-surface porosity described below.

1.4.14.8.2 Variations in unfurling behavior

Returning to the pivotal role of the entrainment defects, it seems clear from practical work in foundries that the highest velocities and maximum turbulence is in the sprue, runner and gates, where Reynolds Numbers of 10^4 to 10^5 are easily attained, indicating severe bulk turbulence. Bifilms entrained in these parts of the running system will therefore be highly compacted, and, at least initially, have reduced damaging consequences for mechanical properties.

The less severe surface turbulence and bulk turbulence in the mold cavity are, paradoxically, a concern. Reynolds Numbers in a nicely filled mold cavity fall to values in the region of 10^3, so that little energy is available to make the films more compact. It follows that any bifilm entrained in the mold cavity itself will suffer minimal furling and will therefore remain substantially open, nearly fully unfurled, maximizing its area. This will cause maximum damaging effect. Additional damaging conditions are ensured by the presence of the entrapped gas in the bifilm. In Al alloys the oxygen and nitrogen are consumed by thickening the walls of the bifilm over a period of a minute or so (Griffiths and Raiszadeh, 2007), the time lengthened on occasion by hydrogen diffusion, inflating the defect, and so further reducing any limited load carrying capacity it may once have enjoyed. It follows that elimination of bifilm-creating conditions in the mold cavity itself is the most important factor to yield reliable properties.

It is easily imagined that a tightly furled bifilm would be reluctant to unfurl compared to a lightly furled bifilm, and there is graphic evidence for large differences in behavior from exactly such a source. Fig. 1.51 shows subsurface porosity which has opened up around a submerged sand core in an Al alloy. The core was bonded with a phenolic urethane resin, which has been thermally broken down at the temperature of liquid aluminum alloy,

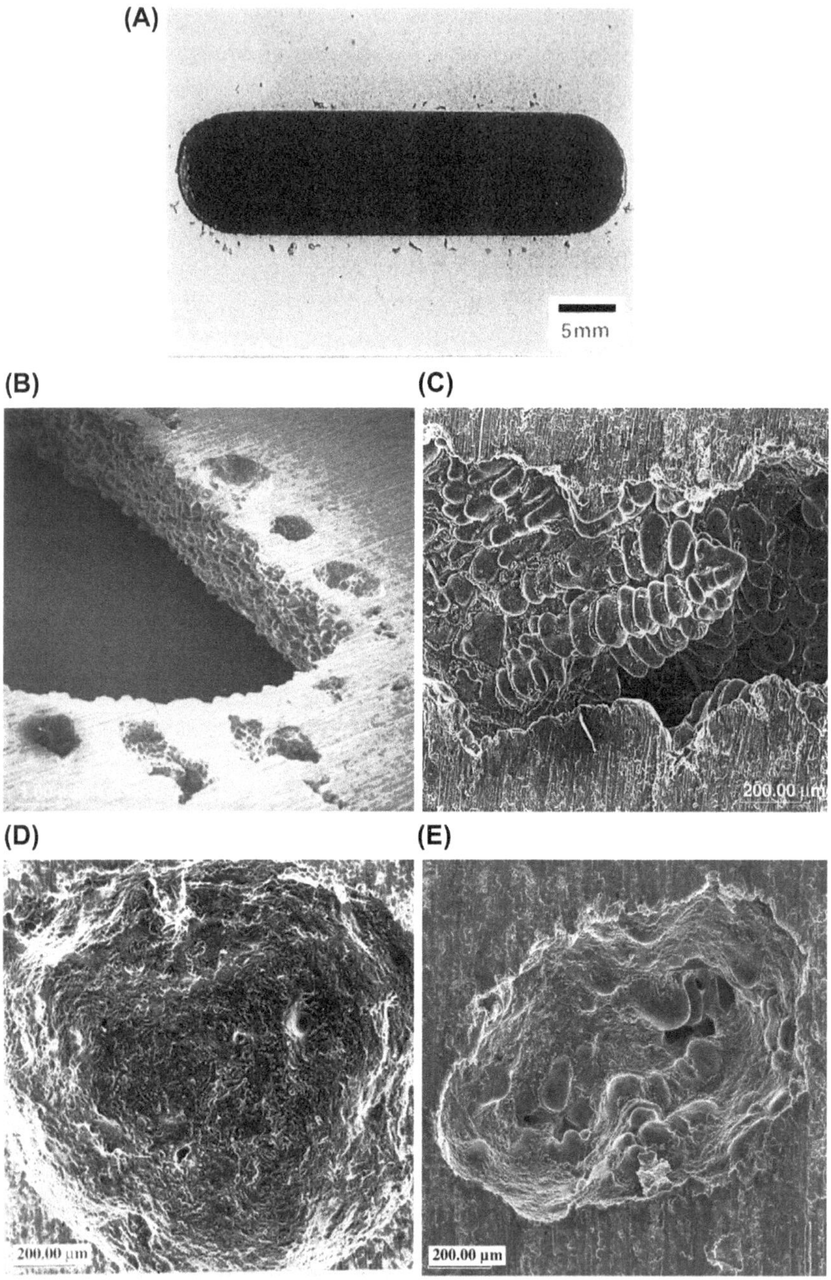

Figure 1.51 (A) A polished section, lightly grit blasted with fine alumina grit, showing subsurface porosity around a sand core in an Al–7Si-0.4 Mg alloy; (B) close up of a thin slice of the cast metal. Closer views of different pores (C) dendritic; (D) round; and (E) mixed.

releasing hydrogen into the surrounding melt. Clearly the flux of hydrogen from the core surface will be expected to diffuse uniformly into the liquid metal. So, all the bifilms will now attempt to unfurl while immersed in the same hydrogen environment. But the pore morphologies are not the same. They vary from spherical to dendritic and all mixtures in between. It is clear that the round pores have formed easily and early, prior to any solidification of dendrites. Their perfect sphericity contrasts with the dendritic pores which must have formed much later, despite the expected uniformity of the hydrogen environment, as a result of some mechanical difficulty to unfurl. When, finally, they have unfurled sufficiently to expand, their growth now has to take place among the already-formed dendrites.

Clearly the interpretation of pore shapes, if round as "gas" and if dendritic as "shrinkage," is quite wrong. In this case both types of pores are clearly hydrogen pores. The two different shapes merely indicate their formation "early" or "late."

It is almost certain that in some circumstance bifilms will be unable to unfurl despite the action of some or all of the above driving forces. This is because they will sometimes be glued shut. Such glues will include partially melted oxides in high temperature liquids such as steels, and liquid chloride fluxes in light metals such as aluminum.

Table 1.4 provides a summary of the logic behind the control of cast defects and cast structure by bifilms. This formal layout illustrates the

Table 1.4 Bifilm control of defect formation and cast microstructure.

	Pouring surface turbulence We > 1 bulk turbulence Re > 2000	**Quiescence We<<1 Re < 1000**	
Example melt	**Entrainment and furling**	**Flotation and unfurling**	**Casting defect state**
Pure liquid gold, Pt, Ir, etc	Film-free surface	Rapid flotation of film-free bubbles	No defects
Al + Ca deoxidized steels B steels and B Ni base alloys high Mn steels	Liquid surface film	Rapid flotation of liquid droplets	Few defects (only small spherical inclusions)

(Continued)

Table 1.4 Bifilm control of defect formation and cast microstructure.—cont'd

	Pouring surface turbulence We > 1 bulk turbulence Re > 2000	Quiescence We<<1 Re < 1000	
Example melt	**Entrainment and furling**	**Flotation and unfurling**	**Casting defect state**
General carbon steels Al alloys + flux	Partially liquid film	Rapid flotation of large sticky bifilms, effectively welded permanently into compact form.	Complex macroinclusions on cope surface
Al alloys Mg alloys Al bronzes Ductile irons	Solid surface film	Rapid flotation of buoyant bubbles	Air bubbles ("gas") Bubble trails (potential leakage defects)
Ni-base superalloys Stainless steels		Suspended furled bifilms with simultaneous unfurling + inflation driven by	General leakage defects.
		1.Gas precipitation	Gas microporosity
		2. Hydrostatic (3D) tensile strain	Shrinkage porosity (macro and micro)
		3. Uniaxial (1D) tensile strain	Hot tears and cold cracks
		4. Intermetallic and second phase crystal growth (unfurling only)	Centerline cracks in particles Matrix decohesion from particles
		5. Dendrite pushing (unfurling only)	Interdendritic and transgranular cracks
		6. Grain pushing (unfurling only)	Grain boundary cracks
Al–Si alloys	Alumina film	Precipitation and unfurl by Si	Alloy with primary (coarse) Si platelets
Lamella cast iron	Silica film	Precipitation and unfurl by graphite	Type A Lamellar graphite cast iron

phenomena for metals (i) without films, such as liquid gold, (ii) with films that are liquid, (iii) with films that are partially solid, and (iv) with films that are fully solid.

Note that the defects on the right of Table 1.4 cannot, in general, be generated without starting from the bifilm defect on the left. The necessity for the bifilm initiator follows from the near impossibility of generating volume defects by other mechanisms in liquid metals. In contrast, of course, the bifilm is seen to possess the potential to initiate cracks and pores with negligible difficulty.

In recapitulation, the bifilm simply opens by the separation of its unbonded halves. Surface tension is not involved (as is usually assumed when nucleating a pore in a liquid) since no liquid is involved. Clearly, the opening of a bifilm is simply a mechanical action between parallel films of (effectively) vanishingly thin oxide separated by a vanishingly thin layer of gas. The opening of a population of bifilms to create porosity in metals occurs with minimal difficulty by the inflation of the bifilm from a closed form, via a lens shape, to fully spherical forms if able to expand freely into the liquid (Fig. 1.36). If solidification is well advanced, the swelling of the pore will then occur between dendrites as illustrated above in Fig. 1.51. These pore initiation and growth mechanisms clearly avoid the nucleation problem which has puzzled researchers for years.

In their book, Dantzig and Rappaz (2009), propose a possible mechanism for pore formation by a bifilm, hinging open at one end, as a crocodile opening its jaws as a result of the expansion of gas in the bifilm. As the bubble reaches a critical size it can detach and float clear, to be followed by the continued expansion of the gas and the detachment of the next bubble and so on. These authors describe this as a kind of nucleation process, which it is. Dispinar and I (2005), using our video X-ray unit, have once watched such a crocodile in an Al alloy. It released bubbles, which we assumed to be hydrogen bubbles, at a rate of one every few seconds as its jaws opened and closed. However, we have observed this type of bubble creation only once. In contrast, we have seen the progressive swelling of bifilms with gas and shrinkage hundreds of times. The avoidance of any kind of nucleation mechanism during porosity formation seems the more general behavior.

1.4.15 Detrainment

The detrainment of bifilms and bubbles from metals can only occur rapidly and easily in the liquid state. The various sedimentation or flotation processes can be relatively rapid, usually occurring to within a tolerable effectiveness within minutes. Detrainment of defects is a huge benefit for the liquid processing route for metals when carried out effectively.

In contrast, of course, the various solid state consolidation routes have no chance to detrain defects.

Powder metallurgy is a typical solid processing example. Bifilms and bubbles entrapped in powder metallurgy products are practically permanent. The application of high pressure will reduce bubble sizes and encourage gases into solution, some of which may escape to the environment by diffusion along inter-particle channels. However, the bifilms trapped between the original powder particles are generally there to stay. Only significant plastic work will be able to effect some bonding across the original double oxide layer and its "air gap." Even so, the original air may have reacted away its oxygen and nitrogen to form oxides or nitrides or be taken into solution in the metal. But the remaining 1% argon gas in the air is insoluble and will therefore remain as a permanent feature. Such residual argon bubbles in metals will always have the potential to initiate volume defects such as pores and cracks. The remnant of bifilms and possibly micro-bubbles is clearly demonstrated in the fineness and density of ductile dimples on a tensile fracture surface of a powder metallurgy product. Each ductile dimple is probably the result of each oxide bifilm formed at each powder contact.

Other consolidation routes such as metal spraying, diffusion bonding, roll bonding and the many varieties of additive manufacture, are all characterized by the presence of oxide bifilms between the additive particles. The bifilms are essentially permanent if they cannot be taken in solution or at least liquefied in some way.

Now moving on to liquid consolidation processes: having briefly looked over the fundamental problems facing solid state consolidation processes, the much more active and positive detrainment processes in liquids constitute a major potential benefit.

However, of course, it has to be acknowledged that in the liquid state the potential for detrainment can be easily undermined by the similar potential for entrainment. This has been the bête noir of casting processes since the beginning of casting. Fortunately, consolidation via liquid metal processes can now understood in terms of the formation of entrainment defects, allowing nonentraining processes to be devised with accuracy for the first time.

Nevertheless, it is salutary that the new nonentraining processes are unable to produce defect-free products if the masses of already entrained defects from prior melting of the metals (much of which is, of course, recycled) are not first detrained. Thus, detrainment must be an essential ingredient of every melting and casting stage of the total liquid process route.

1.4.15.1 Detrainment techniques take a variety of forms

If the oxide surface film is particularly strong, it is possible that even if entrained by a folding action of the surface, the folded film may not be free to be carried off by the flow. It may be firmly attached to some piece of hardware such as the sides of the pouring basin, or the wall of a sprue. Thus the entrained film might be effectively detrained, being held back from joining the general flow, effectively never having been properly entrained. A strong film, strongly attached to the wall of the mold may simply remain hanging in place, flapping in the flow of the melt delivery system, but fortunately never entering the casting, and so remaining harmless.

Such a mechanism appeared at one time to be important in the aluminum industry. Beryllium added at levels of only 0.005% to reduce oxidation losses on Al–Mg alloy probably acted to strengthen the oxide film and so reduce its entrainment during pouring. On attempting to eliminate Be for environmental and health reasons, difficulties have been found in the successful production of wrought Al–Mg alloys by continuous casting (Direct Chill casting, usually abbreviated to DC casting). Instead of hanging up in the supply channels, the surface oxides entered the cast product and greatly increased the incidence of cracking.

Large bubbles in cast metals can detrain easily and quickly. They have sufficient buoyancy to break their own oxide film, and the casting surface oxide skin (once again, the two films constituting a double oxide barrier; the double film phenomenon appearing to be a fundamental and recurring film condition in casting technology) between them and the outside world at the top of the casting. They can thus detrain. But, as we now know, this detrainment is not without trace, because of the legacy of the bubble trail. The bubble trail remains as a serious impairment of the casting.

Small bubbles of up to about 5 mm diameter have more difficulty to detrain. They are commonly trapped immediately under the top skin of the casting, having insufficient buoyancy to break through the double film barrier. They are the bane of the life of the machinist of Al alloy castings, since they lodge just under the oxide skin at the top of a casting and become visible only after the first cut. In iron and steel castings requiring heat treatment, this thin surface oxidizes away to reveal the sub-surface porosity. Similarly, shot blasting will also often reveal such defects.

The most complete and satisfactory detrainment is achieved if the liquid metal has a liquid oxide film. This is because, once entrained, the liquid oxide spherodises to form droplets, and floats out with maximum speed (Fig. 1.34). This important mechanism is dealt with at length in Section 1.4.9 "Liquid Oxide Entrainment."

Detrainment of solid bifilms in suspension in the melt can be encouraged by a number of means:

(i) Filtration. The wide use of filters indicates there are benefits, even though the effectiveness of filters is extremely variable in different situations. The problems are discussed in some detail in the author's Mini Casting Handbook (2018).

(ii) Flotation is important for all the dense engineering metals. It is a natural benefit enjoyed by steel makers in the steel making furnace, where the steel making process takes sufficiently long to encourage a significant cleaning action. Unfortunately, the metal is subsequently ruined by pouring through a fall of several meters into a ladle. Subsequently, thanks to the large density difference between the steel and its oxides, the quality is rescued to some extent by detrainment of many of the entrained inclusions by a fortunate second period of flotation. This detrainment can be so rapid that much entrained material has floated out to the surface of the ladle, adding to the surface slag, during the several minutes that are required for the ladle to be lifted out of the casting pit and transported to the casting station. Unfortunately, the metal is usually greatly damaged for a final time during the pouring to make the casting or ingot. This final degradation can, of course, be avoided by careful design of the pouring system. Such care, although rare at this time, may become more common as more rational casting techniques become more widely known (Campbell, 2018)

The seminal work of Mountford and Calvert mentioned above is worth detailed study. This critical observation of the action of bifilms was so many years ahead of its time that it remained unappreciated until recently. In 1959 at Rolls-Royce, UK, these researchers observed the echoes from ultrasonic waves that they directed into liquid aluminum alloys held in a crucible. What appeared to be an entrapped layer of air in floating debris was observed as a mirror-like reflection of ultra-sound. (Reflections from other fully wetted solid phases would not have been so clear; only a discontinuity like a crack containing a layer of air through which the elastic waves could not penetrate but were completely reflected, could have yielded such strong echoes.) Some larger particles could be seen to rotate, flashing reflections like a beacon when turning face-on after each revolution.

When the temperature of the melt was lowered, iron-rich and Ti-rich compounds appeared to precipitate on the debris, which were discovered to be oxides, causing them to be heavy, and thus falling to the bottom of the crucible where the build-up of sediment could be observed as a second back-wall reflection in front of the original strong back-wall echo. If the melt was stirred, it became opaque with a fog of reflecting particles. However, after a period of 10–20 min the liquid cleared again, with the sediment re-forming a layer seen as a double back-wall echo at the base of the crucible. This process could be repeated.

If the melts were poured after vigorous stirring, they were found to give castings containing oxide debris together with associated porosity. The macroscopic pores observed on their polished sections appear to have grown from traces of micropores observable along the length of the immersed films. Melts that were allowed to settle and then carefully decanted, leaving the sediment undisturbed at the bottom of the crucible, gave castings clear of oxides and porosity.

Other interesting features that they observed included the precipitation of higher-melting-point heavy phases, such as those containing iron and titanium, onto the floating oxides as the temperature was lowered. This caused the oxides to drop rapidly to the bottom of the crucible. Such precipitates were not easy to get back into suspension again. However, they could be poured during the making of a casting if a determined effort was made to disturb the accumulated sludge from the bottom of the container. The resulting defects had a characteristic appearance of large, coarse crystals of the heavy intermetallic phases, together with entrained oxide films and associated porosity. These observations have been confirmed more recently by Cao and Campbell (2000) on other Al alloys.

Over subsequent years there has been significant development of the ultrasonic technique for the study of bifilms in both Al alloys and steels (Mountford and Sommerville, 1993a,b, 1997 respectively). However, the scintillating reflections are not easy to interpret, because a crumpled bifilm tumbling in the liquid could give multiple flashes of reflections as its multiple facets turn perpendicular to the beam. It would be easy to misinterpret such signals as a collection of separate particles. The uncertainty of interpretation seems to have held up the wider uptake of this technique.

Sedimentation of Al8Mn5 particles in AZ91 magnesium alloy has also be studied using real-time X-ray video techniques (Peng et al., 2019). The particles are seen to nucleate and grow on oxide films and sediment to form a sludge layer when the temperature is reduced. The rate of settling of the particles is slower than Stokes law would predict for spherical particles, as is to be expected for particles attached to an oxide film. If the temperature is raised once again, the lower temperature at the bottom of the liquid causes the sludge layer to be slow to remelt.

A wonderfully clear example of detrainment in action is seen in the work of Nadell et al. (2007) in which they used grain refining additions of Ti and B in an attempt to reduce the cracking of high strength aluminum alloy DC ingots. They melted in a furnace and transferred the liquid metal into the continuous casting mold via a sloping launder. When the grain refiner was added to the furnace the ingot became grain refined and free of cracks because of the sedimentation of bifilms weighted down with Ti compounds (although much of the grain refiner now resided at the bottom of the furnace there was still enough Ti and B in solution to provide good grain refinement during solidification). However, when the grain refiner was added to the launder the treated bifilms had no opportunity for sedimentation but were carried immediately into the ingot. The ingot was excellently grain refined, but cracked.

(iii) Aided flotation of bifilms by bubbling treatments. Such treatments are widespread for liquid metals, and generally do seem to be accepted as beneficial, although sometimes the benefit is more probably associated with achieving a more uniform temperature throughout the melt.

Rotary degassing treatment is particularly common for liquid Al alloys. As for filters, rotary degassing is a mixed blessing. Although a key benefit has been the stabilization of melt quality in some foundries, its action appears far from being fully optimized at the time of writing. The relatively small bubbles do seem to be effective in floating out large oxide bifilms efficiently, which is an important benefit, mainly eliminating the primary oxide skins from melted charge materials. However, smaller oxides are more effective in evading contact with bubbles, the oxides moving around the streamlines of an ascending bubble. The breaking of surface bubbles of argon or nitrogen also appears to generate new, small oxides of pure alumina, probably in the waft of air at the instant the bubble opens at the surface. The collapsing of the oxidized surfaces of the bubble naturally create a small bifilm. At this time, it is not known whether this new population of hundreds or thousands of small oxide bifilms is especially damaging or not.

(iv) Sedimentation. The simple holding of melts and doing nothing but wait is an effective means to detrain bifilms in light metals. Bifilms in Mg alloys sink; an action which is encouraged in Mg—Zr alloys by a Zr-rich flux which probably precipitates the heavy Zr on to the MgO bifilms. A similar beneficial action can be employed in Al alloys by the so-called grain refining addition of Ti—B alloy. Alumina bifilms are reluctant to sediment, despite their higher density than the liquid metal, because of the layer of air trapped inside the double films. The air initially favors flotation, but as the air is consumed by reaction with the metal (Raiszadeh and Griffiths, 2008), the bifilm reverts to sinking slowly (The rates given in Fig. 1.52 are an overestimate as a result of assuming the inclusions are spheres and not films). The change of direction from floating to sinking may take minutes or hours depending on whether hydrogen is present to aid buoyancy. The natural tendency to eventual sedimentation can be speeded up by the addition of Ti-rich compounds on the bifilms causing them to sediment like stones, forming a layer of sediment on

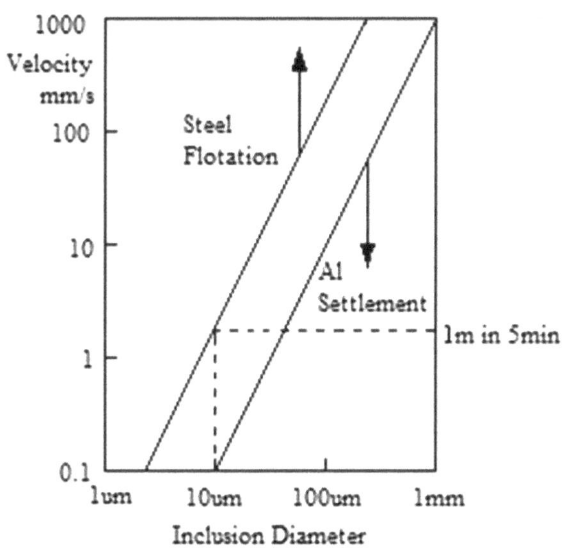

Figure 1.52 Rates of flotation of oxides in steels and sedimentation (unaided by precipitate) in Al alloys.

the bottom of the ladle as directly observed by Mountford and Calvert (1959) using ultrasonics. Cook et al. (1997) produced electrical conductivity grade aluminum by addition of TiB_2 to 99.7Al, although their settling period required 6 h for 90% of the sediment to accumulate at the base of the melt. In the case of more normal casting operations, this melt cleaning technique results automatically if the grain refining addition of Ti and B is carried out properly, even though the lengthy time found in Cook's work suggests that greater potential may be available than is normally achieved in the few minutes normally allowed in foundries. Nevertheless, even for such short sedimentation periods, the effect is valuable and strongly recommended to improve properties (the grain refining effect is practically irrelevant to properties). When designing casting systems for new layouts of Al foundries I avoid all other cleaning techniques, focusing simply on sedimentation. It is simple, highly effective, and minimal cost, if designed correctly into the process.

It is a pleasure to recall memories of the early days (c.1980) of the development of the Cosworth Process, when the melt in the holding furnace had the benefit of days to settle, becoming clear from bifilms, since production at that time did not occupy more than a few production shifts per week. The melt was therefore unusually free from bifilms, and the castings were found to be completely free from porosity, and had outstanding fatigue resistance, reported by one customer to be better than any aerospace foundry in Europe at that time. As the production rate increased during the early years the settling time was progressively reduced to only a few hours, causing a disappointing reappearance of microporosity, and a corresponding reduction of mechanical properties. This link between melt cleanness and freedom from porosity is well known. One of the first demonstrations of this fact was the simple and classic experiment by Brondyke and Hess (1964) that showed that filtered metal exhibited reduced porosity. (When hydrogen was thought to be the only factor in the formation of gas porosity, the elimination of porosity by filtering was thought to be the filtering out of hydrogen! The factor of a mere million in the size of pores compared to the size of hydrogen atom was conveniently overlooked!)

1.4.16 Deactivation of entrained films

The double nature of bifilms, enabling them to initiate cracks and other defects, may be deactivated by (i) the loss of its entrained gas, closing of the bifilm gap thus gaining some shear strength for the bifilm (despite the lack of bonding); (ii) the closing of the bifilm gap by the application of pressure; (iii) the development of some diffusional bonding; (iv) the presence of some kind of binder. Naturally, more than one mechanism can be involved at any one time.

1.4.16.1 Loss of gas

Once submerged, the inside of the bifilm will act as a thin reservoir of air. Air will exist in pockets, in bubbles trapped between the films, and in folds of the films. In addition, most oxides are microscopically rough, so that the opposed roughness of the oxide surfaces will constitute part of the air reservoir and assist to convey the gas to all parts of the inter-layer. A particularly rough oxide that takes a corrugated form is MgO as seen on Al−5Mg alloy (Fig. 1.13B).

Nevertheless, of course, the bifilm is not likely to know that it is submerged. The oxides on its interior surfaces will therefore be expected to continue their growth regardless, as though they were still atop the liquid metal, in contact with the "open air." Thus, the layer of air in the bifilm will be gradually consumed as the oxygen is used up. As soon as all the oxygen has reacted, the nitrogen will then start to be consumed to form nitrides as elegantly demonstrated by Raiszadeh and Griffiths (2008).

In aluminum alloys, although the nitride of aluminum is not so stable as the oxide, and will therefore not form in competition with the oxide, it is nevertheless very stable. It is therefore likely to grow, if only slowly on Al alloys. In contrast, on molten ferrous and Ni-base alloys the kinetics of forming aluminum nitride are rapid at the much higher temperatures. Thus, in a bifilm the layer of air is expected to disappear at different but fast rates depending on the alloy and its impurities. In both Al alloys and steels the 1% remnant of argon in the air would be likely to continue to remain as a gas because of its nearly perfect insolubility in molten metals (Boom et al., 2000).

During the final stages of the disappearance of the gas layer the opposed oxides would be expected to grow together. The closing of the bifilm void will result in the gain of some strength for the bifilm since it will now be resistant to shear stress, as a result of friction between the roughness of the films, and the mechanical interference of jogs and folds. The bifilm will still not be resistant to tensile stress of course. Thus, simply by becoming closed, the bifilm will have gained significant strength, as illustrated schematically in Fig. 1.35.

In addition, on the closure of the central void in the bifilm, there may start to develop some degree of bonding. The bonding may not be particularly strong, but may confer some improvement in tensile strength compared to the layer of air. In an Al–Si alloy, Raiszadeh and Griffiths (2008) observed bridging between the leaves of the bifilm after solution treatment.

If the gas layer is consumed, and if some diffusion bonding has occurred, then the most deleterious effects of the bifilm will have been removed. In effect, the potential of the defect for causing leaks, or nucleating bubbles, cavities, or cracks, will have been greatly reduced. This has been confirmed by Nyahumwa et al. (1998 and 2000) who found that porosity in the fatigue fracture surfaces of Al–7Si-0.4 Mg alloy appeared to be linked often with new oxides, but not old oxides.

Further evidence for the growing together of bifilms with time is provided by the work of Huang et al. (2000) who examined fragments of bifilms detached by ultrasonic vibration from polished metallographic sections. This work was reinterpreted within the framework of the bifilm concept in "Complete Casting Handbook" (2015), revealing that the bifilms of alumina, in the process of transforming to spinel, appeared to be extensively fused together. The interdiffusion of Mg required to transform alumina to spinel appears to drive a kind of low temperature diffusion bonding process. The effect is important in the overall mechanism of the gain in strength following the hot isostatic pressing (HIP) of Al alloys as discussed below.

1.4.16.2 Closing of bifilms by pressure

The deactivation of bifilms is certainly encouraged when the metal is subjected to pressure; mechanical properties of the resulting castings are immediately improved, becoming significantly more uniform and repeatable, and pores and cracks of all types appear to be greatly reduced if not eliminated. Even if the air film in bifilms is not completely eliminated, it seems inevitable that the oxygen and nitrogen will react at a higher rate and so be more completely consumed, and any remaining gases be compressed. In addition, if the pressure is sufficient, liquid metal may be forced through the permeable oxide so as to fuse with, and thus weld to, metal exuding through faults in the far side of the bifilm. The crack-like discontinuity would then have effectively been "stitched" together, or possibly "tack welded." Pressurizing treatments that would promote such assisted deactivation include:

1. The solidification of the casting under an applied pressure. This is well seen in the case of squeeze casting, where pressures of 50–150 MPa (500–1500 atm) are used. However, significant benefits are still reported for sand and investment castings when solidified under pressures obtainable from a normal compressed-air supply of only 0.1–0.7 MPa (1–7 atm). Berry and Taylor (1999) review these attempts.

   It is salutary to reflect that the reduced pressure test, commonly used as a porosity test in aluminum alloy casting, further confirms the importance of this effect but uses exactly the same principle in the opposite direction: the pressure on the solidifying casting is lowered to maximize porosity so that it can be seen more easily (Dasgupta et al., 1998; and Fox and Campbell, 2000). Once again, the effect is visually observed to be of critical importance, despite the relatively small changes in pressure which are involved.

2. Hot isostatic pressing (hipping) is a solid state deactivation process, which, by analogy with the liquid state, appears to offer clues concerning the mechanism of deactivation processes in the liquid. The hipping of (solid) castings is carried out at temperatures close to their melting point to soften the solid as far as possible. Pressures of up to 200 MPa (2000 atm) are applied in an attempt to compress flat all the internal volume defects and facilitate the welding together the walls of the defects by giving time for diffusion bonding. It is clear that in the case of hipping aluminum alloys, the aluminum oxide that encases the gas film will not weld to itself, since the melting point of the oxide is over 2000°C and a temperature close to this will be required to cause any significant diffusion bonding. This closing of bifilms by pressure, but providing no "welding" or bonding of the interfaces, is a typical behavior for a wide variety of metals. There is, as we have seen above, some expectation that the shear strength of bifilms will be increased by the closing of the air gap, but the tensile strength of the bifilm will still be expected to remain close to zero, permitting the possibility of the bifilm crack re-opening in the right conditions.

The lack of bonding is confirmed by the re-inflation of bifilms if the matrix is subjected to a high temperature heat treatment. The residual argon in the bifilm is expanded, creating argon bubbles.

1.4.16.3 Bonding by diffusion reaction

For some aluminum-magnesium alloys bonding between the halves of the bifilm does occur, effectively deactivating the bifilm. This can occur at much lower temperatures than is normally required for diffusion bonding, which would be expected in the range of 1500–1800 C. In contrast, at only approximately 530°C in the case of Al–7Si-0.4 Mg alloy, some bonding has been found, indicating that some additional process is at work. For instance, some diffusion bonding of oxides occurs because of active changes of chemistry and structure of the oxide.

These conditions for bonding are highly specific. In the case of Al alloys that have less than 0.10 wt%Mg, the oxide will be a variety of alumina, and so very stable and unlikely to bond. For higher Mg alloys, above about 1.0 wt%Mg (particularly the Al–5Mg and Al–10Mg alloys), the oxide involved in this case is MgO, and so, like alumina, is stable and unlikely to take part in any bonding action. But for alloys with an intermediate Mg content in the range approximately 0.1–1.0 wt%Mg, after time at temperature, the initially-formed alumina bifilm absorbs Mg by diffusion from the matrix, converting the pure aluminum oxide to a mixed alumina/magnesia mixture, with a different crystal structure known as a spinel (Al_2MgO_4). During the atomic rearrangements required for the reconstruction of this quite different lattice, some diffusion bonding is encouraged.

In an elegant experiment with Al–7Si-0.4 Mg alloy, Aryafar et al. (2010) showed that no bonding across air gap of Al_2O_3 bifilms occurred until all the internal oxygen and nitrogen was consumed. After this, some weak, patchy bonding resulted from the conversion of Al_2O_3 to spinel, $MgAl_2O_4$, in a short time. After longer times, stronger bonding resulted from the conversion of spinel to MgO. Bakhtiarani and Raiszadeh (2010) studied Al-4.5 Mg melts, finding that after entrainment the bifilms continued to grow for several hours, the oxide spinel gradually converting to MgO. The atomic rearrangement resulted in some bonding across impinging regions of the bifilm. The process closely resembles diffusion bonding, but occurring, of course, at temperatures hundreds of degrees lower.

1.4.16.4 Bonding by liquid binder

If the entrained oxide film is partially liquid, some bonding is likely to occur much more readily (Table 1.4). This may occur in light alloy systems where fluxes have been used in cleaning processes, or contamination may have occurred from traces of chloride or fluoride fluxes from the charge materials or from the crucible. Such fluxes will be expected to cause the surfaces of the oxide to adhere by the mechanism of *viscous adhesion* when molten, and as a solid binder when cold. The beneficial action of fluxes in the treatment of liquid Al alloys may therefore consist not only of the elimination of bifilms, aggregating them into a sticky surface dross, but also their partial deactivation by assisting bonding. The fact that flux inclusions can appear commonly on machined surfaces of Al alloy castings is an indication that the rapid detrainment of fluxes by flotation, as would be expected for a liquid inclusion, does not occur. Its flotation has been inhibited by the flux being wrapped in an extensive bifilm.

Many steels have oxides which are partly melted at steel casting temperatures. Such entrained oxides are furled by casting turbulence, but are probably unable to unfurl, having become sticky masses. Their stickiness permits them to agglomerate, adhering to other sticky agglomerates, and so form large inclusions. These are the so-called ceroxide ("ceramic" type oxides) inclusions, which float rapidly as a result of their large size and compact shape. They commonly decorate the upper surfaces of low alloy steel castings.

Other steels have oxides which are completely liquid at their casting temperatures (Table 1.4). Such steels cannot create bifilms and so will be expected to be strong and tough. The deoxidation of steels with calcium and aluminum creates a low melting point eutectic between their two oxides, calcia and alumina, at close to 1400 C, and so is liquid at steel making and casting temperatures. This has now become a standard technique in steel deoxidation prior to casting, although we should note that improved calcium mixtures are now widely used to create other low melting point oxides.

Some steels do not require special deoxidizing additions to create liquid deoxidation products, but produce liquid inclusions naturally. These steels include Hadfield manganese (13%Mn) types explaining the immense toughness of such steels, used for arduous duties such as railway track points and crossings. Boron steels are another special group in which the surface oxide of the liquid steel is a liquid borate, with a melting point close to 1000 C, even though the amount of boron in the steel is impressively low, in the region of only 0.002 to 0.005 weight per cent. These steels are renowned for their strength and toughness, and for their hardenability. Their increased hardenability arises because during the quenching of most normal steels, the rate of heat flow to the quenchant is limited by the air gaps between the halves of the bifilms. In the absence of bifilms, the heat no longer has to take a tortuous path to the outside world, but can simply flow in straight lines.

At multiple instances in his book on Ni-base Superalloys, Reed (2006) mentions boron additions "to strengthen grain boundaries" – compelling evidence to confirm the presence of bifilms in superalloys if any confirmation were needed. It is known from molecular dynamics (MD) computer models of the atomic lattice that grain boundaries are immensely strong, close to the strength of the matrix, so that they do not require strengthening. Weakness is expected only if, of course, the boundaries harbor bifilms. Even so, some confusion remains because there do seem to be some reports of failures, most probably from bifilms, despite the claimed presence of boron in the alloy specification. Perhaps on occasions the very low additions of B are too low to completely prevent bifilm formation, especially in the presence of other reactive elements such as nitrogen, which takes up B as boron nitride, or because of other losses such as evaporation of borane gas, BH_3, especially during melting in vacuum. In such instances some bifilms might be expected, some of which would be likely to be deactivated, with perhaps traces of liquid borates. These uncertainties require careful targeted research to resolve this issue.

1.4.17 Morphological transformations

Most researches of inclusions in metals, particularly steels, assume that all the reactions occur in solution in the steel. This is certainly correct for the important reaction to deoxidize the steel with, for instance, aluminum. Oxygen in solution reacts with aluminum in solution to form solid alumina inclusions.

Unfortunately, excess aluminum (or other deoxidizer) remains in solution. This excess deoxidizer will react with oxygen from the air during the subsequent pouring operations from the furnace into the ladle and from the ladle into the mold or casting machine. This *reoxidation reaction* now occurs *not* in solution but on the liquid surface during the pouring operations. The result is the creation of an oxide film on the surface of the liquid. This surface reaction to create a film has been largely overlooked by researchers.

The surface oxide film is now entrained into the liquid by the turbulence of the pouring action. As we now know, the oxide film enters the melt as a double film.

Throughout its life in the melt, the bifilm now undergoes a series of shape changes. An understanding of these different forms is essential to the understanding of the properties of solidifying metals. The shape changes up to Stage (iii) due to mechanical processes in the life cycle of the bifilm we have already discussed. These are listed below. Stage (iv) is an important diffusion reaction for steels and Ni base alloys which is now added below to complete the list of reactions.

(i) Surface turbulence of the liquid (breaking waves and splashes) lead to mechanical entrainment of the surface oxide film, forming a double film (a bifilm) separated by an "air gap";

(ii) Immediate mechanical furling of the bifilm by bulk (internal) turbulence, typically to one 10th overall size;

(iii) Slow mechanical unfurling in the quiescence of the mold cavity by multiple potential mechanisms, including for instance gas precipitation or shrinkage conditions.

(iv) For high temperature metals such as steels and Ni alloys, a coarsening transformation occurs, driven by surface energy reduction, but controlled by the rate of diffusion. The sequence of morphological changes now is

- **(a)** The thin bifilm develops holes;
- **(b)** The holes grow to form arrays of ligaments;
- **(c)** The ligaments divide into separate particles, retaining the remnant of the "air gap" as an argon "void";
- **(d)** The particles develop crystalline facets (the argon voids remaining attached).

Previous sections of the book are given to the phenomena of entrainment, furling and unfurling. Here we shall examine the final stage in the life of the bifilms in which, given sufficient time and/or temperature, they fragment into particles together with the residue of argon gas.

Fig. 1.53 shows how a stream of liquid ferro-manganese alloy is transformed into droplets. The falling stream of liquid is targeted to impinge on a flat refractory plate, from where it spreads into a thin horizontal film. The shape of the liquid, as a thin film, is now unstable such that any perturbations to thickness will grow under the action of surface tension. Effectively the driving force is the reduction in surface energy of the film by simply reducing its area. For this reason, the film first develops holes,

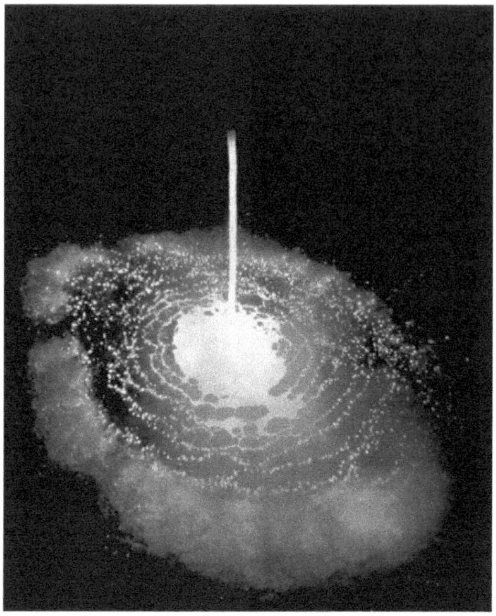

Figure 1.53 Granulation of liquid metal poured on to a ceramic target, showing the characteristic mode of disintegration of a thin film first into holes, which expand to form ligaments, which finally separate into droplets and are quenched in the tank of water. *Courtesy UHT Sweden, 2020.*

which rapidly grow, leading to the formation of ligaments of liquid between the holes. Finally, the ligaments divide into droplets which fall into the quench tank and solidify. The granulation of the ferro-manganese liquid is complete.

In the liquid state all the several changes of shape from thin liquid film to droplet occurs in less than a second.

In the case of a thin film of solid oxide in a liquid metal the rates are not known, but are expected to be at least an order of magnitude slower, possibly measured in minutes. Slower still (add probably another order of magnitude) will be the conditions in which the solid film exists in a solidified metal. We are now perhaps measuring changes in hours. However, interestingly, all these conditions exist in practice.

At the temperatures used for the casting of Al alloys, the temperatures are sufficiently low that entrained bifilms normally retain their double-film morphology indefinitely. However, in the case of Al alloys subjected to very high temperature (900 C) or intensive shearing during freezing, the oxide films are not only mechanically shredded, but it seems probable that diffusion is enhanced by an enhanced vacancy concentration, aiding the evolution of the film into separate particles as has been reported (Wang et al., 2017). Earlier work by Wang et al., (2013) has shown that alumina films appear to fragment into micrometer-sized spinel particles by stirring the melt continuously for several hours at the relatively modest temperature 700 C. However, of course, in more usual casting conditions there remains no doubt that those bifilms which are formed and submerged in milliseconds during turbulent pouring are simply double films and can remain as bifilms throughout their life.

In the case of the common alumina or chromia oxide bifilms in steels, the films are probably solid, but in the very hot liquid their solid forms morph to reduce surface energy, probably mainly by surface diffusion. Fig. 1.62A shows alumina particles which were almost certainly a continuous alumina bifilm prior to the hot rolling deformation. Fig. 1.54 beautifully illustrates a double line of alumina fence posts, a remnant of a bifilm in a Ni superalloy. The two parallel films have transformed as far as two parallel lines of ligaments. Although the lower temperatures of Ni alloys means that bifilms often survive as planar cracks without significant change, in the case of Ni alloy single crystals, the long time at temperature during the growth of the crystal, and the subsequent high temperature heat treatment have provided good conditions for bifilm evolution.

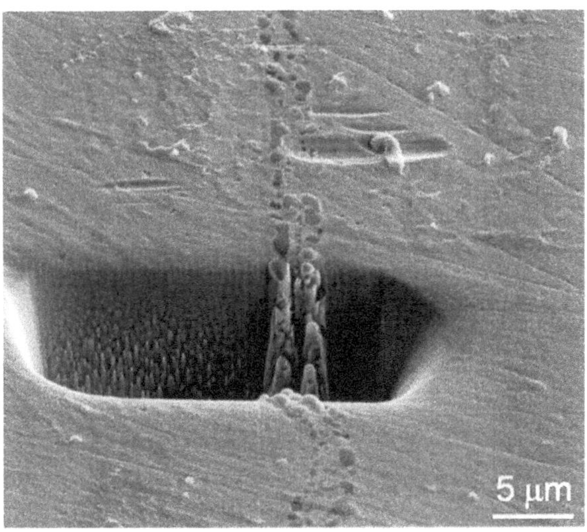

Figure 1.54 An old bifilm in a Ni superalloy morphed perhaps during heat treatment into ligaments (the twin line of fence posts) revealed by FIB (focused ion beam) excavation in the SEM (scanning electron microscope).

In the above case the bifilm was reported to be the result of a crack formed during or post heat treatment and the crack surface was probably oxidized during heat treatment (Keehyun and Withey, 2017). Recalling that the thesis of this book is that all cracks in metals are the result of bifilms, the oxidation of the bifilm during or after heat treatment would merely enhance the thickness of its oxide film. However, in all other respects, it seems safe to conclude that the fence posts denote the site of a very ordinary run-of–the-mill bifilm which may have suffered some thickening of its oxide.

These substantial changes in morphology certainly eliminate the central crack which is by far the most damaging feature of bifilms. By the time that the shape change is well under way, it seems certain that the oxygen and nitrogen of the original air gap of the bifilm will have reacted or dissolved. In Fig. 1.54 the central gap of the bifilm has been eliminated by diffusion of the matrix into the gap. However, the 1% argon from the air will be certain to remain, in this case out of sight. This gaseous volume defect will be expected to remain in contact with some of the fragments of the oxide. It will constitute a small but permanent volume defect capable of initiating a pore or crack if subjected to sufficient stress. Any failure initiated by the argon volume will appear as a decohesion of the oxide from the matrix. (We need to be vigilant in recognizing "real" decohesions as an impossibility because of the immense strength of the oxide/matrix bonding.)

Trials at Warwick University organized by Professor Claire Davis (Mandal, 2018) are revealing. The work involved two microalloyed high strength steels, one treated with 0.07Al and the other with 0.03Ti. These show, respectively, alumina (Al_2O_3) and titanium nitride (TiN), both with their associated argon "voids" (Fig. 1.55). It is important to note that these voids could not have been opened up by the metallographic technique because the matrix/inclusion bond is expected to be inseparable. The voids appear because the original "air gap" in the bifilm can never totally disappear because of the insolubility of the remaining 1% argon in air. This is commonly seen for inclusions, especially those that have been in rolled steels, because the elongated void, partly pulled away from the inclusion, is more easily seen.

Overall, although the damage of entrainment defects is very greatly reduced by the high temperature coarsening mechanism, changing the bifilm into particles, the capability for damage to properties is never totally eliminated because of the lingering presence of the small argon volume defects masquerading as areas of decohesion.

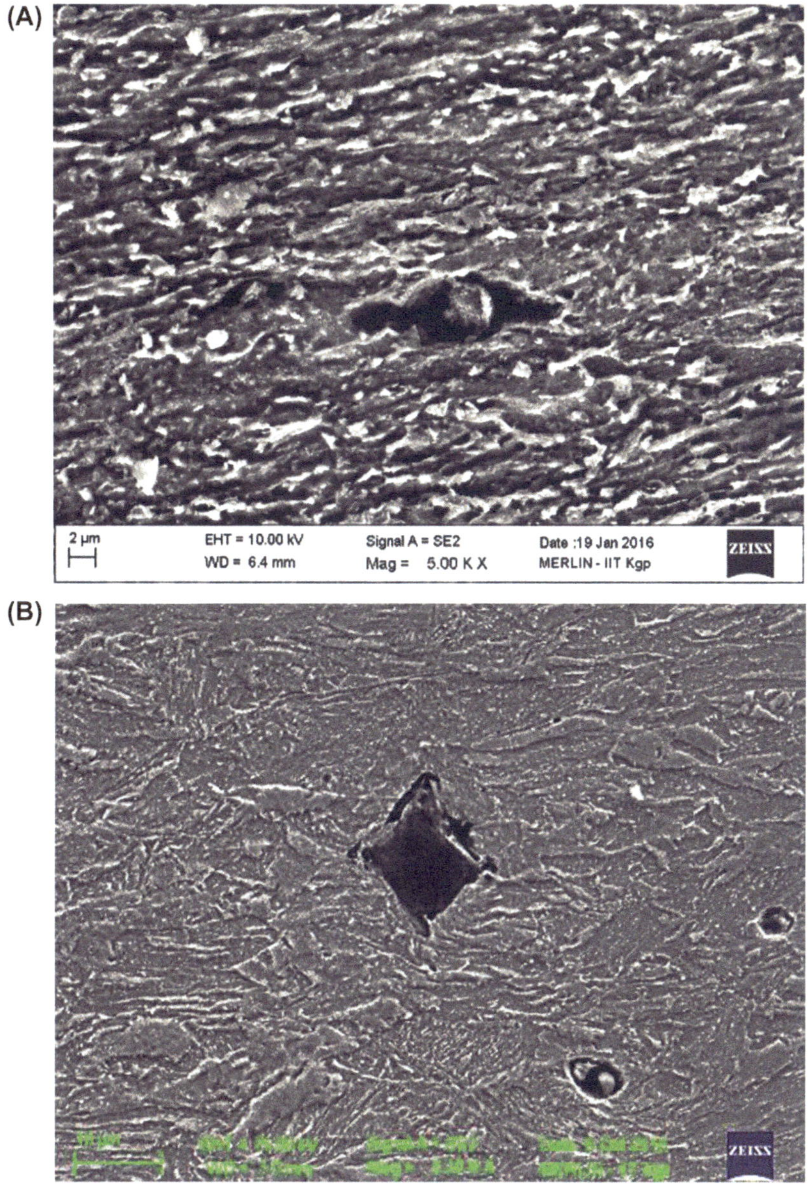

Figure 1.55 A microalloyed high strength steel treated with (A) Al showing alumina inclusion with its associated argon "void"; and (B) Ti treated showing TiN inclusion and argon void (Mandal, 2018).

For the future, for steel casting operations, it would be useful to ascertain whether any argon bubbles were entrained unintentionally from the use of shrouds, submerged distribution nozzles, and stirring devices. Naturally, most of such argon entrained by shrouds will float out and escape within seconds. However, it would be rash to assume that all had escaped, especially the smaller diameter fraction of the bubble population. Such bubbles would be expected to be retained in the liquid metal because of the obstruction to floating posed by masses of bifilms in suspension. Such bubbles would be expected to be of far greater size than those residues from bifilms, and so would be, unfortunately, very counter-productive for properties. It seems essential to evaluate whether much use of argon in steelmaking, particularly in its use with shrouds, is beneficial or otherwise. It is a concern that it may be doing more harm than good. In addition, it does not seem difficult to envisage systems designed to be free from entrainment which require no argon, or at least, require no argon in danger of being entrained.

1.4.18 Soluble, transient films

Although many films such as alumina on aluminum are extremely stable, and insoluble in the liquid metal, there are some alloy/film combinations in which the film is soluble in the metal matrix. This is of course especially true for the higher melting point metals where the higher temperatures and the reduced stability of the films encourage this behavior. This has been assumed to indicate that surface films cannot be present on such metals, which, as we shall see, seems to be a mistake.

Even in those cases where the film-forming elements (such as oxygen or carbon) are soluble in the liquid, transient films are to be expected where the rate of arrival of the elements at the surface of the liquid exceeds the rate at which the elements can diffuse away into the bulk metal. If such a film is then entrained, the film may create a crack, a hot tear, or initiate shrinkage porosity. However, after the initiation of this consequential but possibly permanent "secondary" defect, the originating bifilm then quietly goes into solution, never to be seen again. The perpetrator of the damage evades detection.

It is to be expected that many oxide and nitride films might be soluble in some irons and steels.

Such behavior is commonly seen in gray cast irons. The hydrocarbon gases and vapors which evolve from the molding sand are chemically decomposed on encounter with the advancing hot molten iron.

The hydrogen seems to escape harmlessly, although some may enter the iron. But the carbon is taken into solution in the liquid metal. However, if the arrival of the hydrocarbon gas delivers carbon at a rate at which it cannot be absorbed, a film of carbon is built up on the advancing front of the liquid. The film is known as a "lustrous carbon" film on account of its shiny black appearance.

If a lustrous carbon film is entrained in the melt, it will form a lustrous carbon bifilm in the usual way (Fig. 1.56). However, after some time it will dissolve. The rate of dissolution will be rather slow because the iron is already nearly saturated with carbon. The longer the time available the greater is the chance that the film will disappear. There are various outcomes that can be envisaged:

(i) In a thin section the lustrous carbon will be frozen in place, constituting a bifilm defect to reduce strength as a crack or provide a leak path.

(ii) In an intermediate section the entrained bifilm may last just long enough to initiate some other longer lived and more serious defect, such as a hot tear or shrinkage porosity, prior to its disappearance.

(iii) In heavy section gray iron castings entrained lustrous carbon films are usually never seen, having plenty of time to go into solution.

Titanium alloys, including titanium aluminide alloys, are famous for their high solubility for interstitial elements such as oxygen, carbon and nitrogen. It is expected therefore that oxide films, if they form, are almost certainly soluble and would be expected to disappear. However, as we have noted above, if the rate of arrival of oxygen is greater than its rate of diffusion into the matrix, then a pseudo-permanent film seems inevitable. If folded into the bulk melt it will constitute a bifilm whose lifetime will be expected to be limited. However, it may survive sufficiently long to create a defect, or even long enough to be frozen into the cast product. In practice, early bifilms entrained by casting turbulence will be formed of thin oxides and will rapidly go into solution.

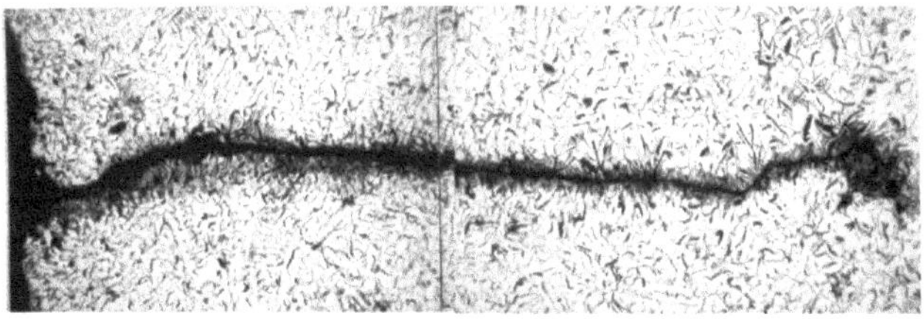

Figure 1.56 A folded-in lustrous carbon film, forming a carbon bifilm in cast iron.

In contrast, thicker oxides, especially if introduced late during casting, may not have time to dissolve before they are sealed into the microstructure by solidification. Despite the expectations of many that oxide films will not exist in Ti alloys, their presence in TiAl alloys has been clearly recorded by Mi et al. (2003).

Nevertheless, it is to be expected that some of these longer-lived bifilms will dissolve during the HIPPING treatment given to all cast titanium alloys; the HIPPING process is therefore expected to be essential for Ti alloys because of the uniformly bad pouring processes the alloys endure in a largely misguided effort to ensure the filling of the mold when casting at low superheats. It is therefore fortunate for Ti alloys that this damage can be healed.

Even so, it seems probable that in some cases not all damage will be eliminated in this way. A few residual "cracks" in some grain boundaries, pinning these particular boundaries, may be the reason for the anomalous grain growth sometimes observed with Ti alloys. Decost and Holm (2017) study model metal systems by Monte Carlo computations to find that abnormal grain growth can be explained by grain boundaries with different mobilities, as would be expected from pinning by residual bifilms. It would be expected that longer heat treatment and/or HIPPING might take more resistant bifilms into solution and result in more uniform behavior.

There does appear to be some indirect evidence for residual oxide bifilms in the form of the development of cracks at the interface between the matrix and boride phases (Campbell CCHB2E). It is also significant that boride particles in TiAl alloys are found to be of widely differing lengths on 2D sections (actually widely differing diameters in 3D, of course) bearing no apparent relation to melting and casting conditions. These features seem essentially uncontrollable, and constitute a strong pointer to their formation on bifilms, which are, of course, not controlled. For the reader interested in the manufacture of titanium alloys the discussion in CCHB2E is strongly recommended.

The presence of 0.4 wt% carbon in TiAl alloys is seen to result in carbides which appear to initiate cracks, and consequently make the alloy brittle. Here is another strong indication that the carbides have formed on bifilms.

In addition, because of the extreme turbulence with which most Ti alloys are cast, the folding and mutual impacting of the liquid during the casting process seems to introduce a weakening effect, with the result that, at this time, all Ti alloys require hot isostatic pressing to develop reliable strength. What effect can cause such weakness other than bifilms?

A nonturbulent melting and casting technique for Ti alloys would revolutionize the Ti casting market; the costly and time-consuming HIP process would not be required.

A class of Ti alloys which are of great interest for their lightness and strength are the titanium aluminides (TiAl). But their single great disadvantage is their lack of room temperature ductility. The behavior of 48Ti—48 Al—2Cr—2Nb (known as 48-2-2 alloy; composition in atomic percent) when machined is revealing: if the cutting tool is not extremely sharp, and the machining cut light, then whole grains are prized out of the surface. Also, when approaching the edge of a machined surface, the edge can disintegrate into fragments. This edge disintegration is a known bifilm effect in badly-cast Al alloys, which is immediately and totally solved by casting without turbulence. There seems to be evidence therefore that TiAl alloys are full of bifilms, probably existing at nearly every grain boundary. A melting and casting technique to avoid the entrainment of bifilms would be predicted to yield an alloy with room temperature ductility, and a potential for much greater metallurgical development.

1.4.19 Liquid Oxide Entrainment (inclusion shape control)

In the case of low melting point metals such as copper-base alloys and below, the surface oxide is invariably solid, so that any surface turbulence results in the entrainment of bifilms and bubbles. Such low melting point alloys based on Mg, Al and Cu (especially alloys such as aluminum bronze) are all seriously impaired by bifilms entrained by poor filling systems.

If the surface oxide on the liquid metal is a liquid, then impingement or folding of the liquid surface will result in liquid oxide — to — liquid oxide contact, so that the colliding liquid interfaces will coalesce, rapidly re-forming into liquid oxide droplets to reduce surface energy. The larger droplets will float out rapidly from the metal. The metal benefits significantly because it contains now only relatively harmless, spherical inclusions which were too small to float out in time. The liquefying of the surface oxide by alloying of the metal or by, in the case of steels, the use of specific deoxidation techniques, is therefore a mechanism of major importance for the reduction of bifilms in metals. Metallurgists proudly call this "Inclusion shape control." The pride is justified, even though the major effect lay in the avoidance of bifilms, which, of course, had not been known. If this had also been realized, the justification for the pride should have been doubled!

The spherical inclusions might be completely "harmless" in the steel, in the sense that they will not be expected to suffer fracture or initiate decohesion from the matrix. However, if they had originated from an entrainment event which also entrained some air, the small quantities of oxygen and nitrogen will have been rapidly be absorbed into the inclusion, or even dissolved in the matrix, but the residual 1% argon will have remained as part of the inclusion. This gaseous volume will assist to lower the energy of formation of a volume failure, such as a crack or decohesion. The spherical inclusion may therefore not always be entirely "harmless."

Steels are complicated. Their high melting points, generally in the region of 1500 C, mean that some oxides will continue to be solid, creating bifilm problems, whereas others will be above their melting points, thereby largely avoiding bifilms. In addition, steels generally require to be deoxidized before they can be cast. The process is necessary to avoid excess oxygen in solution reacting with the carbon in the steel to form carbon monoxide, CO, and so creating porosity, or in a severe case would result in the steel "boiling" during freezing. (Some of us ancient metallurgical engineers remember with fondness the spectacular pyrotechnic displays on days in the steel shop when casting rimming steels.)

As we have noted previously, in the simple case of deoxidation of many carbon steels and low alloy steels, aluminum is used as a highly effective deoxidant. However, after the deoxidation treatment, some excess Al remains, and is now available to react with the air during casting — a process known as "reoxidation." The very high melting point of aluminum oxide alumina (2050 C) ensures that a solid oxide forming bifilms will occur during the pouring of the steel through air, even though the steel contains only about 0.05%Al or less.

However, if a mix of approximately 50% aluminum and 50% calcium used for deoxidation, the mixed oxide (alumina and calcia) has a melting point of only around 1400 C. Entrainment of this liquid oxide results not in bifilms but in liquid films which rapidly spherodise into droplets which tend to float out. On arrival at the upper surface of the casting the droplets are simply assimilated in the surface liquid oxide layer, and disappear. This is the mechanism by which steels finally deoxidized with Ca + Al achieve such high levels of cleanness in comparison with steels deoxidized with the usual Si, Mn and Al.

The beneficial action of Ca deoxidation to form a low melting point CaO-containing eutectic occurs with other oxides, probably most importantly with Cr_2O_3 as in nearly all stainless steels and high temperature Ni alloys.

The addition of boron (B) to steels is another valuable technique for liquefying the surface oxide film on the steel. In this case a borate is formed with an amazingly low melting point close to 1000 C depending on its composition, which probably varies with the steel. Only very low levels, commonly from 0.002 to 0.005, are required to achieve this immense reduction in oxide melting point. Boron steels are noted for their strength and toughness, which must derive from their fortunate absence of bifilms.

The Hadfield Manganese steel (Fe−13Mn) is another extraordinary steel, famous for its toughness. It is used in such punishing applications as railroad points and crossings. Failures in service seem practically unknown. The Mn oxide MnO_2 is liquid at casting temperatures, so preventing the formation of bifilm cracks.

The more recent TWIP (transformation induced plasticity) steels containing over 50%Mn, despite their content of a few percent of Al, typically enjoy 100% elongation, probably contributed at least in part by the absence of bifilms. The lightweight steels Fe−30Mn−9Al continue to exhibit the high elongations, although as Mn lowers and Al rises the elongations fall; perhaps due to the introduction of alumina bifilms at some critical ratio?

We need to be aware that those treatments of minor additions to steels (Ca and B for instance) are vulnerable. In the event of especially turbulent casting conditions the addition can be consumed, effectively overwhelmed by air entrainment, resulting in an overload of oxide formation and a loss of liquefying benefit at some stage during pouring. A top poured ingot (dread the thought!) would therefore require higher Ca or B addition than a contact-poured ingot or continuously cast product. The high Mn steels would not be expected to suffer such a disadvantage during pouring, effectively having an effectively infinite supply of Mn.

Nickel based alloys, especially the so-called superalloys containing Al and Cr (among many other additions) are complicated from the point of view of their oxide during melting and casting. The oxides of Al and Cr are usually solid at the melting points of most Ni-base alloys, as a result of Ni alloys having a lower melting point than most steels. Their lower melting point means that the mix of Al + Ca is now so marginal in its effect that it cannot be certain that it will be effective. Thus air-cast Ni alloys often suffer badly, with the result that cracking during subsequent forging is common. Even when melted and cast in vacuum, cracking on forging provides good evidence to confirm that bifilms are present; vacuum-cast Ni superalloys greatly benefit from turbulence-free casting techniques.

The reader needs to take special note of the metallurgically impossible logic involved in the phenomenon of cracks present in alloys which are otherwise known for their exceptional ductility. Ni-base alloys are ductile and therefore should not fail by cracking. In addition, of course, during solidification the alloys are subjected to only relatively minor stresses, orders of magnitude lower than stresses which could cause cracks. Only the presence of bifilms due to turbulent casting practice can explain the cracking behavior of Ni alloys. Figs. 1.57 and 1.58 illustrate the astonishing cracks which result from poor, turbulent casting practice, in an alloy which should otherwise never crack, and should achieve near-100% reduction in area in a tensile test.

Figure 1.57 Two views of a poorly air-cast Hastelloy showing deep cracks despite its high ductility clearly illustrated by the turned edge of the machined matrix. (Correctly-cast Hastelloy is, of course, free from cracks.)

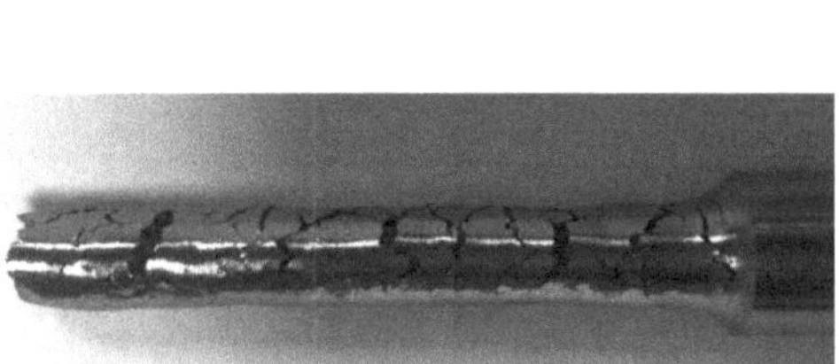

Figure 1.58 A failed tensile test piece in a highly ductile Ni-base alloy CY40, unfortunately full of cracks as a result of poor casting technique.

Moving on from the steels and Ni alloys, the gray cast irons are an exceptionally interesting and complicated case.

Liquid gray iron is clear of oxide film during melting at temperatures in the range of 1550 down to around 1450 C (the precise temperature seems to depend on the composition of the iron). This is due to the preferential reduction of Si and Mn oxides by carbon at these temperatures. The liquid surface has a magical and perfect mirror clearness which is not easy to describe.

Below this temperature a dull, gray film of solid silica (SiO_2) forms. As the temperature continues to fall, finally reaching around 1300 C, the oxidation of Mn in the iron leads to some admixture of MnO_2 in the surface oxide, causing it to melt. At a temperature somewhere below 1200 C the additional oxide, FeO, will reduce the melting point of the oxide further. This complex liquid silicate is one of the reasons for the excellent castability of gray iron, and is probably responsible for the glassy corrosion-resistant sheen on nicely-cast iron castings.

These surface reactions, occurring because of the environment high in oxygen, contrast with the interior of the liquid iron. Below about 1450 C the SiO_2 bifilms entrained into the melt are stable and continue to build up during the turbulent handling of the liquid metal. Thus when the eutectic temperature is reached they are available as the substrates for the precipitation of carbon to form graphite flakes (Campbell 2009). The solid SiO_2 performing a valuable action inside the bulk liquid contrasts with the valuable action of the liquid silicate on the outer surface of the liquid, which is aiding flowability and reducing turbulent defects such as laps.

It is immensely fortunate that cast iron develops the benefit of its liquid oxide silicates on its surface just before it is poured without losing the benefit of its solid interior silica bifilms. This is clearly the reason that, in general, the steam engines were tolerably reliable and the industrial revolution was successful.

1.4.20 Nonprotective and unstable oxides

Most liquid metals enjoy a thin but protective oxide film. For instance, duplex stainless steels are relatively easy to cast as a result of their well-behaved, thin surface film, probably mainly chromium oxide, Cr_2O_3.

Likewise, aluminum oxide (Al_2O_3) occurs as a protective oxide on aluminum and many other alloys even though their aluminum content is sometimes only a fraction of a weight percent. The film is thin and thickens only slowly and uniformly.

These well-behaved thin oxide films contrast with the magnesia (MgO) film on liquid magnesium. It seems the growth of the oxide is disrupted by the magnesium vapor which escapes through fissures in the film and contributes to the building up of the film (Emami and Sohn, 2015). Although the film thickens rapidly, it is fortunate that it still responds well to certain protective atmospheres together with best practice filling conditions, so that good castings can still be produced.

However, there is a Ni alloy, Hastelloy, which contain 6%Mo, which has such a troublesome reputation that many foundries will not attempt to cast it. This alloy has similarities to magnesium, because the MoO_2 decomposes at 1100 C and MoO_3 boils at 1155 C. The outgassing of the Mo-rich vapors disrupts the oxide growth (in passing, alloys of its sister heavy metal, tungsten, do not appear to suffer these problems because the oxide WO_3 boils at 1700 C). The result is an oxide layer which grows to a thick, dry layer resembling pack-ice. As the metal rises in the mold the oxide does not split and move to one side as shown in Fig. 1.7 but simply moves up as a solid cake; a slag-like layer. It takes little imagination therefore to understand that the upper surfaces of castings are completely unacceptable to a customer. There is some recent experience that this unwelcome behavior can be addressed by additions of Ca to the melt prior to pouring. The effect to liquefy the dry, thick slag is immediate, and it seems that enough Ca goes into solution to continue its beneficial action during the filling of the mold. Perhaps this, one of the last remaining intractable alloys, has now been tamed. Tests are being poured as I write. We shall see.

Compared to most of the engineering metals, manganese has an unusually low melting point oxide MnO_2. Its melting point is only 535 C, but, even more unusually, it decomposes at 553 C. This seems to be the reason that the high Mn steels, including the famous Hadfield manganese steel at 13%Mn, and the new TWIP and TRIP steels at higher Mn contents, are favored by low or zero bifilm contents and exhibit such high toughness and ductility. The wind of Mn-rich vapor penetrates the surface of sand molds, where it condenses. A mold coat would greatly reduce this transfer of Mn into the mold.

In vacuum conditions and at sufficiently high temperature, both Al and Cr oxides tend to evaporate. In the case of Al, the effect is believed to occur via the volatile sub-oxide AlO_2. The wind of AlO_2 departing from an Al-rich surface in vacuum seems to leave the surface cleaned of its usual Al_2O_3 film, and prevents new oxygen arriving to oxidize the surface. The surface can remain clean therefore if the temperature and vacuum is good enough (Giuranno, 2006 and Molina, 2007). One would expect therefore that Ni base superalloys containing both Al and Cr should be free from oxide films during casting in vacuum and therefore free from bifilms. It seems a puzzling fact that this does not seem to be true; the Ni and Co superalloys show all the expected failure modes associated with bifilms. From the abundant evidence of defects in vacuum-cast Ni superalloys, it seems certain, if regrettable, that bifilms exist in abundance. Perhaps the vacuum inside the mold is too impaired by the outgassing of the mold, or perhaps the temperature of the surface of most Ni alloys is just too low at the point at which they enter the mold. It would be helpful to know the answer.

1.5 Entrainment avoidance

It is salutary to consider the potential for metals if contamination from bifilms could be avoided.

Gold is one such example. The ancients knew that it could be beaten or pounded into foil (known as gold leaf) thinner than that of any other metal. This is because, since it is uniquely free of oxides, its malleability is not impaired. Thus, thicknesses of 0.1 μm are possible, which are so thin that it is possible to see light through the metal, even though the golden luster of the metal is not lost. This contrasts with metals such as aluminum where foils are generally in the range 200 to 4 μm, but below about 25 μm the foil starts to show pinholes as a result of the presence of inclusions. These would be expected to be mainly oxide bifilms. In highly specialized applications such as ribbon microphones, Al foil might be as thin as about 1 μm, which is still 10 times thicker than the thinnest gold. It would be expected that a tensile test of pure gold would exhibit a reduction of area at failure of 100%.

When constructing Table 1.6, presented later in Section 3, I was expecting to see gold enjoying its pre-eminence at the top of the ductile spectrum of engineering metals. It was a shock to find that lead (Pb) occupied this hallowed site. I realized this was a key message; the only reason that Pb does not enjoy perfect ductility is because it has an oxide, and its oxide is unfortunately usually entrained as a result of poor metal handling practices. Thinking further, this simple truth probably extends to all the ductile metals in the Table. Thinking still further, perhaps ductile behavior extends into the brittle metals if entrained defects are not present? These are fascinating and important questions.

It seems certain therefore that, at the least that can be expected, if other face centered cubic (fcc) metals (aluminum, silver, copper, nickel etc) could be made free from entrained oxides, similar perfect ductility as exhibited by gold would be expected, and foils of thickness similar to that of gold would be achievable.

Naturally, as we have seen, the melting and casting processes are central to the issue of bifilm avoidance. The particulate manufacture of solids, such as powder metallurgy, and perhaps additive manufacture, in general must accept as unavoidable the presence of bifilms which are entrained as part of the consolidation process. The melting and casting route has the benefit of sedimentation and flotation of entrained defects (although remains vulnerable to their re-entrainment as this small book attests.)

Summarizing, clean metals by the casting route require melting processes which encourage detrainment of bifilms, followed by melt handling processes which avoid re-entraining bifilms, followed by casting processes in which entrainment is eliminated or minimized.

In the high temperature metals, steel and titanium for instance, these high aspirations are not easily attained but there is some hope that future designs of melting and casting operations can be vastly improved.

For shaped castings of steel and Ni alloys poured in air, in traditional foundries there are often more employees in the welding shop, repairing defective castings, than there are in the foundry making castings. Casters have not known about the massive problems introduced by the ingress of air into their pouring operations.

With the relatively recent knowledge that turbulence and bubbles create severe bifilm cracks, the development of systems specifically designed to avoid these defects has for the first time become possible (Campbell 2018). The outcome has been greatly improved filling systems, lowering the populations of bifilms by orders of magnitude and resulting in castings of high properties and defect-free appearance when tested by dye penetrant and X-rays. Furthermore, these new techniques are of low or negligible additional cost compared to the traditional casting techniques.

For aluminum alloys, the melting and casting techniques, all have total solutions already available. For instance, the melting of Al alloys is best achieved by a dry hearth tower melting unit, in which the oxide skins on the charge materials are arranged to remain on the "dry" hearth of the furnace. The melted metal escapes, breaking out from these oxide skins, and runs down the hearth into the holding bath of the furnace. In this way the primary oxide skins on the charge materials are separated from the originally melted metal (crucible furnaces and bath type furnaces fail this first requirement as is clear from Fig. 1.8D). The internal population of oxide bifilms can then be separated from the metal by sedimentation in a furnace specially designed to avoid turbulence and convection. The cleaned metal is then pumped counter-gravity into molds at speeds controlled to avoid the re-entrainment of defects.

The area of metallurgy where entrainment avoidance is most sought is in the pursuit of ever cleaner bulk steels by the casting of ingots or the use of continuous casting plants. The target is ever higher performance in terms of strength, toughness and fatigue resistance. This important area will be discussed next.

1.6 The quest for clean steels

1.6.1 Ingot casting

Although much steel is now cast by the continuous casting process, smaller batches for special purposes are produced as ingots. Thus, much of our "special steels" are ingot cast. Ingots vary from a few kg to about 500,000 kg, but the casting techniques are much the same for all.

Fig. 1.59 shows various casting techniques used for ingots. It does not take much imagination to realize that direct top pouring of liquid metal into an ingot mold (Fig. 1.59A) is probably not good for the metal. In fact, after extraction of the ingot from its mold, an inspection of the ingot often reveals laps, cracks, and bifilms clearly visible to the unaided eye from a distance of up to 10 m. Frankly, the structure is awful. The problems of cracks in the faces and sides of ingots, in ingot corners, and vertical cracks up the flutes of fluted ingots (designed with crenelated sides to avoid cracking!) and cracks opening up during subsequent working of the ingot as face cracks, edge cracking, and crocodile cracking, are only to be expected. The failure by cracking of final products made from such ingots are also to be expected. It is relevant to ask, "Could it be worse?" "How can the metallurgical industry have lived with such a catalog of problems for so long?"

The massively destructive turbulence of top pouring also applies to vacuum casting, where inside the vacuum chamber it is not easy to arrange the delivery of the metal in any other way than by tilting the pot of molten metal and emptying its contents into the ingot mold by top pouring. In my

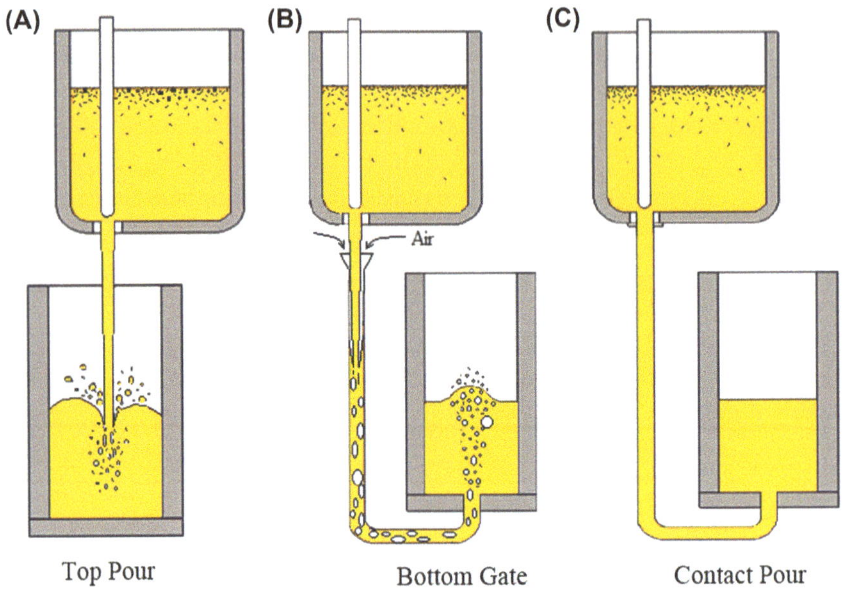

Figure 1.59 Ingot casting by (A) top pouring; (B) uphill teeming; (C) contact pouring.

view this is the Achilles heel of the vacuum induction melting (VIM) process. It is conventionally expected that the vacuum will avoid oxidation problems, allowing cracks and laps to heal. In general, this does not happen. The vacuum in metal melting and casting operations can never be particularly good because of the natural and prolific outgassing of the refractories of the melting unit. As a consequence of pouring in the relatively poor vacuum, the surface of the falling liquid is covered with an oxide film, which naturally becomes entrained as bifilms. In fact, the VIM ingot is probably the wolf in sheep's clothing of the clean steel scene: its bifilms will be so much thinner than air-cast metal, giving the appearance of a good quality product.

Bottom gating (sometimes called "uphill teeming," Fig. 1.59B) is often selected as an improved technique for the casting of ingots. I suspect the real reason is mainly the apparently improved surface finish of the ingot, which may have benefits in reducing some cracking during subsequent working.

In reality, however, the interior of the bottom gated ingot is probably almost as bad as that by top pouring as a result of the huge volume of air which is entrained by the conical trumpet-shaped intake at the entrance to the filling system channels. The cone, with the central jet of liquid, acts as a venturi air pump with awesome efficiency. The consequent mix of air and metal is at least 50/50 by volume, and often probably nearer 80% air and 20% metal. Worse still, the air bubbles will generate long bubble trails of oxide double films. It could hardly be worse. On looking down into the mold as it fills, the rising metal appears to boil. Despite the claims and remonstrations of die-hard traditionalists, it is clearly not possible to make good steel by such an inappropriate process.

The air pump action appears to be by far the most important reason for inclusions in uphill teemed ingot steels. The attempts to deal with this problem by providing an argon shroud are negligibly effective; the shroud gas typically contains 10 or 20% air.

The only effective technique to avoid air entrainment is to exclude air entirely. This is achieved by eliminating the trumpet and eliminating the gap between the ladle nozzle and the filling system intake. The ladle nozzle and filling intake are brought into contact (usually with a compressible seal of ceramic fiber blanket). The technique is simply known as "contact pouring" (Fig. 1.59C). It is totally effective. It makes excellent cast products. It is recommended to be used to the exclusion of all top pouring ingot casting techniques.

The details of ensuring that the filling system channels are not crushed by the weight of the ladle, or the ladle nozzle forced upwards into the ladle, together with the accuracy of a few millimeters required for alignment, are all solvable. Because the bottom of ladles are always rather different (the ladle feet, among other features, get in the way!) every casting operation in which the technique has been introduced has been forced to find a different engineering solution. With regard to accurate positioning, in this digital age, and with laser positioning a mature technology, such challenges should represent minimal difficulty.

The remaining issues with contact pouring of bottom-gated ingots includes the use of pre-formed refractory tubing which is normally used to construct the pouring channels. These are made from fired clay, akin to the making of pottery. They are therefore often referred to as a "pot running system." The tubing has a parallel bore, quite different to the tapering shape of the falling metal as it is accelerated by gravity. Thus, the pot channels contain much redundant air which can become mixed with and degrade the quality of the falling steel. The correct shaping of filling system channels for castings is now understood with considerable sophistication (Campbell, 2018). It seems probable that a filling system correctly molded to shape in sand would give a superior quality ingot (sand inclusions are never observed if the channels are correctly shaped) and would be significantly more economic and environmentally friendly. The filling system design is more fully described in Section 1.5.5.

A contact pour ingot may be 10 to 100 times cleaner than a standard ingot, but this may on occasions not be good enough for some special products. In the quest for perfection therefore, the remaining impairment of a contact poured ingot resides in the priming process. The air already in the filling system channels mixes with the initial fall of metal, so that a clutch of bubbles and oxidized metal is released into the ingot mold during the first few seconds of the pour. A design of filling system in which this first metal is discarded, and bubbles diverted by the use of a "filter" (actually a ceramic block permeated by pores usually around 1 mm diameter) into a centrifugal trap, is a relatively low cost sophistication which may be sometimes worthwhile (Campbell 2018).

1.6.2 Continuous casting

Continuous casting is a far more efficient casting technique for bulk steel as a result of its continuous operation, reducing discard from the top and tail of rolled ingots, and its saving of significant amount of rolling by providing a semi-finished shape.

In addition, however, continuous casting is often quoted as delivering a better quality of steel (fewer inclusions) than ingot steel. Part of this enhanced quality will derive from the greater length of time it takes to deliver metal from the ladle into the casting machine. This extended time will greatly help the floating out of damage introduced by pouring from the melting furnace.

The wide adoption of continuous casting eliminates, in principle, many of the problems of the batch disadvantages of ingot casting. Even so, there remain a number of fundamental issues with continuous casting which hamper the development of its full potential.

First of all, it is not a continuous process. It has to be started. This necessarily involves the pouring of liquid metal from a height of at least a meter into an empty mold, sealed at its base with the dummy bar in position. The fall of the melt into the mold is necessarily turbulent and splashy. For most steels, this will create masses of bifilms. Although these will be concentrated in the early part of the ingot, of which typically 2 or 3 m must be cut off and discarded, the continuous addition to the liquid metal in the mold during casting will continuously dilute the originally contaminated metal, distributing bifilms far along the length of the cast strand. Naturally, even though the numbers of defects will reduce with distance from the start, the original generous quantities of bifilms will never be entirely diluted to zero. This unfortunate start-up damage is a major issue limiting the production of clean steels by continuous casting.

This traumatic start to what is otherwise a tolerably low-turbulence casting system requires to be eliminated if possible. It is not suggested that this will be easy. However, this is the certainly a major contributor to inclusions in commercial steels and must be addressed if clean steels are to be targeted.

A potential solution is shown in Fig. 1.60 in which all launders, particularly the final launder which empties into the mold, is fitted not only with a submerged entry nozzle (SEN) but with a temporary extension nozzle (TEN). The TEN is internally tapered to follow the contour of the falling stream. Both nozzles are initially closed. The TEN opens first to deliver the metal stream to the base of the launder or mold. The details of the design of this tube and the geometry of its impact with the launder base or dummy bar require to be carefully designed to reduce surface turbulence to a minimum; in the figure the TEN targets a simple steeply walled depression to confer some control over surface turbulence during filling. The placing of a ceramic foam filter at the base of this depression will significantly aid the control of splashing and turbulence during the initial seconds of filling. When the mold is filled to the level at which the SEN can take over, the TEN is closed and the SEN opened. The TEN is at that point superfluous, and unwanted. If made of steel it may melt sufficiently (while

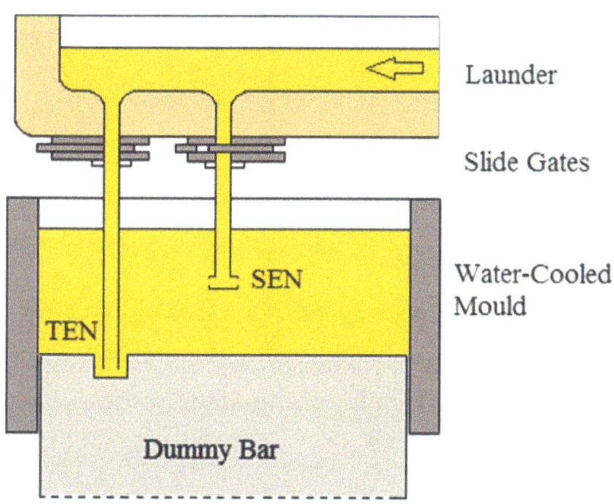

Figure 1.60 A suggested scheme to reduce start up turbulence in continuous casting.

introducing a few relatively minor defects such as its own oxide skin!). If made of ceramic it would have to be detached and lifted clear. Tilting the TEN will facilitate its withdrawal from the mold. Naturally, there is plenty of room in a slab mold, but some smaller square or round casters may have problems to withdraw the TEN. Although slide gates are shown in Fig. 1.60, other control features such as mechanical stoppers or fusible discs may work sufficiently well, particularly if the discs were teamed with a controlled heat source such as an induction coil.

Clearly, this suggestion has not been tried, and may not work, or may not be practical. The arrangement in Fig. 1.60 is suggested at this point to emphasize that some kind of solution is required. It is not acceptable to continue to live with the seriously damaging starting trauma of this important process.

In the early days of continuous casting the withdrawal of the casting from the mold was intensively studied, resulting in a system in which the cast strand was held stationary for a short period in the mold to build up and strengthen its solidified skin to avoid break-outs. This was followed by a withdrawal stroke of varying distance and rate. These efforts seemed to deliver mixed results. It was the arrival of a casting flux, a synthetic low melting point slag, added as powder to the mold, and which melted to protect the surface of the liquid metal and lubricate the withdrawal action, which had good results. However, the deep "oscillation marks" on continuously cast steel were a problem for subsequent working, especially rolling. Although the casting flux helped, the oscillation marks have never quite gone away.

Looking back on the reasons for the early experiences with break-outs of liquid metal as the cast strand emerged from the mold, it seems certain this was a bifilm problem. If a bifilm of sufficient size happened to find itself near to the edge of the mold, and if the mold friction on the withdrawal stroke was high, the bifilm would open, forming a fracture for the escape of metal.

If bifilms were better controlled by better launder and mold filling, break-outs would probably disappear. The intermittent withdrawal techniques now used for steels might usefully be abandoned, reverting to simple continuous withdrawal as used for the continuous casting of Al alloys. The regular laps, the oscillation marks, which are a feature of the surface of continuous cast steel might then be eliminated. Smooth, continuously cast metal would be welcomed by all.

1.6.2.1 *Nozzle design*

The stop-valve systems for the transfer of liquid metals consist of a ceramic or refractory nozzle which can be sealed by a mating stopper. They are provided at the base of ladles and launders, so that metal can be released from the ladle or launder when the stopper is raised.

Most nozzles which the author has seen are provided with a parallel bore. This must present difficulties for manufacture, because it would seem probable that a tapered bore could more easily be formed. Moreover, the widespread use of parallel bores is perversely unhelpful. Such nozzles in the base of ladles are designed to block by the build up of oxide because they will never be able to fill properly. The fast jet of metal, moving at least at 5 m/s down the center of the nozzle, will drag air down with it in the annular unwetted gap formed by the metal accelerating and drawing away from the walls of the nozzle as it falls. As air is accelerated downwards inside the nozzle, air is necessarily sucked upwards into the remainder of the annular gap to replace it (Fig. 1.61). This upward drawing of air into the nozzle continuously fuels the oxidation inside the nozzle, eventually leading to its blockage. Most researchers have assumed that blockage occurs by the accretion of alumina inclusions, carried in suspension in the melt, on to the walls of the nozzle during the passage of the metal through the

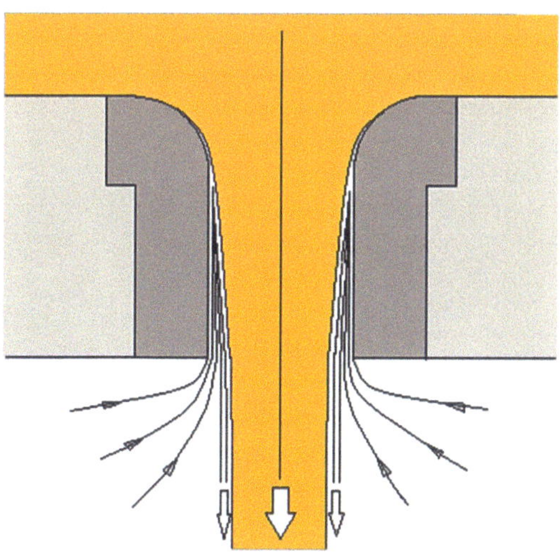

Figure 1.61 The entrainment of air up into a parallel bore nozzle, to replace air dragged down by the metal stream. The enhanced oxidation leads to blockage of the nozzle by oxide.

nozzle. This is not likely to be true because the liquid metal is out of contact with most of the nozzle. The fact that build up of oxide is reduced if a well-fitting argon shroud is used to keep air away from the nozzle, indicates air entrainment driven by the zero-taper of the nozzle is at fault.

Observations of Way (2001) that show the blockage of nozzles of steels in continuous casting plants to be the result of arrays of separate granular alumina inclusions (Fig. 1.62). Separate inclusions somehow separated from the fast flowing jet and somehow adhering to the nozzle seem hardly conceivable. Much more probably, such inclusions were originally alumina films formed by the annular back-circulation of oxidizing air inside the nozzle where the flow detaches from the nozzle wall (Fig. 1.59).

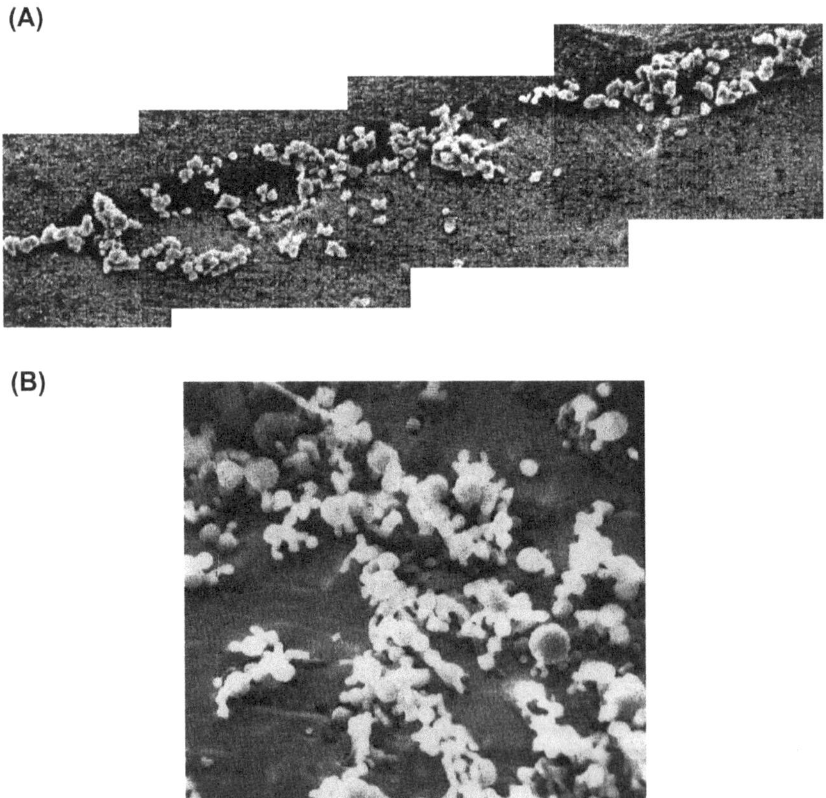

Figure 1.62 (A) Alumina stringers in deep-etched rolled steel; (B) alumina particles out of a blocked nozzle from a continuous casting plant for steel (Way, 2001).

The battering of the films in the energetic turbulence of the air and metal interface, plus the action of coarsening to reduce surface energy, would finally result in the films transforming into particles.

The continuous casting process generally uses a succession of metal-handling features, all of which require nozzles with stoppers. These are (i) from the ladle, (ii) controlling distribution into a cascade of launders, and finally (iii) into the mold by the SEN. The SEN is a rather special variety of nozzle, extended in length, and probably also has a parallel bore, making its own contribution to air entrainment and oxide creation. Like the SEN, all nozzles would benefit from being correctly tapered and having their exits submerged.

Traditionally, an attempt is made to counter air entrainment by argon infiltration into the metal stream. This is another threatening technology. Instead of a smoothly continuous flow, the flow is now intermittent, separated by bubbles which enter the mold. Not all the bubbles will detrain successfully. The trapping of argon bubbles in the steel is known to occur. These are nontrivial defects which, as a result of the insolubility of argon, will survive working operations and constitute potential sites for the initiation of pores and cracks. The avoidance of argon by correctly tapering all the nozzles and the SEN would avoid argon entrainment defects.

Although the use of argon to reduce the oxidation of the surface of the liquid metal in launders is essential, the enthusiasm for bubbling through the liquid is probably counter-productive: the disturbance of the surface entrains defects, and the contaminants in the argon, which will be lucky to be as pure as 80% by the time it is entrained, will react to create inclusions in suspension.

Much research on continuous casting has been plowed into the design of launders and the action of the SEN in the melt pool to encourage the *detrainment* of inclusions. However, the prevention of *entrainment* has been largely ignored, but is, of course, of immense importance; no detrainment from launders would be necessary if attention was paid to avoiding entrainment in the beginning!

It remains, therefore, the prevention of entrainment is likely to be far more effective in reducing inclusion content of steels, because entrainment is fast, certain, occurring in a fraction of a second, but can be controlled; detrainment is slow and uncertain, and will therefore always be limited in what it can achieve.

1.6.3 Secondary remelting processes

Having made our ingot, or length of continuous cast product, its relatively fine dispersion of inclusions continues to degrade properties to the extent that for many applications a further melting and refining process is beneficial and cost effective.

For steels, the secondary remelting industry is widespread and active. It employs vacuum induction remelting (VIM), vacuum arc remelting, sometimes referred to as vacuum arc refining (VAR) and electroslag remelting (ESR).

The ESR and VAR processes are the most widely accepted refining processes. They are based on the concept of taking a steel electrode of ordinarily "good quality," or possibly a VIM melted and cast electrode, and remelting the steel drop by drop using electrical energy. The fairly rapid solidification of the liquid metal in a water-cooled mold ensures impurities are redistributed more finely, making an outstandingly clean product. The secondary remelted steels are used for products requiring the highest strengths and reliability.

The mechanisms for the improvement in properties are rather different in each process, but we shall start with the VIM process which is common to both, because both remelting operations often use VIM ingots as consumable electrodes.

1.6.3.1 Vacuum induction melting (VIM)

The VIM process is shown at its simplest in Fig. 1.63 in which a melt of special steel or Ni alloy is melted under vacuum and top poured under vacuum, dropping heights of at least a meter, into an ingot mold. Practically all research installations worldwide are of this simple design, so that practically all research on steels and Ni-base alloys has been carried out using this process.

In industrial practice, VIM installations are nothing like this simple cylindrical furnace. The industrial equivalent can be awesome, monumental in size and complex and design, occupying prodigious areas of the foundry. The furnace assembly is multi-chambered, allowing charging under vacuum and pouring into a long, vacuum launder system which can deliver metal under vacuum to many ingot molds simultaneously. This is truly daunting technology.

However, whether a small simple research furnace, or a vast industrial unit, the central problem with all VIM installations is the top pouring into the mold. The technique creates bifilms, the reactive alloys in the liquid

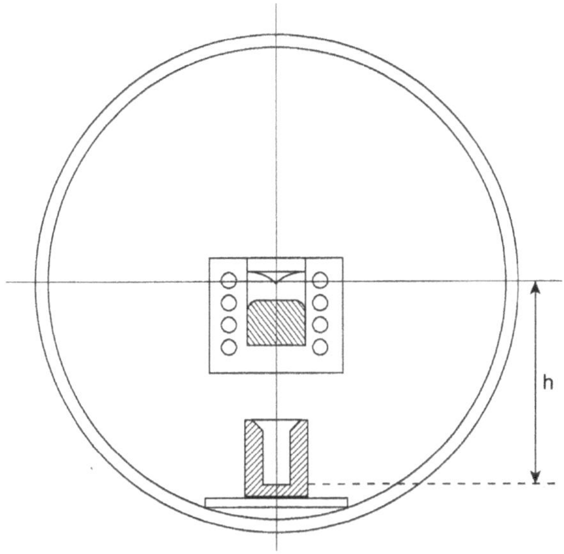

Figure 1.63 The problem of the fall distance "h" during top pouring in a conventional vacuum induction furnace.

metal "gettering" the oxygen available in the vacuum. As the reader will by now appreciate, bifilms are already difficult to see when casting in air, and now in vacuum are immensely thinner, making them practically impossible to detect. The VIM steel does give the strong impression of being cleaner.

The fact that the oxygen content of the steel is reduced has been widely accepted as proof that the steel is cleaner, which, of course, in a way it is. However, there is no evidence that its population of cracks has been reduced. It is probable that the crack population will be largely unchanged, being representative of what would be expected from the fall height for that particular steel, whether poured in air or vacuum. Careful thought indicates it is reasonable to suppose that the type of atmosphere surrounding the pour would not be expected to reduce the bifilm population in any significant way. Dilute air would produce thinner bifilms than undiluted air, but their area would not necessarily be changed.

The true potential of VIM might be realized if its top pouring mode could be avoided in some way. By way of suggestion, Fig. 1.64 is offered. It is, as yet, totally unproven and quite possibly an unworkable concept. The bottom drain, whether slide gate or fusible disc, will almost certainly require a supplementary heating technique such as a small separate

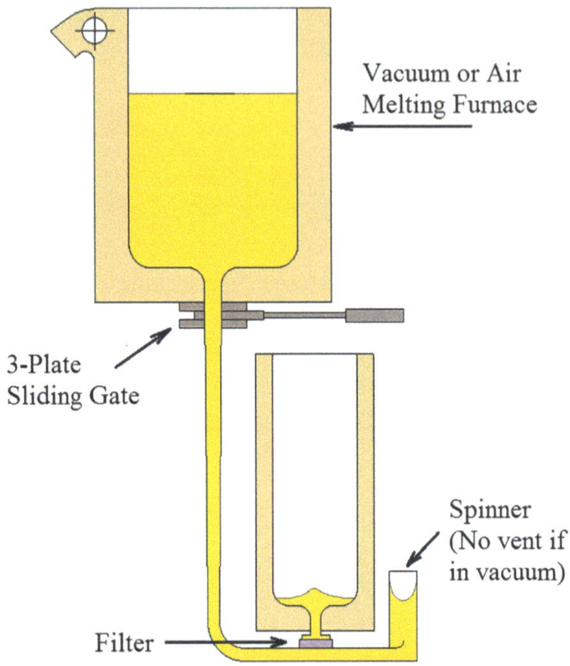

Figure 1.64 A potential but untested suggestion for bifilm-free VIM ingots avoiding top pouring.

induction coil – possibly disposable, and preferably avoiding water cooling to avoid the possibility of accidents.

For existing VIM installations, the reader is challenged to either test this suggestion, or think of some better solution to avoid pouring. Counter-gravity could be an attractive alternative, and probably subject to fewer problematic features such as uncontrolled leaks or blockages due to freezing. Furthermore, for future installations, counter-gravity would almost certainly be more easily engineered.

Ingots cast by VIM are commonly specified as consumable electrode material for the secondary refining processes, VAR or ESR (Fig. 1.65).

Both VAR and ESR remelting processes use heavy electrical currents passed through the electrode to melt the tip; the VAR process using an arc as its source of heat, and the ESR process the resistance heating of a depth of liquid slag. The ingot is built up drop by drop. Both processes produce ingots which are regarded as ultimate achievers of the cleanest steels. It is true that their microstructures are impressively clean, with fine inclusions finely distributed. Thus, when small parts of the ingots are sampled for

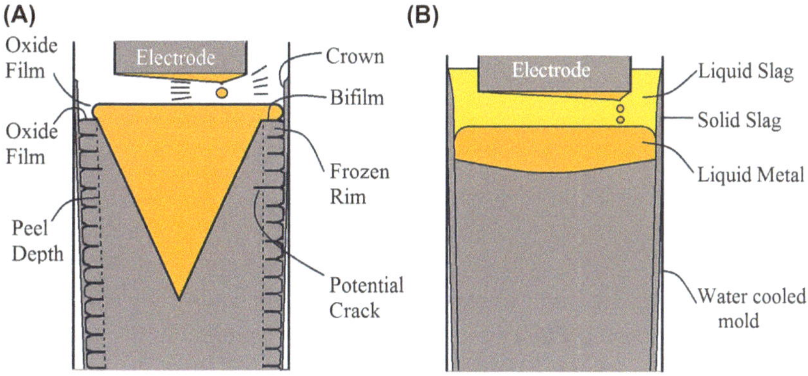

Figure 1.65 (A) VAR and (B) ESR processes.

such tests as tensile or fatigue, these small samples of metal display excellent results. Such small samples, however, usually cannot alert the user to macroscopic faults. This is the subject of concern in this account.

It is regrettable therefore that the bifilm defects in the top-poured air cast or VIM electrode are a threat to the success of these subsequent refining processes. This is because the bifilm cracks are sometimes extensive, occupying a large percentage of the area of the electrode. When the electrode is gradually melted, on occasions, a large rock-sized chunk of the electrode can detach and fall into the growing VAR or ESR ingot, crashing through the slag layer in the case of ESR and splashing into the metal pool. The introduction of such large, unmelted fragments, together with the oxide film covering its surface, means that a massive bifilm can be introduced into both VAR and ESR ingots, wrapped around a large chunk of unmelted and unrefined metal. The unmelted chunk will also preserve its own interior population of bifilms unchanged. I have personally seen such a defect larger than my fist in the center of an otherwise excellent ESR ingot.

At the present time therefore, electrodes for secondary remelting processes are not themselves reliable, and threaten the remelted ingots with uncontrollable and very serious defects.

1.6.3.2 Vacuum arc remelting (VAR)

"Vacuum arc remelting" is often referred to by the producers as "vacuum arc refining." This is not altogether untrue, but the words "vacuum" and "refining" give an aura of quality, conferring a distinct commercial advantage to this material. For VAR we shall stay with the descriptive name "remelting" rather than "refining," and shall attempt to give a more objective assessment.

(The more prosaic name for its competitor, electroslag remelting (ESR) has overtones of "slag" as an unwanted nonmetallic phase, representing a distinct commercial disadvantage. Although in its early years some effort was made to re-naming the processes "Electoflux Remelting" the name was never widely adopted. It is a great pity that ESR labors under such a commercially unattractive name, when the process and the product offer many benefits.)

Remelting by VAR occurs by the passing of droplets of metal through an electric arc under vacuum, ensuring that hydrogen and other volatiles are reduced to a low level. The very high temperature gradient in the solidifying steel provides a fine dendrite arm spacing, so that inclusions, if re-precipitated at all, are formed extremely finely and well dispersed. So far so good.

The splatter of droplets and condensation of volatiles against the water-cooled walls of the vacuum chamber above the slag layer produces a thin deposit known as "The Crown." This is gradually overtaken by the ingot as it forms. However, the outside meniscus of the metal against the wall of the mold appears to be substantially solid, and certainly oxidized as a result of the relatively poor vacuum. The result is that the low melting constituents of the crown melt but do not mix with the bulk of the ingot. The melted crown fills the gap between the contracting ingot and the water-cooled mold wall, forming a smooth outer skin for the ingot, covering the laps shown in Fig. 1.66.

Figure 1.66 The build-up of a VAR ingot in layers separated by circumferential bifilm cracks.

VAR suffers occasional falling in and partial remelting of its "crown." This material will be to some degree oxidized and so will result in the creation a large bifilm as the oxide surface on the melt pool wraps around the oxide on the submerging crown fragment (Fig. 1.8). I have been unable to find any reports about this possibility. It is possible that it is not important. However, it lurks as an uncontrolled and uncontrollable feature of the process, and its behavior appears to be unknown.

By far the most important feature of the VAR process appears to be generally ignored. It is the way in which liquid metal dribbling from the electrode builds up the ingot. The newly arriving metal forms a central pool, but the outlying regions of the surface close to the water-cooled mold may be already solidified or partly solidified. The arriving liquid spreads, overflowing those partly solid outlying regions which are strong enough to support the liquid rolling over its surface. The spreading action of the liquid may occur either in irregular "stops and starts," or continuously spiraling around the circumference. (Spiral advance is the norm in the continuous casting of aluminum alloy cylindrical ingots known as "logs"). The rolling over of the spreading metal, laying down its new oxide on the already oxidized surface of the ingot forms extensive, horizontal bifilm cracks as has been described in Section 1.4.1, and is seen here in Fig. 1.66.

The depth of cracks will be variable because the melting region of the electrode wanders over the base of the electrode. This migration of the source of heat and source of hot metal will occur partly because of the natural action of the melting regime, in some way transversely progressive, which will be characteristic of the melting geometry on a planar surface. This would be an interesting study. Of more concern, it seems probable that the electrode, having been turbulently cast, will contain large bifilms. These layers of oxide and air will be insulating barriers, re-directing and concentrating the flow of electrical current in different parts of the electrode, and thereby causing the arc to take large erratic migrations. The surface of the growing ingot will therefore suffer irregular changes to the site of arrival of hot metal, so that large parts of the surface will cool, possibly starting to freeze sufficiently to support the weight of the arrival of fresh metal. Most seriously, the cooling metal will oxidize. New hot metal will eventually arrive to roll over the cool oxide films, laying down the fresh thin oxide on the older thicker oxide to create asymmetrical bifilms as extensive horizontal cracks of variable area and depth. The cracks will decorate the exterior of the ingot, at vertical spacings in steps of approximately 8 mm which is the height of the sessile ("sitting") drop of most steels and high temperature alloys.

The cracks are not obvious on the as-cast ingot when it is taken out of the mold. This is because the crown, with its spatter and lower melting point constituents, high in condensed Mn-rich vapor, melts to cover the ingot surface. In the domestic parlance of home decorating, the appearance of the ingot benefits from a kind of "papering over the cracks."

The cracks are well known to continue to be effective despite their covering of melted crown. Personal experience can be recalled such as attempting to forge a VAR ingot of Ni alloy "Waspalloy." When not first peeled to remove the surface cracks, it cracked on the first stroke of the forge – in contrast to an otherwise identical as-cast, un-peeled, ESR ingot in its as-cast skin, which forged like butter.

For a sufficiently large ingot diameter, and sufficiently slow rate of melting, the formation of cracks at every 8 mm height of the ingot, around the circumference of a 300 mm diameter ingot 1 m high would lead to a total crack length of over 100 m. Although, of course, this outrageous length appears ridiculous, the huge figure is still huge even if reduced by an order of magnitude. It is highly probable that cracks will be present, and of highly variable depth. It is significant that Qu's team (2019) studied the structure of a triple-melted superalloy VAR ingot, finding that oxide inclusions reduced, and properties increased, up to a depth of approximately 48 mm.

The presence of the cracks becomes obvious if the ingot is forged because the ingot simply cracks. Cracking on forging seems, in general, to be reduced by machining away the ingot surface. The machining of steel VAR ingots is carried out by grinding. For more expensive Ni materials, the surface is machined by turning on a lathe, commonly referred to as "peeling." This yields "peelings" (turnings) which are more easily remelted.

In practice, the surface removal seems usually to be carried out to a depth of approximately 5–10 mm. This is clearly a serious concern in view of Qu's findings which suggest cracks may be close to 50 mm deep. Despite the fact that the ingots, in general, seem to forge satisfactorily, it is impossible to believe that all the cracks are completely removed. The compressive stress in the surface layers of the ingot, as in all cast products (with corresponding balancing tensile stresses in the center) will ensure that, after solidification and cooling to room temperature, all surface cracks will remain firmly closed. In view of the very modest working of secondary remelted ingots by forging, any cracks remaining after peeling are certain to survive and be present in the final forged product.

It is to be expected therefore that VAR steels will be cracked, and that at least a proportion of the cracks, as thin bifilms under compression, are probably not detectable.

The presence of such cracks has been reported in commercially available material; Cambridge researchers (Peet et al., 2017) specially selected VAR material as the starting material for their research to have the best chance of a steel as clean as possible. They were greatly surprised to discover the presence of a 60 mm long crack (Fig. 1.67) parallel to the peeling cuts. This is not expected to be an exceptional event.

More seriously, tragic engineering accidents have been reported in detail, particularly for helicopter failures, for which it seems that no other explanation can be found other than the probability that bifilm cracks in VAR steels are responsible (Campbell, 2016).

1.6.3.3 Electroslag remelting (ESR)

In contrast to the several problems of VAR, ESR suffers from few of these faults. It remains vulnerable to the falling in of electrode fragments, but this is hardly the fault of the ESR process; a good electrode production route would eliminate this potential danger. If ESR electrodes were produced using *contact pouring in air* it would yield a reliability and perfection of secondary remelted steel which would probably be unbeatable. This appears never to have been tested. The ingot itself is fundamentally crack-free and can therefore be forged in its as-cast state without problem.

In the ESR process the heat for melting is provided by Joule heating in a liquid slag layer. The electrode melts under the slag surface, transferring its

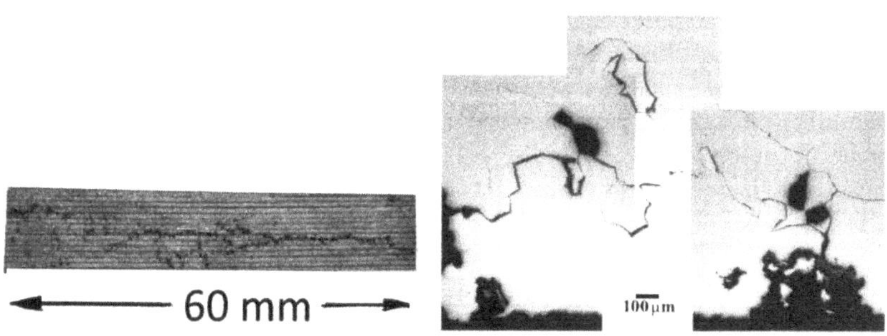

Figure 1.67 A crack in a VAR ingot revealed after a salt-bath heat treatment (Peet et al., 2017).

liquid metal drop by drop into the molten pool. The size of each droplet of liquid metal is approximately 10 mm. If a bifilm from the electrode was present in the droplet, and if it projected through the surface of the droplet into the slag phase, it would be wetted by the slag and sucked out of the steel by capillary forces. Thus, we can probably set an upper limit to the size of bifilms which can be present in an ESR ingot. This simplistic approach neglects any furling of the bifilm, of course. However, we can refine our estimate of maximum bifilm size further. The bifilm must first travel through the molten layer covering the base of the electrode, before it reaches the location where droplets are detaching. The thickness of layer of melted metal will be only of the order of 1 mm, placing a narrow limit on the maximum possible size of bifilm which can be present. Furthermore, of course, any surviving bifilms have additional time to float out from the metal pool prior to solidification.

The whole surface of the molten metal at every stage of the process is protected by the slag, which, as a high temperature liquid oxide mixture, will take into solution any oxide films or bifilms, not only because of thermodynamic considerations, but because of the high surface energy of these thin features encourages their mechanical extraction by the slag if they intersect the metal interface. The liquid slag serves the additional role of providing a uniform and steady heat source, and a reservoir of heat, so that freezing of the ingot occurs under steady and controllable conditions. In addition, electromagnetic forces arising naturally from the flow of electrical current, stir the slag layer, assisting the redistribution of heat. Thus, the process does not suffer the problems of the migration off axis of a concentrated heat source.

It seems probable that the ESR process is successful to eliminate bifilms larger than a few millimeters (even though it is probably that a large population of smaller bifilms remains) which may explain the mystery reported by Hoyle in his book on the ESR process (1983). Hoyle reports the universal observation of improved properties for steels and Ni alloys following ESR treatment, but has to admit the apparent absence of any detectable metallurgical reason.

In conclusion,

(i) With regret, VAR steel cannot be trusted. As much as its successes are recognized and promoted, its failures appear so far to have been overlooked.

(ii) The use of ESR with a standard VIM electrode is, at this time, an unknown risk, but is definitely far less of a risk than VAR. The very thin bifilms created during the top pour under vacuum may pose the danger of parts of the VIM electrode detaching and falling into the molten pool. This would require to be investigated and proven to be safe, prior to accepting the use of VIM/ESR as a safe production route.

(iii) The use of ESR with VIM using an improved vacuum casting process, for instance, Vacuum Contact Pour or "Vacuum Counter-Gravity Filling" may represent an ultimate in cleanness and reliability if it can be achieved. However, perhaps perversely, it may be "a bridge too far," being unnecessarily clean if Contact Pour in air followed by ESR proves to have, as might be expected, unbeatable mechanical properties, and the absence of major faults. Unbeatable properties would include generous ductility and toughness, and practically infinite fatigue life.

(iv) "Contact Pour in Air"/ESR is achievable immediately, and may represent an unbeatable if not ultimate combination for perfection in reliability.

Despite its low cost, CPA/ESR, in addition to its improved reliability, might equal the expected cleanness and mechanical property performance of VIM/ESR, and might, or might very nearly, equal the expected achievements and reliability of "VIM with improved vacuum casting process"/ESR if such an option was ever to become practical.

1.6.3.4 Commercial thoughts

A final note on the difficulties which can be envisaged for producers of VAR metals seems appropriate. It is hoped that it is not regarded as ignorant and unwelcome intrusion on top of unwarranted criticism. It seems inevitable that the criticisms leveled here at current ESR and especially at VAR processes will, with great regret, present the secondary remelting industry with major challenges. Even so, the industry must admit that it has an impressively strong customer base in aerospace and defense, and every VAR producer I know already has the ESR process working alongside. Thus, the industry would be in an excellent position to expand and upgrade the reliability of its ESR production, taking up a new marketing stance to promote reliability and promote enhanced ESR in place of VAR. In addition to an increased reliability of electrode material, there are well known techniques for improving ESR materials, such as using collar molds to enhance the cooling rate of the ingot by direct water cooling. The application of an inert gas environment over the melt also tends to reduce dissolved gases etc. The other marketing aspects of ESR should be emphasized such as the flexibility of ingot shape, no longer limited to cylindrical forms, but changeable to the degree that even some shaped castings can be produced.

In the meantime, improved inspection and/or significantly greater peeling depths would temporarily allow the upgrading of VAR reliability at some cost, but would permit, if necessary, a controlled retreat from this market. Without doubt there will be other possible strategies. These major changes will not be easy and not painless. However, the elimination of the promotion and provision of cracked steels is, of course, necessary. Our duty of care as engineers leaves us no option.

1.6.4 Shaped castings

Shaped castings have an important place in the world of engineering; so important that it is difficult to list the numerous applications. Steel gears and Al alloy gearbox housings for helicopters, turbine housings in steel for power generation, rotor shafts and turbine shafts, valves and fittings in stainless steels or titanium alloys for the petrochemical industries, etc. The list may not be endless, but is awfully long.

It must be admitted that the final product quality is often not up to standard that either the caster or the customer would like. Most castings will be crammed with bifilms, some so large that the castings may leak or crack in service. The training which the caster may have received is likely to have at least paid lip service to the concept of the importance of turbulence-free filling systems. However, turbulence was never brought under control by early filling system designs, and the overwhelming poor performance of cast products was not realized. My first book, Castings 1991, contained only an outline review of approaches existing at that time, which in general were unsuccessful to address the problem.

The presence of major crack-like defects in Al alloy castings is usually easy to demonstrate. If a needed feeder head is not provided for a casting, the resulting nonfed condition will result, unsurprisingly, in shrinkage porosity. However, much worse are the cracks which now proceed to tear the casting apart. The bifilms remained closed and unsuspected while the feeder continued to pressurize the solidifying metal in the mold. As soon as this pressure is removed, the bifilms start to open up under the action of shrinkage. The finished casting resembles a heap of fragments, possibly still hanging together by a few connections. (The doubting reader is encouraged to carry out this exercise; a good example is a permanent mold cast automotive cylinder head produced by a poorly designed filling system, and normally covered in a mass of feeders weighing more than the casting.)

This serious issue has only gradually been understood, and solutions gradually worked out since about 2000. Successive publications have provided improvements, culminating only recently (Campbell 2018), when I finally realized that bubble entrainment, causing bubble trails, a kind of long bifilm, was probably the major damage-producing mechanism in most cast metals (see Section 1.4.2). In addition, most filling systems used conical pouring basins which acted as air pumps, vastly increasing the content of air introduced together with the liquid metal, causing massive problems from bubble damage.

An early experiment to compare a traditional filling system design for turbine blades was compared in an air pour for a Ni alloy (Li et al., 2004). Although the detailed mechanism of defect generation was not specifically targeted at that time, a pouring design which was predicted to be an improvement over the traditional system reduced defects by a factor of 10.

It is a relief to report that the performance of gravity filling systems has now been improved by at least a further factor of 10. For the first time, techniques have now been devised to address both bifilm and bubble damage in ingots and shaped castings. Better melting conditions and the employment of counter-gravity filling would realize even greater potential (JC 2012 and 2018).

Fig. 1.68 shows one example of the new thinking. It contrasts with the traditional filling from a bottom poured ladle into a conical trumpet and parallel pot runner system, which, despite its bottom gating, results in a seriously faulty product.

The new technology includes contact pouring to exclude air at the system intake, plus a hyperbolically tapered down-sprue, a smoothly ducted transition to the runner, maintaining the velocity of the melt. The first metal bypasses the casting ingate, protected by the novel flush filter, and proceeds to a spin trap. This early metal unavoidably contains air, so bubbles are necessarily entrained during the priming of the filling system. The centrifugal trap holds the bubbles, centrifuging them into the center of the trap so they cannot escape back into the runner. As the metal in the spinner rises, it provides a gradually increasing back-pressure to the flush filter, encouraging metal through the filter sufficiently gently that no jetting or splashing occurs in the mold cavity. After 1 or 2 s, the spinner is filled, at which instant the full pressure from the down-sprue becomes available on the filter and ingate to fill the mold at its designed speed (Fig. 1.68).

Steel and Ni-base castings poured in air by the new gravity-filling systems are immensely improved, being cleaner by factors of 10 or 100 or more. Much of the limit to the cleanness of cast metals now rests with the prior metal handling out of the melting furnace and into a ladle. For the future, counter-gravity casting of steel direct from the melting unit will clearly solve most of these issues.

For aluminum alloy castings, the counter-gravity Cosworth Process, built in 1980, was far from perfect in this respect, but showed that aerospace-quality of toughness and fatigue performance could be achieved immediately. Further developments from those early days have ensured that most of the problems are now solved. Castings of extreme cleanness and with significantly higher properties are to be expected as such processes become commercially available.

The above are summaries of exciting and substantial developments in gravity and counter-gravity casting in air.

Turning now to vacuum casting, the situation is rather less satisfactory. The provision of a filling system which can fill successfully to suppress bifilm formation is not easy in most vacuum environments, and nearly all vacuum casting systems fail in this respect.

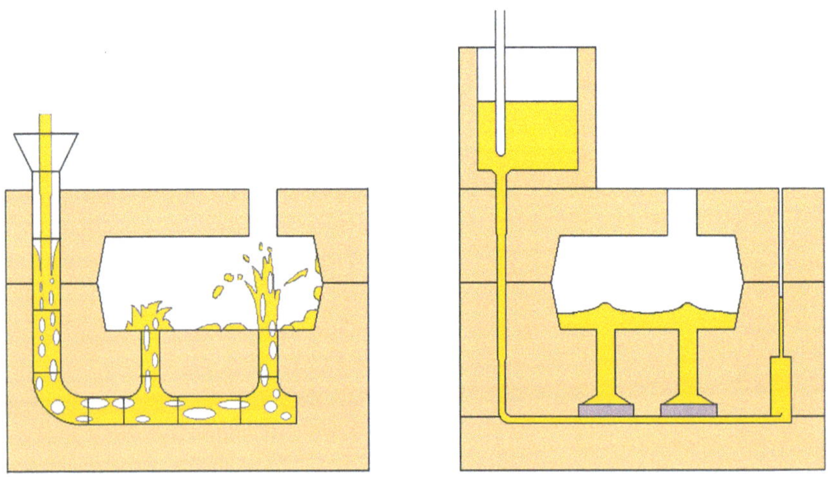

Figure 1.68 The traditional filling technique for castings using preformed conical pour basin and refractory tube system, compared with the new filling technology using contact pour, sand molded channels, flush filters and spinners (Mini Casting Handbook, 2018).

In Ni-base alloys, the provision of melting stock by Ni alloy producers using top pouring into long steel tubes ensures that the metal is degraded even before the turbine blade foundry receive it. Whether the metal improves or degrades further during the melting process in the foundry is not clear – if the metal quality changes during melting, then the dwell time in the melting crucible prior to pouring will probably be important. However, in nearly all cases of pouring in vacuum, the pouring geometry is primitive. It seems so difficult to rectify this situation that it may be that the only remedy is counter-gravity casting. If it could be achieved by both alloy makers and casting producers, it would be expected to solve quality problems in this industry at a stroke.

Titanium alloy shaped castings are melted and cast under vacuum, but, but with regret, under such grievously turbulent conditions they require to be HIPed to gain some reliability. Once again, it seems to me that the only way forward that has any chance of success is counter-gravity casting.

1.7 Potential for quality assurance

Standing back for a moment to view the larger scene of the commercial supply of castings, it is particularly sobering to see the proliferation of standards and procedures throughout the world to control a number of features of castings in an attempt to control their properties. Although once widely known as "Quality Control" (QC) the procedures to maintain the quality of products are now more appropriately named "Quality Assurance" (QA).

For instance, one of the central feature of all standards for steels and other alloys is their chemistry. The chemistry is under extreme control with upper and lower limits cited for all major elements. In terms of the "quality" of the final product, the observable defects such as gas porosity and shrinkage porosity in castings are usually closely limited or forbidden.

Naturally, although I admit to important exceptions, in general, I personally am skeptical about the slavish devotion to accurate chemistry; much of it appears to be to be capable of significant relaxation – a move which would greatly benefit the world by reducing the vastly over-generous number of alloys we have. In any case, the likely presence of bifilms can easily outweigh chemistry excursions. In addition, of course, they are difficult, if not impossible, to detect with any degree of reliability, and are likely to create more serious consequences for properties.

The significance of bifilms is clear, and worth repeating. They are often not detectable, or on the limit of detectability by the common nondestructive testing technique dye penetrant testing, but can be more important than clearly observable defects. They are often so numerous and/or so large that they can control the properties of castings, often outweighing the effects of alloying and heat treatment.

Bifilms which happen to intersect the surfaces of castings or their machined surfaces are sometimes detectable by dye penetrant or other techniques. However, as a result of the general situation in which the interior regions of castings experience tensile stress, and the surface regions are in compression, it follows that bifilms are often tightly closed, making them difficult, if not impossible, to detect.

The much rarer situation of a very clean material with only a few large bifilms is illustrated in the vacuum arc remelted (VAR) material. A VAR ingot with only one 25 mm deep bifilm crack, if totally cut up to make 100 tensile test bars, is most likely to reveal all 100 bars with perfect properties. At worst, the result would be 99 bars acceptable and 1 failure. In practice, of course, the whole ingot is not destructively tested, but three or four test pieces are taken which are assumed to be representative of the whole of the ingot. These few samples will guarantee that the ingot will be accepted even though it contains a major defect which will probably survive processing, but which may destroy the helicopter. Clearly, the reliance on sampling techniques in the case of an unsatisfactory casting technique is seriously misguided.

In stark contrast, it is salutary to think that a similar sampling technique in the case of an electroslag remelted (ESR) ingot would most probably have been completely representative and therefore reliable.

The conclusion is inescapable: it is more important to specify and control the casting manufacturing *process* to *avoid* the formation of bifilms than to employ apparently rigorous QA procedures, searching retrospectively (and probably without success) for defects.

I have a clear memory of one sunny day, on emerging from the blackness of the foundry. I was complaining to QA experts from Boeing, Dale McLellan and Fred Feiertag, that although many specifications were in place to control the chemistry and other features of casting products, specifications to ensure effective manufacturing techniques to ensure bifilm-free castings did not exist. Clearly, Dale became impatient with my complaints and turned to me, saying "Write one!" That stopped me in my tracks. He knew how to issue a challenge. The result, about 18 months later, in

2004, was the publication of "Casting Practice." In spite of my best efforts, I found it beyond my ability to write it as a specification. It was merely an explanation of casting problems together with some recipes for good casting practice as was understood at the time. Its recommendations are, of course, now rather dated.

Time has passed, but a manufacturing specification has never appeared. The later "Complete Casting Handbook' 2011 and revised in 2015 contained updated recommendations for good practice. The distilled and down-sized version of this book "Mini Casting Handbook" was designed specifically for casting producers in 2017. This in turn was updated in 2018 (Campbell 2018) to contain the first effectively guaranteed design rules for the achievement of defect-free cast products. It remains the nearest approach so far to a manufacturing specification for cast products.

However, these latest recommendations are effectively guidelines, not a specification. But they do, for the first time, appear to offer solutions for defect-free castings. Such products would be expected to be significantly stronger, totally reliable, and ultimately lower in cost than those made by our current production techniques. The first steps are being demonstrated for a number of alloy systems in a number of casting operations world-wide as I write. These are early days.

The early results are encouraging, but we remain without a manufacturing specification. Therein lies a challenge for someone or some organization. Dale has now long been retired, but I am sure he would be pleased to see it.

CHAPTER 2

Fracture in the liquid/solid state

Contents

We shall consider the detailed mechanisms of failure of partly solidified metals under tensile strain. Such failures are known as hot tears, or hot cracks, or described as a "hot-short" condition. We shall use these terms interchangeably.

The hot tear or crack is one of the most serious defects in cast products. It is the cause of breakouts of liquid steel from continuous casting machines and thereby limits the rate of production of many steels. In addition, the need for major weld repair and upgrading costs tends often to double the cost of many steel castings. Some shaped steel castings, large turbine housings for instance, sometimes resemble hedgehogs, bristling with "tear bridges" to chill the locality and to hold the casting together and thus prevent the opening of hot tears. The hot tear is a nontrivial tensile failure to be respected.

The importance of the hot tear has attracted many researchers, practically all adopting their own new design of test technique, but in every case exhibiting a shocking absence of any knowledge of filling system design. All these lamentable pouring designs will have provided generous quantities of

The Mechanisms of Metallurgical Failure
ISBN: 978-0-12-822411-3
https://doi.org/10.1016/B978-0-12-822411-3.00002-2

bifilms to ensure hot tearing. This is probably appropriate when investigating hot tear susceptibility, but beyond this, the literature on hot tearing suffers from a dearth of well-designed systematic work. In this short chapter, therefore, we shall gather together what seem to be the relatively few results which shed light on failure in the partly solidified condition.

It is widely appreciated that the tensile failure in the partly solidified pasty condition happens principally in long freezing range alloys when the grains are soft and highly ductile, with mechanical properties more nearly like butter. At a later stage during solidification, the grains are separated by thin films of residual liquid which are said to confer a degree of brittle behavior because the tensile failure is usually accompanied by little deformation, although the matrix is in its weak and plastic condition.

The negligible stress required for fracture, as confirmed by the lack of plastic deformation, is to be expected if the failure is the result of the tensile opening of one or more bifilms. In these conditions of a highly ductile matrix, a crack cannot propagate, strongly suggesting that the observed crack *is* the bifilm and not the result of mere initiation and significant subsequent growth.

However, before launching into the detail of the tensile failure mode, it is instructive to consider first the flow of residual liquid in the pasty zone. Much traditional thinking about hot tearing involves such concepts.

2.1 Interdendritic flow

There have been beautiful computer simulations of the flow of the residual liquid through partially solid alloys, notably from the school in Lausanne (Vernede et al., 2006). These models show the liquid to flow in an elegant organic pattern, as though branching like the tree, or the circulation of the blood via multiple branching capillaries. The models for metals and alloys quantify the expected permeability of the dendrite mesh as a function of solid fraction and temperature in clean metallic liquids.

In the case of real metals, however, the situation seems quite different. A little-known experiment from Brazil (Fuoco et al., 1998) reveals the real extent of flow in the pasty zone for the widely used Al−7Si−0.4Mg alloy is practically zero. Clearly, there is a major divide between the predictions of elegant, classical, physical metallurgy and real metals. This commonly used commercial Al alloy of Fuoco's work has a relatively short freezing range and so would be expected to permit interdendritic flow with minimal difficulty.

In his experiment, the alloy was poured into a channel formed in a sand mold. At the end of the channel, a 0.3 mm thick steel barrier held back the flow until the temperature in the metal indicated a target percentage fraction of solid, at which point the barrier was removed. Fig. 2.1 shows the flow of material after the removal of the barrier at different solid fractions.

The results by Fuoco and co-workers are a shock. They reveal that even at very low percentages of solid, in an alloy with a relatively short freezing range, in which the liquid would be expected to flow practically unhindered, only a minute volume of liquid permeates the dendrite mesh under the head of pressure from the feeder. Interestingly, if grain-refined, some mixed solid/liquid flow takes place (sometimes referred to in the literature as "mass flow or mass feeding") but this also is small even at only 20% solid. At 55% solid, all flow is stopped.

This amazing impermeability of the pasty zone would be difficult to understand if it were not expected that a dense population of bifilms is likely to be present. The bifilms will easily bridge the channels between dendrites and so hinder the flow. The population of bifilms is to be expected despite the metal having passed through a ceramic foam filter before entering the mold. Such filters typically have a pore size of close to 1 mm. We can conclude that the metal has a high population of bifilms smaller than 1 mm and possibly smaller than 0.1 mm, although, of course, possibly in the furled condition. This content of fine barriers in suspension clearly and effectively suppresses movement in the pasty zone (Fig. 2.2).

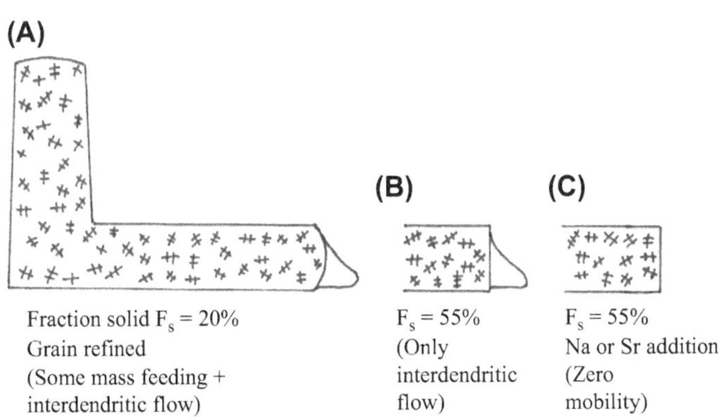

Figure 2.1 Results by Fuoco et al. (1998) showing (A) flow of liquid/solid (mass flow) at $f_L = 20\%$ if grain refined; (B) only interdendritic liquid flow at $f_L = 55\%$ if no additions; (C) zero flow of either liquid or solid phases at $f_L = 55\%$ if modified with Na or Sr.

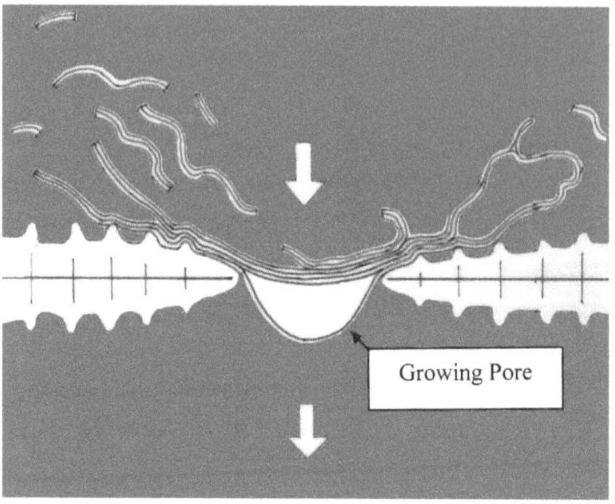

Figure 2.2 Bifilms in suspension in the liquid will be expected to block interdendritic flow and lead to the downstream initiation of porosity by the opening of the last bifilm in the blockage.

Fuoco investigated the modification of the Al–Si alloy by both Na and Sr additions, finding that both additions reduced flow even further. The explanation for this is rather lengthy, but seems logical. Before modification, because bifilms are known to be substrates for primary Si particles (Campbell and Tiryakioglu, 2010), a large fraction of the bifilm population would have already been effectively removed from the liquid by having been incorporated into Si particles. However, after Na or Sr addition, the AlP nuclei on the bifilms are deactivated so that Si can no longer nucleate on bifilms and grows instead as a classical coupled eutectic at a lower temperature. The bifilms, no longer imprisoned in primary Si particles, are now released in profusion, becoming free to jam interdendritic pathways and block the flow of liquid.

Steinbach and his international team (2017) confirms the impairment of interdendritic flow by only small additions of Fe to Al–Si alloys, as expected from the effect of the beta-Fe phase flattening bifilms and strengthening them, creating even more effective barriers.

It is important to remember that such effective arrest of interdendritic flow does not always occur. For instance, inverse (dendritic) segregation is an excellent demonstration of successful flow of residual liquid through a pasty zone. Perhaps it is dependent on the bifilm density. It seems therefore that we need to be cautious regarding both computation and experimentally based predictions of flow in the pasty zone; dense populations of bifilm will not always be present, but if present will be expected to inhibit flow.

2.2 The presence of residual interdendritic eutectic liquid

Fuoco's experiment demonstrates that long-range flow in the pasty zone is not to be relied on. However, in certain conditions, long-range flow is not necessary to prevent local interdendritic starvation when subjected to tensile strain. There are many experiments for binary alloy systems in which the second alloying element is added in gradually increasing amounts, testing the susceptibility to hot tearing at each stage. The susceptibility increases progressively as the alloy addition increases as a result of the progressive increase of the freezing range (the difference in temperature between the liquidus and solidus temperatures). When the addition reaches a critical value at which the nonequilibrium eutectic appears, hot tearing is at a peak. Additional alloy causes a sudden and precipitous fall in susceptibility as interdendritic liquid becomes plentiful and is available everywhere throughout the structure—it does not have to flow from a distant location (Fig. 2.3).

Recent hot tearing measurements by Chinese researchers (Bai et al., 2016) on some high-strength 7000 series Al alloys assessed by dog-bone tensile specimens agree impressively with a much earlier study by

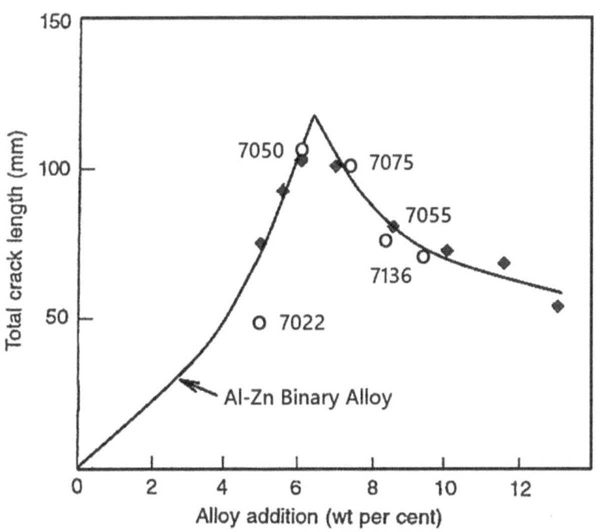

Figure 2.3 Ring test results for hot tearing susceptibility for binary Al–Zn alloys (Pumphrey and Moore, 1949) compared with comparative crack lengths for 7000 series alloys (Bai et al., 2016).

Pumphrey and Moore (1949) on Al−Zn alloys using the simple ring test (Fig. 2.3). This interesting agreement confirms the appropriateness of the ring test in its simplicity and reliability over the myriad of complex, but generally unsatisfactory hot tearing tests. The results confirm the interesting fact that the hot tearing susceptibility of the 7000 series alloys, which are among the most complex chemistry of all the cast Al alloys, is mainly controlled only by its zinc content. However, even this alloying control will be expected to be overridden, hot tearing falling to zero for all compositions for bifilm-free conditions. Experiments to demonstrate that hot tearing can be eliminated in the ring test if the 7000 alloys are sufficiently clean and have yet to be performed, but the experiment by Chadwick, described later, is reassuring.

Turning now to the Al−Cu system, it is surprising to many that the eutectic appears so much earlier than is expected from the equilibrium phase diagram. Instead of forming at the equilibrium value 5.7Cu, as a result of nonequilibrium microsegregation, it forms at approximately only 0.7Cu (Fig. 2.4). The subsequent steep fall in susceptibility occurs because feeding liquid, in the form of the nonequilibrium eutectic, now suddenly occurs everywhere in the microstructure, including, of course, at the very point of the separating grains where it is needed; no long-range transport of feed liquid is required. The resulting characteristic cusped curves, illustrating a maximum hot tearing susceptibility at the nonequilibrium eutectic, are observed for many binary systems including Mg−Al, Mg−Zn, Al−Cu, Al−Mg, and Al−Zn (JC, 2015 in CCHB).

The susceptibility to hot tearing falls to low, practically negligible levels after the cusp-like peak, as predicted in Fig. 2.4, so that at around 5%−6% Cu in the Al−Cu system, no hot tearing should be expected. This interesting prediction is, however, quite wrong. The aluminum alloys with this content of copper are among some of the strongest cast aluminum alloys and so are popular with aerospace designers, but they have a reputation for being extraordinarily difficult to cast because of tearing and cracking failures. The hot tear problems are so severe that these alloys are avoided by many foundries.

Once again, traditional metallurgy fails to predict these tensile failures in this important region of the liquid/solid zone. It is interestingly perverse to reflect that this second major divergence between theory and fact has been overlooked for many years. We might be forgiven for concluding that the fact of tear failures in this range of

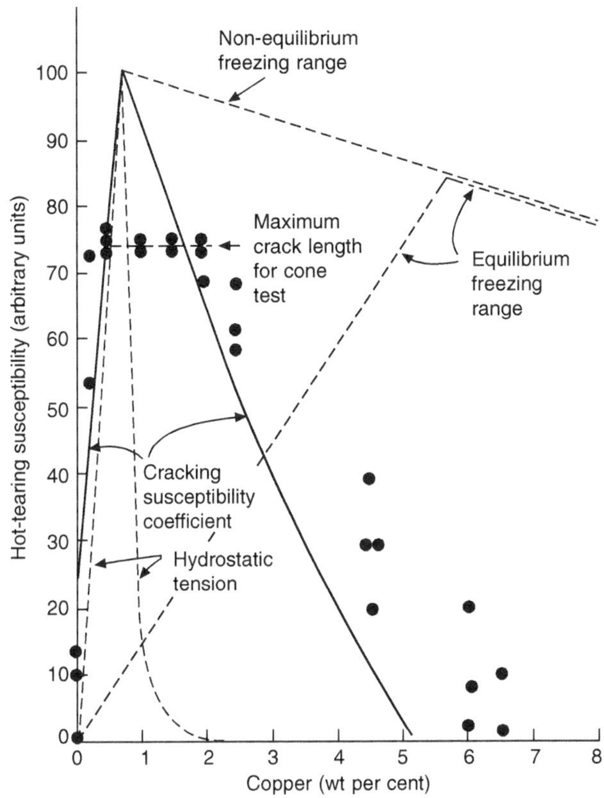

Figure 2.4 Results by Warrington and McCartney (1989) for hot tearing response of Al–Cu alloys showing a peak (necessarily extrapolated somewhat; the results being limited by the limited height of their conical test ring) compared with theoretical models. Freezing ranges and hydrostatic tensions by JC (1989); cracking susceptibility by Clyne and Davies (1977).

composition underlines the inadequacy of theories which neglect bifilms. This conclusion is reinforced below.

In contrast with the great volume of work on Al alloys, there are comparatively few results on the hot tearing susceptibility of steels. The information on Fe–C alloys from an early student practical exercise on porosity (Campbell, 1969) reflects a similar behavior to tearing and is of interest because it shows that peak susceptibility occurs close to 2%C, the maximum width of the pasty zone as predicted by the equilibrium diagram (Fig. 2.5). This closeness to equilibrium behavior occurs simply because carbon diffuses so quickly in Fe that, unusually, the predictions of the Fe–C equilibrium diagram remain accurate for solidifying alloys.

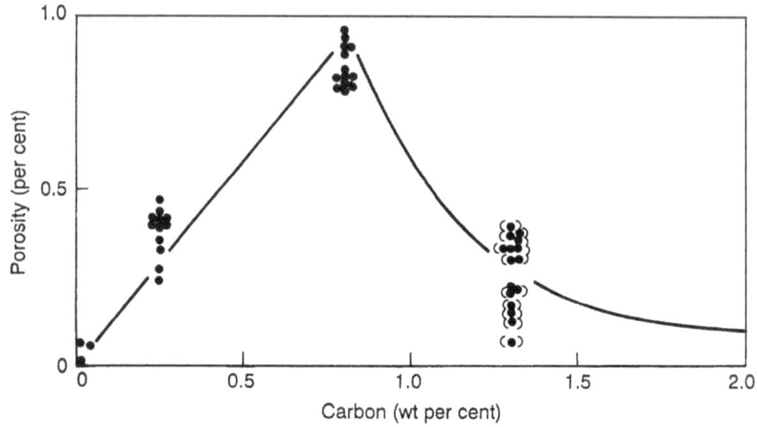

Figure 2.5 Fe–C alloys showing a peak in porosity at close to 1%C, corresponding to a maximum in the freezing range. Data in brackets are those microstructures displaying eutectic.

2.3 Hot cracking driven by transformation strain

Fig. 2.6 is an example from industrial experience (Gray et al., 1977) on steels of much lower carbon content, in which the peak cracking is close to 0.1%C. The cracks will almost certainly have initiated on bifilms, but, this time, the increased width of the pasty zone will not have been expected to have caused a significant driving force for their opening. The major driving force is likely to be the delta ferrite to austenite phase change. Grill and Brimacombe (1976) report that the linear contraction when transforming from delta to gamma is 0.38%. The austenite would tend to form at delta ferrite grain boundaries where the bifilms happened to be located. The strain of transformation would be accommodated by the bifilm which would be forced open a little by the formation of the austenite. The modest additional stresses from solidification would have further assisted the opening.

The cracking of steel ingots of many types during forging and rolling remains a serious issue even to this day, with reasons for cracking still appearing mysterious to researchers (Bombac et al., 2018).

2.4 Intergranular effects driven by thermal strain

Returning once again to see what we might learn from simple theoretical models (Campbell, 2015), hot tears can be envisaged to open as a gradual rearrangement of grains and residual liquid as illustrated in the

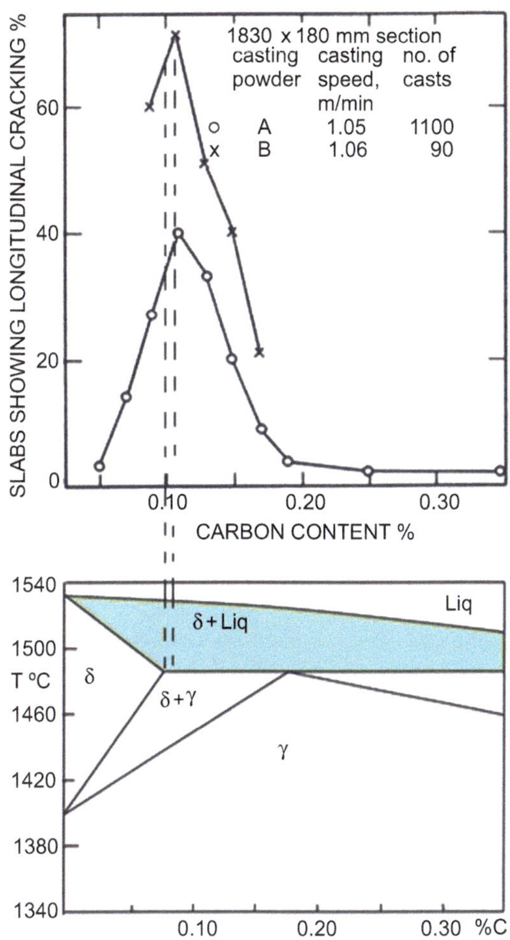

Figure 2.6 Fe–C alloys showing a peak in porosity at close to 1%C, corresponding to a maximum in the freezing range. Data in brackets are those microstructures displaying eutectic.

two-dimensional (2D) hexagonal model shown in Fig. 2.7. At stage 1, the initial unstrained distribution of liquid and grains is assumed to be uniform. In stage 2, on suffering some unidirectional strain, the grains rearrange. The model shows the development of intergranular pools which are readily seen in grain-refined alloys, as seen in Fig. 2.8.

The 2D model describes nicely the opening of the segregate pools in grain-refined alloys and the extent of "pretear" extension. However, beyond this stage, it is interesting to note that it fails to explain convincingly

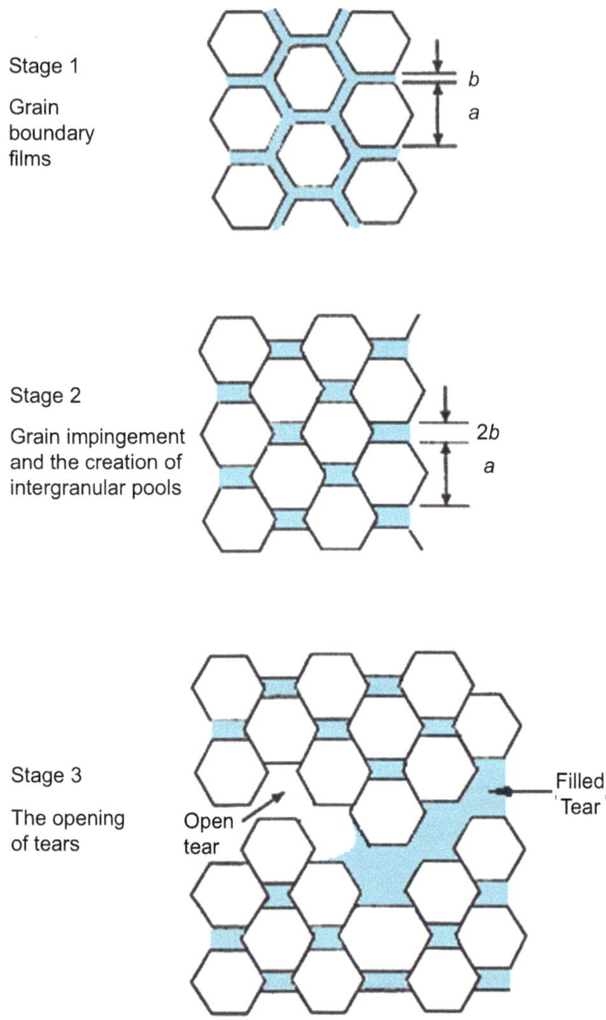

Figure 2.7 Stages of hot tearing using a 2D hexagonal model. Stage 1 is the unstrained structure; Stage 2 shows isolated regions of segregate. Stage 3 shows opening tears including so-called "filled" tears.

the arrival and propagation of a tear. Stage 3 illustrates the opening of a hot tear from the surface of the casting. The model implicitly assumes that grains on either side of the tear could continue to open despite the effect of friction between grains which would tend to hold the whole structure together.

The initiation of a tear, progressing inward from the surface, as envisaged in Fig. 2.7, is probably uncommon. The surface oxide film plus surface

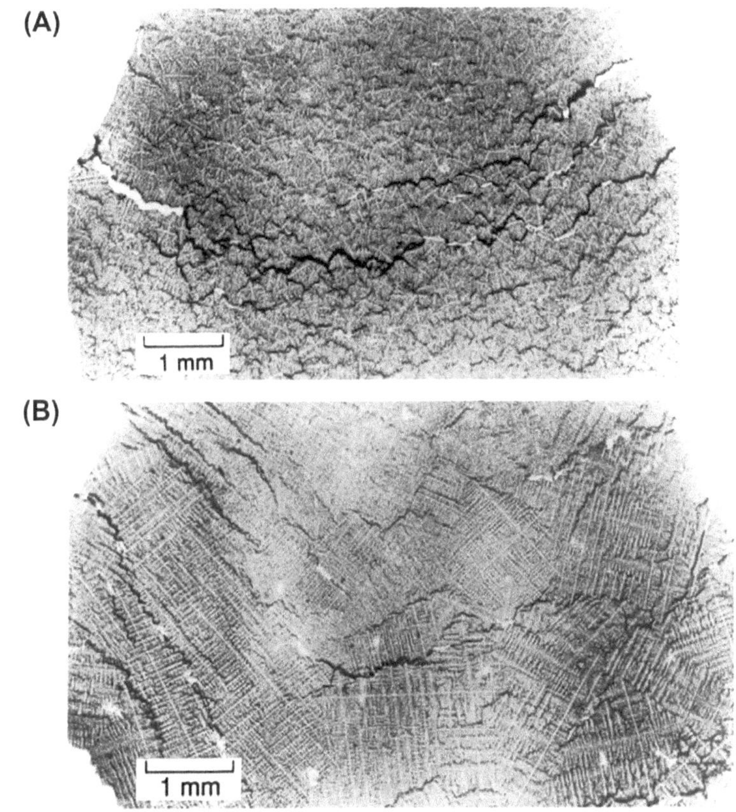

Figure 2.8 (A) Radiograph of a hot tear in an Al-6.6Cu grain refined alloy. Dark regions are copper-rich; white regions are open tears; (B) radiograph of hot tears in an Al−10Cu alloy not grain refined (Rosenberg et al., 1960). *Courtesy M C Flemings.*

solidified layer plus internal pressurization from the hot top feeder all would cooperate to discourage tear initiation at the surface. A tear will therefore rather tend to occur internally, in which tear initiation can only be envisaged to occur at an entrainment defect such as a bifilm or bubble. In Fig. 2.7, further internal propagation of the tear can only be accompanied by plastic flow of grains. Interestingly, this is completely counter to observations of the fracture surface of tears, in which arrays of dendrites are clearly seen to be devoid of any signs of stress or strain as is clear in Fig. 2.9. Clearly, our model in Fig. 2.7 is unsatisfactory: it can only describe pretear extension and the accompanying development of intergranular segregated pools; it cannot describe tear propagation.

(A)
(B)
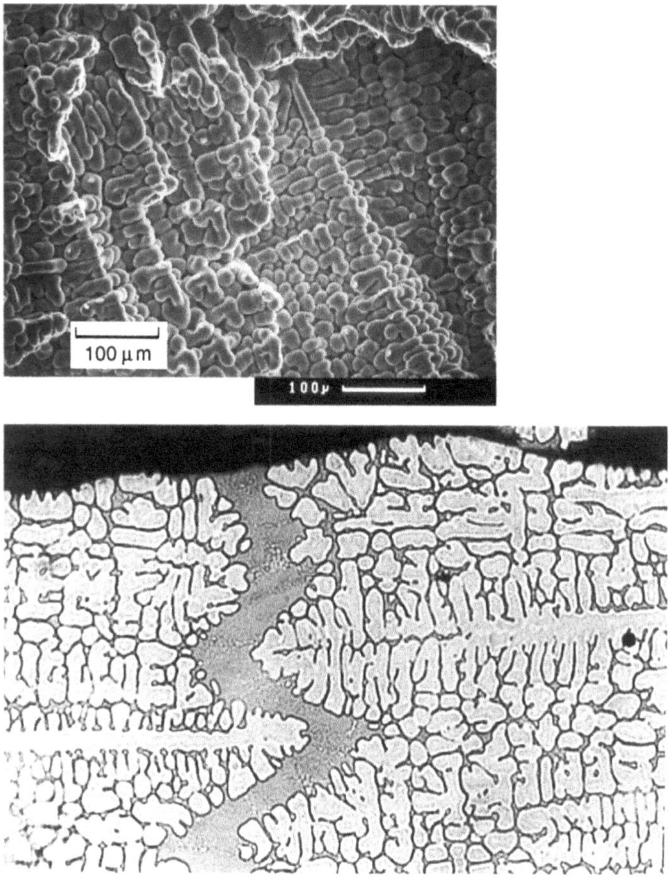

Figure 2.9 (A) A scanning electron microscope image of the surface of a hot tear in Al–7Si–0.4Mg alloy sand casting (the half of the oxide bifilm covering the dendrites is invisible in this image); (B) a so-called "filled" hot tear in an Al–10Cu alloy by Spittle and Cushway (1983) (No oxide bifilm is involved in this relatively harmless phenomenon).

2.5 Practical hot tearing phenomena

We might surmise that the additional opening of the tear in Fig. 2.7 was facilitated by the presence of a bifilm between the grains. This would open easily, i.e., with minimal stress, under tensile strain. This suggestion is convincingly corroborated by a unique piece of evidence. When a hot tear is observed in a shaped casting, the immediate remedial action by foundry personnel is, understandably, usually concentrated in the tear area by attempts to reduce the stress on the cooling casting. Typically, this is achieved by reducing the strength of the mold and/or cores in that area.

My personal experience over the past 40 years is that for every alloy in every casting that I have tackled, totaling dozens of examples, I have never bothered to reduce the cooling stresses on the casting. Invariably, hot tearing was eliminated at a stroke simply by improving the pouring conditions for the liquid metal. I admit this is hardly a scientific experiment which might be commended by a scientist! The control of tensile failure in the liquid/solid state by controlling the pouring conditions during casting seems, nevertheless, to be an important observation whose explanation is discussed below.

When making a new cast product by pouring into a mold, the liquid metal will be expected to contain a wide variety of bifilms. Many will be "old," having originated early in the melting process or perhaps even from the distant history of the metal which may have undergone many recycles. These aging bifilms will have lost much of their original air layer separating the two films so that friction between the films, or even some bonding, may have occurred. These will therefore not constitute easy-opening bifilms which would be likely to initiate a tear.

However, a number of bifilms will be formed during the action of pouring. These freshly entrained bifilms will have their original supply of entrained air to "lubricate" the two halves by tending to keep them apart. Such freshly entrained defects will have a minimum strength, conferring on them a greatly enhanced potential for the creation of cracks. In addition, bifilms formed *inside* the mold cavity (as distinct from inside the running system) will tend to retain their full size and will not be furled significantly because of the reduced turbulence—the metal enters the much larger volume where metal velocities fall and conditions become tranquil. The bifilms created inside the mold cavity are special: they will therefore be expected to be the ideal initiators of failure by cracking in the liquid/solid state.

Good filling design will eliminate these large, easily opened bifilms. Although the casting will still contain its population of older bifilms, these are not so easily opened. Thus, the good filling system design seems always to eliminate hot tears (although it clearly cannot eliminate all bifilms).

It seems, therefore, clearly necessary to take account of the differences in the various populations of bifilms. For instance, those originally present in the melting furnace as a result of prior recycling, or even poor control of turbulence in the melt, will still be present. But this older population, as we have seen, seems comparatively poorly influential in creating hot tears because, despite the high strains, stresses remain extremely low at these high temperatures.

Digressing for a moment to consider how this older population of bifilms might affect the tensile fracture of the room temperature product; when applying far greater stresses during tensile testing to failure, such bifilms would be expected to become active and visible in the form of the ductile dimple morphology of the fracture surface. They will also, of course, reduce the tensile properties. The examination of the fracture surface from a tensile test is probably the optimally sensitive test for the presence of bifilms in the metal.

The room temperature tensile test contrasts with the reduced pressure test (RPT) used for Al alloys, in which the tensile stress is merely a fraction of an atmosphere, so that only relatively young and large bifilms can be detected. The RPT technique should therefore be a useful guide to the hot tearing potential of a batch of liquid metal.

As a further aside, the fracture surface of the hot tear (Fig. 2.9A) shows dendrites with hardly a trace of deformation, confirming the negligible stress involved in the tearing process—although an oxide film should be found covering the whole surface, revealing minute folds and creases to confirm its origin as a film originally on a liquid (and not merely an oxide grown on the solidified fracture surface while still cooling). The negligible stress required for hot tearing questions the relevance of those researches which have attempted to measure stress, while the sample is subjected to overwhelming forces from friction with the mold. The hot tear process is mainly driven by strain; the stress is negligible and largely irrelevant.

A further issue which we should note is shown in Fig. 2.9B. The "filled hot tear" requires careful definition. The name suggests it was once open but has been subsequently filled. In general, this seems to be not true. The image by Spittle and Cushway shows arrays of dendrites forming grains which have been pulled apart, opening up a region of residual liquid. These features are common in castings. They have never been cracks or tears. This conclusion is corroborated by our simple 2D model in Fig. 2.7. When tested for mechanical strength, such "filled tears" in solidified metals usually exhibit no loss of strength or ductility. Even so, the filled tear, if continued to suffer tensile strain, may develop into a hot tear as its liquid drains away. The filled tear is probably an early warning of modest cooling strain which may develop into an unwelcome failure.

Finally, in this section, I cannot resist the retelling of an experiment in 1997 by one of my mature students, Herbert Chadwick. Herbert was investigating the hot cracking of A201 alloy, a high-strength aluminum alloy notorious for its tendency to hot tear (it contains approximately 5.5%Cu). Herbert carried out a major factorial study of the effects of varying all its chemical elements within the specification limits of the alloy. He used the simple ring test, in which the alloy is poured into an annular gap around a cylindrical steel core. As the ring casting cools and contracts onto the steel core, it develops cracks. The total length of cracks exhibited by the casting can be measured; total crack lengths up to 100 mm or more are typical.

On completion of the study, we were amazed to find that the only clear influence on the hot cracking tendency was the size of the spoon from which the casting was poured! Having got over the shock of this unlooked-for prosaic result, it was realized that the result strongly suggested that the pouring conditions were important. I therefore suggested to Herbert that he changed the casting procedure from top pouring to bottom gating via a filter.

Herbert arrived in my office a few days later with the new castings; four rings had been cast and each was contracted on to the steel core so tightly that they could not be extracted. Furthermore, the cast rings exhibit not a single crack. I was shocked into silence. This was probably the world's most hot tear prone alloy cast on to the world's most severe tear test, but with zero cracks. It was an amazing result. While I was gathering my thoughts to comment appropriately on the immense significance of this outcome, Herbert leaned back in his chair and stared despondently into space.

"Yes" he said, "it was a complete failure."

I was aghast at this remark and could only manage to get out the word

"What?!"

He went on sadly, resigned to his disappointment,

"What was the use of a hot tear test that did not hot tear?"

As a postscript to this amazing piece of work, I wrote up a formal paper in the usual way and submitted it to a scientific journal specializing in solidification and casting. The submission was rejected, together with an admonishment by the Editor not to submit such patently unbelievable material in future. Herbert's work was never published.

2.6 Incipient melting

Widespread experience teaches that if a high-temperature heat treatment exceeds a critical temperature at which grain boundary phases start to melt—a condition known as incipient melting—on subsequent cooling to ambient temperature, the mechanical properties of the material are found to be impaired. Accidental overheating of heat treatment furnaces can therefore be a serious matter. Furthermore, having to limit temperatures to avoid incipient melting deprives the metallurgist of the benefit of extra rapid, efficient heat treatments. With heat treatments, I am always looking to increase temperatures as far as possible. (Each 10°C rise doubles the reaction rate, thus 20°C rise would be 4 times faster; 30°C rise would be 8 times faster, etc. A better gift for improved productivity from heat treatment facilities could hardly be imagined!) Incipient melting was an irritating limitation.

The loss of properties by the melting of the boundaries is a curious fact because, after the overheating, when the boundaries freeze once again as the alloy cools, the mechanism might be expected to go into reverse, regaining the soundness, so that no permanent damage should have been suffered. This unexpected nonreversal of behavior appears never to have been explained.

As an instance of the practical problems, in Al–7Si–0.4Mg alloy, high-temperature heat treatments are generally limited to 545°C because iron-rich and other grain boundary phases melt at 555°C. Shivkumar et al. (1990) reported that despite improvements to the structure in the form of the spheroidizing of the Si particles, which would have been expected to improve properties, in fact, properties fell dramatically at and above 560°C.

A bifilm mechanism to explain the property loss on overheating has been proposed (Campbell, 2009). It is assumed that the alloy will normally contain a large population of bifilms residing mainly at grain boundaries. The grain boundary location is a result of (i) dendrite pushing in the liquid state and (ii) grain boundary migration arrested at such discontinuities in the solid state. As is common, intermetallics and second phases precipitate on the outer interfaces of bifilms because the oxide usually appears to be a favourable substrate.

When suffering overheating, the grain boundary segregates melt, with (usually) an expansion accompanying the phase change from solid to liquid. This is around 3% for many metals, corresponding, of course, to a 1% linear expansion in the three orthogonal directions. As both the solid and liquid phases are effectively incompressible, this expansion must be accommodated somehow. For those regions of the grain boundary without bifilms, the expansion strain, being approximately 10 times greater than the yield strain of around 0.1%, is necessarily to be accommodated by plastic flow of the matrix. The high strain energy involved discourages this transformation. For those regions of the grain boundary where there is a bifilm, however, the strain energy is largely avoided because expansion will be easily accommodated by the "air gap." At the same time, of course, the penetration of part of the new precipitate into the air gap will tend to force open the bifilm to become a visible crack.

On subsequent cooling of the material, the melted regions now freeze. The reversal of the phase change is accompanied by a reversed contraction. In the absence of an associated bifilm, the solid matrix would have to reverse the plastic strain. However, in the presence of the bifilm, the strain is not reversed; instead, having been opened, the bifilm now simply remains open because there is negligible bonding across its internal interface and no other mechanism to encourage it to close. The cooling alloy now exhibits its bifilms as open cracks, sometimes reported as porosity. It is the nonreversal of the bifilm opening, which permanently impairs the tensile properties.

It follows that material that has been carefully manufactured in the liquid state so as to avoid the entrainment of surface oxide would be capable of avoiding the penalties of incipient melting. Such material would be able to enjoy higher heat treatment temperatures, with a significant reduction in heat treatment times and/or an improvement in properties.

The author can recall some experience in support of the expectation of this benefit from the casting of Al alloy with the Cosworth casting process (unfortunately details could not be released at the time for commercial reasons). The lengthy quiescent dwell of the liquid in the holding furnace followed by countergravity filling of the mold via an electromagnetic pump reduced the turbulence of mold filling and routinely gave a cast alloy with low levels of bifilm defects. For normal casting conditions, as discussed above, the Al–7Si–0.4Mg–0.4Fe alloy would be limited to a solution temperature of approximately 550°C because of the melting of grain-boundary iron-rich phases at 555°C. The Cosworth material was able to be solution treated at 560°C. Properties were only observed to fall at a temperature approaching 570°C. The extra 10°C from 550 to 560°C meant that heat treatment times could be reduced by a factor of 2.

In interesting related work, Gauthier and Samuel (1995) studied 319 alloy (Al–6Si–3.5Cu–0.5Fe), finding a maximum solution treatment temperature 515°C. If the 515°C was exceeded, melting the $CuAl_2$ grain boundary phase, porosity was seen to form, reducing properties. However, if the $CuAl_2$ phase was taken into solution by a prior treatment at 515°C for 12 h, they were able to subsequently raise the solution temperature to 540°C, resulting in excellent properties. This approach clearly constitutes a lengthy, complex, and expensive treatment. It is proposed that a substantial reduction in bifilm population would achieve a similar result more easily and more effectively because the reduction in bifilms often costs nothing other than care of the liquid alloy during melting and casting.

The hypothesis proposed in this section would require that alloys whose residual liquid contracted on melting and expanded on freezing should not suffer from pores or loss of properties as a result of incipient melting; the bifilm would expand to accommodate the melting but close again on freezing. This prediction appears to be fulfilled by observations of Nili-Ahmadabadi and Feghahati (2007) on ductile iron. In detail, the explanation of this result is probably associated with some small additional precipitation of graphite, easing itself into a closer equilibrium condition, the accompanying expansion on freezing retaining the pressurization of the microstructure.

2.7 Slurry processing

It has become fashionable to criticize the traditional approaches to the production of castings by the filling of a mold with liquid metal. With regret, the liquid casting approach probably deserves such criticism, earned by its abject and repeated failures over the years to deliver castings of respectable quality.

It is understandable therefore that the casting of liquid metal has been thought to be the fundamental problem and has been the impetus for a huge expenditure and effort on the casting of various partially solid alloys and liquid/solid mixtures, widely known as "semisolid" mixtures. I have to say I try to avoid such inaccurate language. The mixtures are rarely 50/50 liquid/solid. Imprecision in our vocabulary can become wooliness in our thinking. I agree the accurate name, "partially solid alloys," does not alliterate to help it slip off the lazy tongue, but it is correct, and in my view therefore worth the effort.

Even so, as this book records, the author remains convinced that despite all the well-known difficulties with the casting of liquid metals, the difficulties are worth living with, and ultimately, good technology, delivering benefits to properties hard to beat. For this reason, this writer remains strongly committed to the production of castings by the casting of liquid metal and convinced that the new improved solutions to this production route (Campbell, 2018) will provide the major benefits to the greater part of the engineering industry for the foreseeable future.

This is not to deny the special features and advantages enjoyed by some nonliquid, or perhaps we should say, partially solid routes to casting manufacture.

Most liquid/solid mixtures are created by various shearing techniques in which the rigid dendrite arrays are fragmented, allowing the mix to flow as a slurry. The shearing processes are often stirring processes of different kinds. During the mixing processes, it is unfortunate, and usually overlooked, that most techniques tend to stir air and oxides into the mixture. The result is, of course, a mix that now contains a dispersion of oxide bifilms that may contribute somewhat toward strength, but in addition, will also act counter-beneficially because of the presence of the associated "voids" of air or argon entrained together with the oxides. The oxides will strengthen, but the voids will provide initiators for failure.

The contribution of oxides and "voids" to the properties of slurries is almost always overlooked, but has to be significant. The good news is that the dispersion of mixed defects will be reasonably uniform, so that the properties may not be optimum, but the product will probably be uniformly reliable (or unreliable).

The advantages of casting partially solid mixtures are well-known and include their use in high-pressure die casting (HPDC), in which the flowing slurry seems more resistant to disintegration and thus avoids the worst excesses of entrainment usually suffered by the HPDC technique when applied to liquid metals. Furthermore, because a significant fraction of the heat content has been removed from the mixture before casting, the metal dies run cooler. The dies therefore suffer reduced crazing because of thermal shock or thermal fatigue and so last longer. Freezing is also faster, thus increasing productivity. Although reduced shrinkage porosity is also cited as an advantage because, once again, a significant fraction of the cast material has already solidified; this is almost certainly more apparent than real; the so-called shrinkage porosity in pressure die castings was mainly misdiagnosed entrainment problems including air and oxides. However, of course, there has to be some small component of reduced shrinkage problems, but in such thin section castings this will usually be negligible. Anyway, the undoubted overall improved soundness confers improved and more reliable mechanical properties and improved leak tightness.

The disadvantages of the processing of partially solid mixtures relates to the increased cost of processing of the mixtures, and for some approaches, the increased control necessary to form a correct mix. (Some approaches have automatic, intrinsic, built-in control, such as the approach by Rheometal (Granath et al., 2008), using not temperature control, but a method of mixtures based on enthalpy equilibration, melting a known amount of solid in a known amount of liquid. The oxide skin from the melted solid addition appears not to have raised any problems so far as is known to date, but probably requires further investigation.)

2.7.1 Rheocasting

Rheology is the science for the study of the flow properties of glues and goos. Metallic liquid/solid mixes tend to be flowable while being sheared, but develop strength if allowed to stand. This behavior, well deserving of its fame, was discovered by Spenser, Mehrabian, and Flemings at MIT in 1972 while investigating the fluidity of metals.

From this discovery, "rheocasting," as it became known, was developed as a casting process. A mechanical stirring technique applied to a cooling alloy in its liquid + solid range, breaking up the dendrites as they formed, to become compact, near-spherical shapes, forming a flowable mixture which could be poured or injected into a mold. The process continues to find niches in the industrial and commercial world in the second millennium.

The detailed behavior science behind the flow behavior continues to be investigated (for instance Atkinson and Favier, 2016). Practical developments also continue (Dybalska et al., 2017) to optimize processing, finding minimum shearing times to be only a few minutes. Other approaches to generate a useful slurry by simultaneous cooling and shearing have employed the simple act of pouring the liquid alloy down a sloping plane (Das et at 2016), a process with no moving parts which seems surprisingly effective!

2.7.2 Thixocasting

Thixocasting was a variant of the Rheocasting approach in which the partially solid mixture was *precast* in the form of semicontinuously cast "logs." The logs required to be sawn to correct length, reheated to the casting temperature at which they would retain their shape, but be easily sheared to convert to a flowable slurry, and finally injected or poured into a die. This approach has always suffered from the cost of the separate melting and Rheocasting process steps. Furthermore, the foundry returns (runners, feeders, and scrap castings) present a problem for recycling. The just-in-time processes for the production of partially solid mixes, described below, avoid both of these issues.

2.7.3 Just-in-time slurry production

New concepts require agonizingly long development periods before they become accepted by users. Companies such as Alcan Canada (Doutre et al., 2000 and Lashkari, 2007), and Ube Japan (Lee and Kim, 2007), Idra Italy (Yurco, 2003) and Rheometal, Sweden (Granath et al., 2008), among others (Midson, 2008), have continued to develop novel just-in-time slurry production techniques to counter one of the challenges of the process, the production of controlled qualities of slurry at commercially attractive production rates, and commercially attractive costs at just the right moment.

2.7.4 Mg alloy slurry production from granules

The work by Fan (2007), Zerwinski (2008), and others has demonstrated the feasibility of producing Rheocastings in Mg alloys by the processing of alloy granules through a twin screw injection molding machine as part of a pressure die casting process. The processing of Mg granules via a route more normally associated with the processing of plastics is possible partly because of their low strength and hardness and partly because Mg reacts with the steel screws to only a limited extent, compared, for instance, to Al alloys which dissolve iron readily. Once again, the uniform dispersion and shredding of the oxide skins from the original granules is seen to deliver a product that has uniformly reliable, if not optimum, properties. In fact, the precipitation of the so-called β-$Mg_{17}Al_{12}$ phase at grain boundaries in thixotropic Mg alloys containing over 7%Al has been cited (D'Errico et al., 2008) as limiting ductility. It seems almost certain that this phase is precipitating on bifilms generated by the mixing process and pushed ahead of the growing grains, so effectively precracking the grain boundaries.

The shearing of aluminum alloys by ceramic twin screws has now been abandoned for practical reasons, finding the simple rotary stirring of a partially solidified melt appearing to be equally effective.

2.7.5 Strain-induced melt activation

Strain-induced melt activation (SIMA) produces a thixotropic condition by first deforming the solid alloy followed by a heat treatment into the mushy zone. The approach is unique in avoiding the entrainment of air and oxides suffered by competitive techniques for slurry production. Chen et al. (2010) describe its use for magnesium alloy AZ91, using equal channel angular pressing (ECAP) to produce a high but uniform strained condition, and thereby achieve impressive results. Even so, it seems likely its potential for commercial exploitation will be limited by its challenging economics, in common with other semisolid routes.

2.7.6 Metal/matrix composites

Solid/liquid mixtures can be made by the addition of particles of solid into the melt. The exploration of such mixtures has been pioneered by Pradeep Rohatgi. Since his time in the International Nickel Company Laboratories, New York, in 1965, he has explored mixtures of practically everything with everything (see his Rohatgi Symposium reported by Gupta, 2006). Most matrices studied so far are those of Al alloys, but other work includes Zn, Mg, and, later, superalloys (Rohatgi, 1990). The particle additions include graphite, SiC, Al_2O_3, and low-cost additives such as fly ash, a waste product from power plants. Thousands of tones of such metal/matrix composites (MMCs) are now processed for use in the transport industries.

The process most used to make mixtures is the so-called vortex method in which the solid additions are fed into a central vortex made in a rapidly stirred liquid metal. Some work has also been carried out using particles in suspension in a carrier gas, introduced under the melt surface from a lance.

All processes that introduce particles from the outside into the interior of the melt suffer the problem of having to penetrate the oxide skin on the melt. This means that large particles usually enter the matrix wrapped in the oxide film from the surface of the liquid. However, fine particles such as powders cannot easily penetrate the liquid surface because of the mechanical strength of the surface oxide and the resistance of its surface tension. The result is that fine particles enter the liquid as clumps, each clump being enclosed by a wrapping of the oxide film necessarily entrained from the surface (Fig. 1.8). The problem with such clumps is that the reservoir of air they contain continues to thicken and strengthen the surrounding oxide envelope. The vigorous stirring required to break these large agglomerates down into smaller agglomerates automatically introduces yet more oxides into the melt via the central vortex. Tian et al. (1999) study clusters of alumina particles in liquid Al, finding clusters from 1 particle of 10 μm diameter to clusters of particles 100 μm diameter. In each case, the clusters are enclosed by amorphous alumina films.

The MMC studies by Emamy et al. (2009) also illustrate particle clusters inside oxide "paper bags." There is evidence from this work that the loss of fluidity with higher additions of particles is not primarily the result of the higher percentage of particles but the result of increased oxide bifilm content from the vortex stirring process. The huge increase in bifilm population in such vortex-produced MMCs is also seen to degrade ductility, although, paradoxically, the oxide appears to confer some slight increase in strength.

Badia (1971) explores the production of MMCs in Al and Zn matrices by additions of a variety of solids including SiC, Al_2O_3, SiO_2, and graphite. He found that coating of the particles with Ni was essential to achieve some kind of wetting so that the additions would remain in suspension; otherwise the additions appeared to surface once again, rejected from the melt as a dry powder. The action of the Ni is not completely clear and may depend not only on its metallic nature but also on its powerful exothermic reaction with Al to form nickel aluminides. Any benefit to the MMC of the introduction of a lightweight phase is lost, however, because the 5 μm Ni coat on particles 5 μm diameter corresponds to an addition containing well over 50% Ni. Badia used the vortex addition technique and subsurface injection from a lance but preferred the vortex approach because continued stirring with the rotor could maintain the addition in suspension.

Practically all of the above studies have focused on the use of particle additions to a melt. Occasional exploratory studies have included a variety of fibers (Rohatgi, 1990) and SiC whiskers (Das, 1981) for which surprising gains in strength have been recorded for levels of only 0.5 volume percentage addition.

In contrast to these laboratory studies, in which the introduction of particles was attempted with relatively inefficient protective atmospheres, Alcan instigated a major production initiative in which they produced an Al alloy—SiC MMC in tonnage quantities in a high vacuum environment (Hammond, 1989)—although at the time the production conditions were a commercial secret. The uniform dispersion of the particulate phase in Duralcan, as it came to be known, indicated its relative freedom from oxide films, although work by Emamy and the author did reveal some residual bonding problems, although these were much less than those commonly seen elsewhere. Later work (Hoover, 1991) confirmed the excellent specific stiffness, good tensile properties including fatigue, and good retention of properties at elevated temperature.

2.7.7 In situ metal/matrix composites

The formation of MMCs by the precipitation of the strong, hard particles in situ in the matrix by a metallurgical reaction during solidification is an attractive technique. In this case, the strengthening particles suffer no poor cohesion with the matrix because of an intervening oxide film and its associated thin layer of air; the particles are bonded to the matrix with atomic perfection, having grown atom by atom from the matrix.

The difference in the bonding of an in situ generated MMC based on TiB_2 particles in an Al alloy matrix, and Duralcan (a good quality extrinsically generated MMC based on SiC particles) was shown clearly in work by Emamy and Campbell (1995). A nonfed casting designed to create a reduced internal pressure (possibly a slightly negative pressure corresponding to a hydrostatic tensile stress) was seen to cause significant dispersed microporosity in Duralcan, but very little in the in situ MMC. The difference between the relatively poor bond and the perfect bond was clear.

Hadian (2009) studied the system Al–Mg_2Si and the refinement of its Mg_2Si particles by more rapid freezing and by the addition of Li. Although the Mg_2Si particles are usually regarded as brittle, this study is notable for its image of a Mg_2Si particle cracked (one assumes by the presence of a central bifilm on which it formed) but plastically peeled open, revealing the particle to be impressively ductile. This is expected to be a universal feature of in situ MMCs; all the hard, strong particles will probably deposit on a bifilm and thus all will be expected to possess an in situ crack, in apparent conflict with the intrinsic strength and crack resistance that they would be assumed to possess.

The common Al–Si system similarly behaves as an in situ MMC in which the silicon particles act as the hard strengthening phase in the ductile Al matrix. Once again, the presence of central cracks in the Si is attributed to the nucleation of the Si particles on bifilms in suspension in the melt. It is known that clean melts cannot form Si particles in this way, with the result that the Si is forced to precipitate at a lower temperature and grow continuously in the form of fine ribbons as a finely spaced eutectic, which we call a "modified" Al–Si eutectic. The addition of such elements as Sr and Na similarly act to inhibit nucleation on the oxide bifilms, thus encouraging the modified structure (Campbell, 2009). The Si now grows in a continuous branching form known as a coral morphology which can no longer be regarded as separate particles, even though, of course, it takes on the appearance of separate particles on a polished micrographic section. This form of the Al–Si MMC can be particularly strong and tough if bifilms are somehow removed by careful processing because the development of the Al–Si MMC structure now no longer depends on their presence.

2.8 Conclusion to Part II

The assumption of the near-universal presence of bifilms in metals appears to be consistent with both the microscopic and macroscopic failure modes of all solid/liquid mixtures examined so far. In the absence of bifilms, such mixtures do not fail by cracking.

CHAPTER 3

Fracture of solids

Contents

The Mechanisms of Metallurgical Failure
ISBN: 978-0-12-822411-3
https://doi.org/10.1016/B978-0-12-822411-3.00003-4

3.1 The concepts of ductility and brittleness

In our short account of the failure of *metals*, we shall consider only the failure modes resulting from uniaxial tensile stress: brittle fracture, and ductile flow. In the case of perfect brittleness, fracture would occur without plastic deformation, like a glass. In the case of perfect ductility, cracks could not form, but the tensile failure would occur by the necking down of the metal to a point, being 100% reduction in area (100%RA). In practice of course, these extreme modes are usually mixed, but the extremes are nevertheless important concepts and are very nearly exemplified in interesting cases.

Whole books have been devoted to these subjects, so we are entering a mature field. Even so, the bifilm has, up to the present, been largely unknown but usually playing a major role. We shall endeavor to explore the wide scope of the changes to our thinking that are to be expected, taking account of the presence of bifilms in metals.

However, it seems helpful to start clarifying the concepts of intrinsic brittleness and intrinsic ductility.

It was Weertman (1996) who published an interesting list of materials, mostly metals, in which from the theory of dislocations he records the ratio of the theoretical tensile strength to the theoretical shear strength (Table 3.1). For those metals with ratios above approximately 7, he concludes the metals are intrinsically ductile because a crack tip will put out dislocations before any advance of the crack. For those metals and compounds with a ratio less than 7, he concludes that they are intrinsically brittle because a crack tip will move before any dislocation movement.

A rather more complete table of our common engineering metals is presented below (Table 3.2) based on their elastic properties, from the most ductile to the most brittle of our metals.

It is important to keep in mind that both of these interesting spectra of failure behavior are based on the assumption that a crack preexists. If a crack does not exist, there may be no operable mechanism available for crack

Table 3.1 Ductile/brittle behavior of selected materials.

Material	**Ratio**	**Strain behavior**
Au	36	Intrinsically ductile
Ag	31	
Cu	28	
Ni	15	
Ir	10	
Fe	7	
W	3.7	Intrinsically brittle
Al_2O_3	2.7	
Si	2.3	
NaCl	2.2	
Diamond	1.2	

Table 3.2 The ductile/brittle behavior of our common engineering metals.

Element	**μ/B**	**ν**	
Liquid	0	0.50	Ductile
Pb	0.12	0.44	
Au	0.15	0.42	
Nb	0.22	0.40	
Pt	0.22	0.39	
d	0.23	0.39	
Hf	0.27	0.37	
Ag	0.29	0.37	
Al	0.35	0.34	
Cu	0.35	0.34	
Zr	0.39	0.33	
Ti	0.42	0.32	
Ni	0.43	0.31	
Co	0.45	0.30	Transitional
Fe	0.48	0.29	
Mg	0.49	0.29	
Mo	0.48	0.29	
Nd	0.50	0.28	Brittle
W	0.52	0.28	
Re	0.54	0.26	
Ir	0.56	0.26	
Cr	0.72	0.21	
Be	1.42	0.02	

μ, shear modulus; B, bulk modulus; ν, Poisson's ratio.

initiation, in which case even the intrinsically "brittle" materials will be unable to fracture, and strain will necessarily be achieved in some other way, possibly by plastic flow or twinning.

The famous observations on common table salt, NaCl crystals (Mendelson, 1962), are a perfect example. These crystals are brittle when crushed in air but can deform in a ductile mode when compressed under water as surface flaws are encouraged to dissolve away.

Such behavior is analogous to the condition of an intermetallic or second phase, traditionally considered to be "brittle," formed in situ in a melt or in a solid matrix. Having grown atom by atom by a diffusion-controlled process, the particle necessarily enjoys perfect atomic contact with the matrix whether liquid or solid; there can be no pores, cracks, or delaminations. The interphase interface will be atomically perfect. The new phase will therefore be expected to exhibit extremely strong resistance to decohering or fracturing. In the extreme, if sufficient stress could be exerted on it by the matrix (which in general is unlikely), it would be expected to deform plastically.

This important conclusion is undermined repeatedly in traditional metallurgical thinking. When studying an as-cast metal microstructure, the cracking of intermetallics and second phases is often attributed to cooling stresses in the solidifying casting, despite the well-known properties of many grain boundary phases such as carbides being immensely strong and crack-resistant, and stresses during solidification usually modest because of the weakness of the solid near its melting point.

It is a concern that most of our studies on the strength of inclusions are in this category, regrettably groomed into habit by generations of wrong thinking. If the inclusion had been formed in situ in the matrix (whether liquid or solid), it will not usually be expected to fracture even if the matrix is highly stressed. This is partly because inclusions are often immensely strong and because they are extremely strongly bonded to their matrix. If the stresses are enormously high, it is possible that the inclusion will deform plastically. There is plenty of evidence for this. For instance, consider the Guinier–Preston (GP) precipitates in Al–Cu alloys. These nanometer-sized precipitates are, of course, perfectly bonded to the matrix. It is not conceivable that they could decohere or fracture, whereas at high stresses they are commonly intersected by dislocations which result, naturally, in some deformation. For more massive inclusions such as carbides in steels, generally assumed to be brittle, plastic deformation can be commonly identified on fracture surfaces. Similarly, Mg_2Si intermetallics, also widely assumed to be brittle, have at times been observed on fracture surfaces to undergo massive bending deformation without fracture (Hadian et al., 2009).

Tensile failure by ductile flow is perfectly well understood. When operating to perfection, the metal necks down to a point, representing 100% reduction in area. Examples of such failure are frequent. However, this simple and straightforward process is not discussed further in this work.

We shall concentrate on failure modes involving cracking. These modes of failure are important, and in my view, not properly understood. We need to tread carefully, making a new sifting of the evidence.

3.2 Theories for the theoretical strength of the solid

We shall attempt to review the theories explaining the initiation and propagation of brittle fracture, starting from those theories which propose that there are intrinsic mechanisms built into metals which can initiate and propagate brittle cracks. Because of the immense success of physical metallurgy in explaining the properties of metals in terms of the atomic lattice, with excellent descriptions of conductivity, elasticity, and ductility, etc., it has been understandable that the discipline of physical metallurgy might expect brittle fracture to be similarly explicable.

However, the probable failure of *every* theory for the initiation of fracture is a situation which seems to be not admitted and perhaps not realized. We shall take a critical look at our current theories put forward to explain the initiation of cracks and find them wanting. We shall find that we eliminate, or at least probably eliminate, all the possibilities, forcing the conclusion that the initiation of cracks depends on the introduction of unbonded defects entrained during consolidation of our metals, recalling that there appears to be no shortage of such defects in our current metals. Thus, perhaps to the reader's surprise, we shall find ourselves with little option but to conclude that most if not all cracks originate in the liquid state.

3.2.1 Dislocation mechanisms

A traditional approach to estimating the theoretical strength of a solid includes consideration of the critical stress for slip. Alan Cottrell (1964) gives a clear account using the definition of the shear modulus G as the ratio of stress to strain. Assuming a critical value for strain of about 0.1 to initiate general slip of planes of atoms over other planes of atoms and taking the values G = 26.1 GPa for aluminum and 81.6 GPa for iron, we derive the critical stress for slip 2.61 GPa for aluminum and 8.16 GPa for iron. These are, of course, high stresses. However, they are almost certainly not stresses sufficient for fracture by decohesion of planes of atoms from other planes of atoms. In this case the estimate relates to the initiation of slip, in which the atoms have moved one atomic spacing but have not separated. Thus, although these impressively high values are estimates of the yield strength of solids, they are not estimates of the fracture stress.

A mechanism for initiating failure by the generation of microcracks in the lattice was put forward later by Cottrell in 1958. This was an elegant dislocation model assuming the junction of two intersecting slip planes. Smallman and Bishop (1999) give the details. A dislocation gliding in the (101) plane coalesces with one gliding in the (001) plane according to the reaction

$$a/2[111] + a/2[111] \rightarrow a[001]$$

The new dislocation which has a Burgers vector a[001] is a pure edge dislocation and acts as a wedge, one lattice constant thick, inserted between the faces of the (001) planes. The crack grows, a lattice spacing at a time, by the arrival of other dislocations running into it. The authors comment that although the mechanism accounts for the observed (100) cleavage plane of bcc metals, examples have never been directly observed.

Somehow, Cottrell's elegant model has fallen into oblivion. Researchers into fracture now seem to have forgotten Cottrell's theory and have generally concluded that fracture of steels initiates from cracked carbides in grain boundaries. We shall have more to say about this conclusion in due course.

3.2.2 Dislocation pile-up

The pile-up of dislocations against an obstacle is a widely accepted mechanism assumed to be capable of creating a crack in the metal lattice. The stress concentrates the dislocations against the obstacle, so that coalescence of their half-planes opens a crack in the lattice. It is interesting, however, that for the past decade and more, hundreds or thousands of pile-ups have been observed in the electron microscope and have been studied in simulations by molecular dynamics (MD). Pile-ups are commonly seen and studied, but, once again, the development of a crack has never been observed. In a way, this is perhaps only to be expected. Once again, it is the result of the immense strength of the bond between atoms; in normal circumstances it is impossible to pull atoms apart.

3.2.3 Vacancy condensation

From time to time, the nucleation of pores in metals has been postulated to occur as a result of the condensation of vacancies. This appears a reasonable assumption because throughout metallurgy, the formation of second phases is observed to occur in conditions in which the solid solution is supersaturated in the elements of the second phase. A well-known example is the formation of GP zones in Al alloys which eventually grow to form the $CuAl_2$ precipitates.

It is clear from thermodynamic arguments that immediately after solidification the metal is briefly supersaturated with vacancies. It has been natural therefore to assume that they would aggregate and precipitate to form voids.

However, there has been direct evidence for over 40 years that cracks and pores never form by condensation of vacancies (Kuhlmann-Wilsdorf, 1965). Transmission electron microscope (TEM) observations of the condensation of a supersaturation of vacancies in a lattice might be expected to form cavities in the same way that condensation of supersaturated solutes can form second phases. However, for all metals studied so far, this is not true. TEM observations of quenched fcc metals, including Al, Ag, and Au, indicate that condensation of vacancies does occur, but instead of the formation of vacancy discs or three-dimensional voids, the lattice collapses completely. Under its own interatomic forces, the metal lattice consolidates to full density, creating dislocation loops or stacking fault tetrahedra depending on the stacking fault energy, not pores or cracks (Uberuaga et al., 2007).

In more recent electron radiation studies of vacancy condensation in Fe (Moser et al., 1987), Mo (Sen Gupta et al., 1987), Zr (Hellio et al., 1987), and U (Weiberg and Quere, 1987), voids were never reported; only dislocation loops were observed.

There have been suggestions that hydrogen atoms, diffusing into vacancy condensations, might stabilize the void as, effectively, a hydrogen bubble. Tiryakioglu (2019) has put forward a strong rebuttal of this suggestion for aluminum, drawing attention to the effect of pressure on the solubility of hydrogen which strongly encourages the hydrogen to escape from a shrinking bubble and return into solution. There seems a possibility that similar behavior should be expected with other metals.

Overall, however, the apparent absence of void nucleation by vacancy condensation in most circumstances investigated so far indicates of the strength of the forces between atoms. The strength of the atomic bonds wins every time so far, preventing void formation.

3.2.4 Nucleation of a void

The nucleation of a void by the imposition of a high hydrostatic tension in the liquid state has been discussed in Section 1.1. The elegant classical treatment by Fisher (1948) indicated strengths of 7.0 GPa for liquid Fe and

3.1 GPa for liquid Al. Tiryakioglu (2018) agrees tolerably closely with the Fisher result, finding 4.0 GPa for liquid Al in his astonishing compilation of fracture strengths. The theoretical strengths of solid metals are expected to be somewhat higher because of the closer interatomic distances. Tiryakioglu confirms this once again for solid aluminum in Fig. 1.2.

In this text so far, we are starting to build up a case that both liquid and solid metals are expected, with good reason, to have high strengths out of reach of normal conditions of the stressing of metals and certainly far higher than can be expected in solidifying metals.

As discussed in Section 1.1, the extremely high stresses for the "homogeneous" nucleation of pores or cracks might, of course, be reduced in the presence of a poorly wetted substrate that would allow "heterogeneous" nucleation. However, although the nucleation stress is predicted to be reduced by a factor of 20, the fracture stresses remain sufficiently high to be generally out of reach for normal conditions (Campbell, 1968).

Significantly, the ability of a solidifying metal to withstand any high level of internal tensile stress is very limited as every foundry person knows; a poorly fed casting can collapse, its external surfaces dishing inward under only atmospheric pressure (0.1 MPa). Even with the benefit of reduced stress requirement for heterogeneous nucleation on an extremely unwettable surface, the stress to decohere from the surface to create a pore remains in the range 10^3-10^4 times higher than can be met during solidification. In summary, the tensile stresses sufficient to create volume defects in solidifying metals cannot be generated, simply because interatomic forces are too high to allow pores to open, and metals near their melting points are in general too weak to generate and support such stresses.

Moving on to consider the unfortunate illogical identification of heterogeneous nucleation, based on the pore sitting at the base of the crevice in the solid inclusion (Fig. 1.5). This pore cannot have been formed by solidification processes, but must have arrived from outside the liquid metal: it must have been introduced or "entrained." Once entrained, it can easily expand to create new pores at almost negligible applied stress or gas content in solution. However, once again, if such an entrained inclusion were somehow avoided or eliminated, we return to the situation that a pore becomes impossibly difficult to form.

Finally, in efforts to understand void nucleation in metals, an impressive and welcome development over recent years has been the ability to model large clusters of atoms to allow a study of their behavior simulated on the atomic scale. The computer-based approach is known as MD. It has been used not only for metals but also for more complex aggregates of atoms such as those in complex polymers and organic molecules. A short web search will reveal many descriptions of this exciting new window into events on the atomic scale.

When applied to metals, an accurate picture of lattice defects such as dislocations, vacancies, and interstitial atoms can be generated; their movements and reactions between such defects as vacancy pairs and higher vacancy clusters can be studied in detail.

During studies in which many MD simulations were carried out to study the *growth* of cracks in metals subjected to high stresses in the GPa range, the *nucleation* of a crack has so far not been observed (Yamakov, 2011). It seems reasonable to conclude that nucleation of cracks in metals can only occur at extremely high stresses.

In confirmation, in those studies in which tensile conditions are applied to stabilize or grow cavities, MD studies by Milstein et al. (2004) indicate that a tensile stress of over 15 GPa was required to stabilize a void in Ni, causing it to grow explosively to promote failure.

MD simulations (Sabochick et al., 1987; Milstein et al., 2004; Traiviratana et al., 2008) show how clusters of up to 45 vacancies are unstable and collapse within a few nanoseconds to form stacking fault tetrahedra. Such rapid elimination of voids is perhaps not surprising in view of the natural thermal vibration of atoms throughout the lattice, buzzing about their average positions at a rate of around 10^{13} Hz.

Thus, the MD studies confirm the TEM observations that voids are not easily formed and will collapse rapidly and disappear unless the hydrostatic tensile stress exceeds some large value of the order of the ultimate tensile strength. Furthermore, the various estimates of the ultimate theoretical strength of metals appear to agree as well as might be expected, at least to indicate that the strengths are immensely high, given the very different theoretical approaches.

3.2.5 Decohesion of a weak interface

Turning our attention now to grain boundaries in metals, it is a widespread but unfortunately mistaken belief among metallurgists that grain boundaries are weak. This conclusion is reached from the observations of apparent decohesion of boundaries in some conditions. The mismatch between assumption and fact starts to be apparent when it is recalled that virtually all of our major engineering alloys are full of grain boundaries: it is inconceivable that such features are weak; in fact, strength is enhanced in alloys by increasing the density of boundaries (i.e., making grains smaller).

We shall therefore examine the evidence to suggest that those boundaries which appear to decohere actually contain bifilms; they are effectively pre-cracked. Clearly, boundaries that do not contain bifilms will be expected to be strong as a result of the close atomic spacing over a large percentage of the boundary area. These expectations are corroborated by a number of studies.

Void growth studies by Meyers et al., (2009) used reflected shock waves in Cu, finding that voids formed at grain boundaries, but only when the tensile stress exceeded 37 GPa, confirming the immense strength of grain boundaries in copper.

MD calculations by Yamakov (2011) for boundaries in Al confirm their high strength for this metal also. He concludes that for single crystal Al its strength is 7.5 GPa whereas the grain boundary has a strength generally in the range 6−7 GPa depending on the type of boundary. For instance, for those boundaries with a high percentage of coincident sites (atom sites common to both adjacent grains) such as a coherent twin boundary, the boundary has strength equal to that of the matrix.

In addition, it is worth spelling out clearly that decoherence will not be expected at phase boundaries in which the phases have grown in situ by precipitation in the matrix. Such "home-grown" interfaces between the matrix and inclusions and second phases such as carbides, nitrides, and sulfides in steels would never decohere. Similarly, the Si/matrix interface in Al−Si alloys would never decohere.

In relation to this last example, there is an interesting MD study by American researchers led by Horstemeyer (Gall et al., 2000) at Tuscaloosa, Alabama, in which they calculated the tensile strength of all the main interfaces of the Al-Si alloy. The Al−Si interface was 22 GPa, the Al−Al interface 29 GPa, and the Si−Si interface 49 GPa. They conclude that the decoherence would occur with increasing ease from Si to Al to the Al−Si interface. With due respect to this illustrious team, this is quite the wrong conclusion. The more accurate conclusion is "Within the realms of day-to-day engineering, *none* of these interfaces would decohere—they are all massively too strong."

Exactly the same reasoning applies to MD results by Liu et al. (2019) who calculate the strengths of various interfaces in a simulated bifilm, concluding that for some oxide compositions, failure could occur in the Al matrix rather than at a (perfect) oxide interface in the center of a bifilm. Once again, the common sense conclusion is that the strengths of all the interfaces they study are far too high to fail by fracture in any practical situation. Further still, it is of great importance to consider the real structure of the bifilm: the two films are expected to hardly contact, except in a few scattered points. Although the MD study probably models the contact points accurately, these locations represent perhaps less than 1% of the total area of the bifilm.

Viewing now decohesion in metals as generally observed, those phases associated with an unbonded interface, dragged into the melt from or through the liquid surface (Fig. 1.8), are likely to have almost zero contact with the surrounding matrix and thus easily decohere in practically any direction, thus easily initiating a volume defect such as a crack.

Decohesion appears also to occur from those phases which happen to have nucleated and grown on only one side of a bifilm. The phase is "glued" to this particular film of the bifilm, but the other unbonded film on the unoccupied side of the bifilm is, of course, easily separated, giving the appearance of the matrix having decohered from the particle. This apparent decohesion is, of course, the natural consequence of the interface never having been cohered in the first place. However, such apparent decohesions are certainly capable of initiating pores or cracks.

However, there are numerous experimental observations that support the common sense prediction that decohesion does not occur from those phases formed by precipitation from solution in the melt or precipitation from solution in the matrix because of the necessarily perfect atomic bonding.

The ability of a surface to resist decoherence from a matrix in perfect molecular contact was nicely demonstrated as long ago as 1867 by Gernez. He showed that crystalline solids which had been grown in the liquid and which had never been allowed to contact air were incapable of inducing effervescence in a liquid supersaturated with gas. Otherwise identical solids which had been allowed to dry always caused effervescence. In this experiment the decohering forces were relatively weak, but the principle is sound. By analogy, but operating at higher stress, the contact between intermetallics and the matrix from which they were formed will be atomically perfect and thus will be strong.

The experiment by Emamy and Campbell (1995) cited earlier, comparing the bonding between second phases and the matrix in two commercial metal matrix composites, confirms the resistance to decohesion of the interface in atomic contact compared to the interface made by entrainment.

In general, therefore, it seems decohesion is only to be expected from boundaries which are already decohered as a result of the interposition of a bifilm.

3.2.6 Griffith/Barenblatt preexisting cracks

From all of the above considerations in this section, it seems we cannot nucleate a crack or other volume defect. This in itself is an important conclusion.

Even so, of course, nucleation of a crack thus becomes unnecessary if a microcrack already exists, so that initiation of a macrocrack becomes possible. Griffith's elegant and farsighted proposal as early as 1920 proposed that every crack required a preexisting crack for its initiation. This has become the accepted model, much quoted and elaborated (see Cottrell, 1963, 1993).

We have, it seems, tended to think of the initiating Griffith crack and the subsequent macroscopic crack in the singular: one microscopic preexisting Griffith crack initiates just one observed macroscopic crack.

However, when it is realized that a Griffith crack appears always to be conveniently present wherever a metal is stressed sufficiently to form an observable crack, it follows that a large population of Griffith cracks must exist in metals. Minimal additional consideration is needed to suggest that the population of Griffith cracks is not confined to the surface because any newly created surface revealed by cutting or machining could provide similar initiating cracks.

Effectively, we arrive at the realization, if this were not obvious already, that Griffith's widely accepted theory proposes that our metals generally contain a dense population of cracks. Our straightjacket of traditional unthinking has led many of us into the limitation of thinking of singular Griffith cracks and singular resulting macrocracks. Griffith was not limited by such blinkered thinking. He proposed that metals were full of microcracks of different sizes and orientation and that initiation of failure by cracking was the result of the application of stress exceeding a critical stress for the largest, favorably oriented microcrack. In his words from 1920, "the general conclusion may be drawn that the weakness of isotropic solids, as ordinarily met with, is due to the presence of discontinuities, or flaws, as they may be more correctly called, whose ruling dimensions are large compared with molecular distances. The effective strength of technical materials might be increased 10 or 20 times at least if the flaws could be eliminated."

Clearly, he was under no illusions about the immense importance of his theory, and the accuracy of his foresight is awesome. It is therefore all the more sobering that his approach tended to be ignored or treated with skepticism and even suppressed by those in charge of research in the laboratory where he worked (Gordon, 1988).

Griffith formulated an energy criterion in which if the elastic energy released by extension of the crack by a distance 2c was greater than the creation of additional surface energy of the crack, then the crack would propagate. The well-known relation is

$$\sigma = \sqrt{(\gamma E/2c)} \tag{3.1}$$

where σ is the stress, γ is the surface energy, and E is the elastic modulus. Because the equation is describing the stress required to separate atoms by the propagation of an existing crack, we may obtain an estimate of the theoretical fracture stress of the metal. The famous equation, modifying the format Eq. (3.1), is

$$\text{Fracture stress } \sigma_f = \sqrt{(4\gamma G/\alpha\pi a)} \tag{3.2}$$

where $\alpha = (1-\nu)$ where ν = Poisson's ratio and "a" the half crack length. Clearly where a starting crack does not exist (a = zero) the fracture stress is infinite. However, taking a = one atom radius, equivalent to a crack length of an atomic diameter, and Poisson' ratios as 0.345 and 0.293, respectively, we obtain critical fracture stresses of 13.3 for Al and 34.2 GPa for Fe.

These estimates of ultimate fracture strength for the solid are of the same order of magnitude as other estimates, giving some confidence that they are a reasonable guide. Clearly, there seems good reason to accept that both liquid and solid metals have extremely high strengths which are simply a fundamental reflection of the strength of atomic bonds.

Griffith's simple and elegant energy model has been universally accepted. As an energy model, Griffith's approach required no specific geometrical form for the defect. Barenblatt (1959 and 1962) proposed a somewhat different approach, still accepting the concept that the population of microcracks preexisted, but adopted a specific geometrical form. The new model assumed a defect in the solid matrix consisting of a planar area of any shape and size but zero coherency surrounded by matrix with full coherency. Willis (1967) showed that Barenblatt's approach gave a solution for fracture stress as an equation of the same form as that of Griffith, underlining the basic agreement between the entirely different formulations.

From the time of its conception, no-one has had any idea of how an area of the metal matrix with zero coherence could exist. However, as a description of a bifilm, Barenblatt's model could hardly have targeted the description of a bifilm with greater precision. His prophetic theory gives an underlying theoretical basis for the bifilm as an effective crack.

3.3 Summary of the theories for the ultimate strength of solids

From our above examination of all the accepted theories for the initiation of cracks based on atomic or lattice mechanisms in our engineering metals, it seems we must conclude that metals have no built-in mechanism for the initiation of failure by cracking; all the currently proposed intrinsic mechanisms for crack nucleation in crack-free metals are inadequate.

It should not be surprising to us that there are no operational fracture mechanisms in a lattice. All the atoms are in close contact and held by enormously strong interatomic bonds. It would be surprising if there were fracture mechanisms capable of breaking the metallic bonds in this close and regular assembly.

The only crack initiation mechanism which appears to be viable is that in which cracks preexist, as correctly proposed as cracked carbides in grain boundaries of steels, and as Griffith (and Barenblatt) propose as preexisting in the matrix.

The Griffith/Barenblatt proposals for preexisting cracks avoid the difficulties of the alternative proposals which attempt to explain the initiation of cracks, although, of course, the question of formation of the cracks is now neatly side-stepped. The widespread acceptance of Griffiths/Barenblatt is a widespread acceptance of the existence of a population of preexisting cracks in our metals; it is intriguing that, over the decades since 1920, no-one appears to have questioned how such cracks could arise, and certainly no-one questioned how they might be controlled.

As will now be clear, we propose here the preexisting population of cracks to be an expected natural consequence of the consolidation processes for metals. All the cracks are bifilms. The mechanism creating the double oxide films is simple and clear and already experimentally demonstrated in many ways. The widespread early doubts about a bifilm theory are therefore not easy to understand. At this time there seems to be no alternative explanation for the presence of populations of preexisting cracks in metals.

Moving on, a new fascinating and important question arises. Can fracture occur if the preexisting Griffiths crack, or Barenblatt crack, or the preexisting population of pores, did not exist?

This is not a trivial question because even a cursory overview of solidification (and of other bulk forming processes such as precipitation from solution or condensation from a vapor) gives strong pointers that such defects as microcracks are not to be expected. If this is true, we are immediately forced to conclude that fracture in the absence of preexisting microcracks is probably not possible at stresses less than the ultimate theoretical fracture stress.

This surprising conclusion requires to be tested carefully. The conclusion is in line with the prediction of immensely high theoretical fracture stresses.

In his wonderful book, Gordon (1988) describes experiments by David Marsh in 1956 in which microscopic whiskers of various materials including metals were subjected to tension. Their natural absence of defects resulted in their failure at stresses at or near to their ultimate fracture stress. These stresses were so high that the elastic energy stored in the whisker at that stress approximates to that of the total chemical bond energy as would be stored in a conventional chemical explosive such as dynamite. Thus, when the whisker fractures, there is not the usual sharp "bang" which accompanies the fracture of a normal tensile test piece into two pieces. Instead, "there is an explosion followed by a little cloud of dust or smoke. Otherwise, there are no visible remains of the test specimen, which has simply vanished. Only the very small size of such test specimens prevents them from doing serious damage to both the equipment and the experimenter." March's experiments clearly illustrate the potential strengths of metals in the absence of defects.

In the absence of fracture initiation mechanisms when approaching the ultimate tensile strengths of metals, we would probably expect extensive plastic flow of macroscopic test pieces, necking down to 100% reduction in area (RA). We should routinely expect that all metals should have 100% RA, and if not, something was wrong! Even for very brittle metals such as beryllium, explosive failure of macroscopic specimens is difficult to envisage.

Continuing this exercise of the imagination for other metals, it follows that we would predict that, in the absence of entrained defects, an Al–Si eutectic alloy undergoing a tensile test would not exhibit a failure of a single Si particle. The hard and strong Si particles would be expected to flow in a ductile manner, together with the ductile aluminum matrix, until the whole specimen finally necked down to 100% RA.

Analogous images can be imagined for steels with grain boundary carbides. Not one carbide should crack. If the matrix were sufficiently strong, the carbides would deform plastically with the matrix during tensile deformation, eventually the whole matrix-plus-carbide composite would neck down to failure at 100% RA.

We have reached the conclusion that metals contain microcracks, often in abundance. If neither pores nor cracks preexisted in a reasonably brittle material, it would be problematic to explain fracture. However, as we have noted earlier, the presence of a defect that could constitute a microcrack or a pore is not necessarily to be expected from traditional physical metallurgy, but, of course, *is* to be expected from practical metallurgical processing.

For instance, there is much vague belief in traditional metallurgy that solidification can create defects. This is not true. Solidification cannot be expected to create either cracks or pores because this simple phase transformation involves the movement of atoms by merely fractions of an interatomic distance, from a randomly close-packed site to a regular close-packed site. Effectively, the atoms never come apart. Solidification can only form interfaces which remain in perfect atomic contact. It can never form cracks. (Take care to note that all those defects, I think without exception, which appear to be solidification defects, such as porosity of various forms, appear to be initiated by, or actually are, entrained defects.)

Other scenarios can be envisaged such as embrittlement by a dense population of small bifilms. The ductile/dimple fracture mode falls into this category. There is a widely accepted convention, regarding a fracture surface displaying an array of ductile dimples as a "ductile failure." This is neither accurate nor helpful. Its *preexisting* population of pores or cracks is usually provided by inclusions or second phases precipitated on bifilms. The presence of the bifilm causes the inclusions or second phases to appear to be brittle (Fig. 3.1). The macroscopic failure can be seen therefore to be strictly a mixed fracture mode, in which the population of apparently brittle particles is cracking or apparently decohering, thus initiating the failure at the base of

Figure 3.1 Ductile dimples on the fracture surface of an Al–Si alloy showing apparently brittle Si particles at the base of each dimple (Zhang, 2007).

each ductile dimple, but the matrix is failing by plastic shear. The plastic shear of the matrix is seen to form knife-edge arêtes, proving its ductility and ability to neck down to 100% RA. The formal descriptions of this mixed failure mode by Thomason (1990) and Knott (1997) are recommended.

In summary, from the above considerations, conditions for failure appear to be clear and logical: in general, fracture cannot initiate in a perfect metal. This sounds awfully tautological, but the description "perfect" in this case merely implies that the matrix contains no microdefects such as pores or cracks. Furthermore, we conclude that pores or cracks can initiate in a metal only from interfaces that are unbonded (because atomic bonds are too strong to be broken). As unbonded surfaces cannot be formed by phase change processes such as solidification or vapor phase deposition, etc., such interfaces must be introduced from *outside* the metal. These are necessarily *extrinsic* features and necessarily have to be *entrained*.

We have therefore three main extrinsic defects: pores, bifilms, and extrinsic (external, exogenous, or entrained) inclusions. All are effectively introduced into the matrix by the impingement of surfaces during consolidation. Thus, in powder metallurgy (PM) routes, such defects are necessarily unavoidable; they are *intrinsic*. In casting they are unfortunately common, but unnecessary and avoidable, as has been discussed in Part 1.

Overall, it seems likely therefore that the near-universal presence of bifilms in our metals is one of the most probable, if not *the* most probable, reasons for most of the phenomena listed in this chapter. By accepting the Griffith/Barenblatt theory, we have effectively been accepting the fact of a dense distribution of cracks in metals since about 1920. The contribution of this book is to identify the nature and source of the cracks.

Most importantly, by appropriate handling and casting of liquid metals, it is now becoming possible to produce metals free, or nearly free, from cracks (bifilms). In the near future, we can hope for critical tests of the proposals made here.

3.4 Consolidation

Consolidation is a simple concept, but one which has far-reaching implications. It simply draws attention to the (rather obvious) fact that in the creation of engineering structures, the metals we use have been produced by collecting together smaller pieces and bonding or welding these to make workable engineering bulk shapes such as girders, plates, and sheet (Campbell, 2016).

There are one or more steps in the production of metals in which the metal is gathered and joined together ("consolidated") from smaller fragments to make larger pieces for engineering purposes. In almost every case the gathering together raises difficulties. The fragments almost invariably possess an oxide skin, so that on consolidation, the oxides "get in the way," obstructing the creation of a "weld" type of contact; in fact, oxide impinges on oxide to create an unbonded interface between the two oxide ceramic layers. We have created a bifilm. The dense population of bifilms in many metals is a natural and fundamental consequence of the necessary mechanism of consolidation.

Furthermore, of course, some production routes involve multiple consolidation processes, so that the generation of bifilm-type defects is likely to grow at each phase. A typical example is given by Salehi and colleagues (2014) who cast melt spun fibers which required a further hot consolidation stage. Similarly, the Al recycling market ensures that the "secondary" alloys in the recycling loop have undergone repeated reconsolidations, therefore having the greatest chance to accumulate multiple populations of entrained defects.

It is salutary to survey below the broad spectrum of engineering metals, from those with perhaps zero or few bifilms, through processes which control the size and distribution, to cast metals with often uncontrolled random sizes and random distributions of defects.

3.4.1 Atomically dense clean metals

Probably metals grown slowly on a planar front by vapor condensation, or electrodeposition from solution on to a planar growing front, can be free from bifilm defects. Although engineering chunks of such materials have been produced to demonstrate the feasibility of these processes, the glacial production rates do not encourage prospects for large-scale production.

Nevertheless, it is salutary to reflect that it is not easy to conceive of such atomically uniform materials suffering from stress corrosion cracking (SCC) or exhibiting ductile dimple fracture surfaces.

3.4.2 Powder metallurgy

The archetypal solid-state particulate consolidation route is powder metallurgy (PM), in which minute particles of metal are brought together to be consolidated by various mechanisms such as pressing and sintering.

Metal powder particles are commonly covered with an oxide film. As an example, Schaffer et al., (2004) measured the thickness of the oxide on his aluminum particles to be only 5–15 nm. The film is clearly in perfect atomic contact with the powder particle because it has grown atom by atom from the particle. However, when the powder particles are consolidated, the particles are compressed together oxide film-to-oxide film. This impingement of ceramics (such as extremely stable alumina or chromia with melting points over 2000 C) will result in a powder compact which will be resistant to the development of strength, having a dense population of oxide-to-oxide interfaces which are resistant to bonding, but will act as cracks.

As, of course, is well known, PM succeeds by special efforts to achieve consolidation of PM compacts. Some oxides can be taken into solution, and others can be melted, especially if trace elements are present to lower the melting point of the pure oxide phases. The plastic deformation of the compact is likely to fracture bifilms and encourage some metal-to-metal contact between powder particles. These techniques can be successful to the point at which the residual oxides can act as strengthening features of the microstructure. Efforts by Rao et al. (2001) with Inconel 718 powder are typical, in which they found HIP alone inadequate to achieve significant ductility, but a solution treatment at 1270 C, involving some incipient melting, disrupted boundary oxides and delivered good properties. It is common for properties of PM products to be enhanced over those of conventional alloys, almost certainly as a result of the continued presence of the finely dispersed residual oxides in the structure.

Even so, not all efforts are successful to eliminate the oxide bifilms inherited from this solid-state consolidation process. The light alloys do not enjoy the benefit of sufficiently high temperatures to generate low melting point oxides to aid consolidation. Narayanasamy et al. (2005) found that when cold forging compacts of pure aluminum powder, which would be expected to consolidate well as a result of its extreme softness, the resulting properties were found to exhibit a scatter similar to that of castings.

Mechanical alloying occupies a unique place in this list of consolidation techniques. Its starting point is the mixing of powders, so it has some commonality with PM discussed below. But the powders are then subjected to the mechanical alloying process, usually in a ball mill, in which the powders are hammered, sheared, and smeared to such an extent that they are atomically mixed. The original oxides will have been to some degree separated into their component elements, as in solid solution. The resulting fairly homogeneous powder has then to be consolidated using familiar PM techniques, at which point defects may become entrained.

3.4.3 Spray forming

It is to be expected that sprayed metals will exhibit generous quantities of oxide double films, as a result of the work piece itself oxidizing to create a surface oxide film before the arrival of the next liquid or partially liquid particle together with its oxide film. The oxide films naturally impinge to form bifilms.

The separation of splats by double oxides is wonderfully clear in the work by Aug et al. (2015) in which they study plasma spraying of high entropy alloys including AlCoCrFeNi (Fig. 3.2). Studies by Kong et al. (2003) have shown similar clear images of oxides separating splats of oxyfuel thermally sprayed Stellite 6.

The formation of bifilms between deposits of spray droplets of 7000 series Al alloy is strongly suggested in the work of Salamei and Cochrane (2003). Subsequent examination by TEM material consolidated by extrusion showed grain boundaries containing voids together with coarse precipitates of $MgZn_2$ particles. When subjected to tensile failure, the microstructure near to the fracture surfaces revealed the $MgZn_2$ particles appeared to have decohered from the matrix.

The formation of porosity in spray samples is merely seen to be the noncontacting of oxide films in random locations. This obvious mechanism is in fact precisely analogous to the formation of porosity in cast metals.

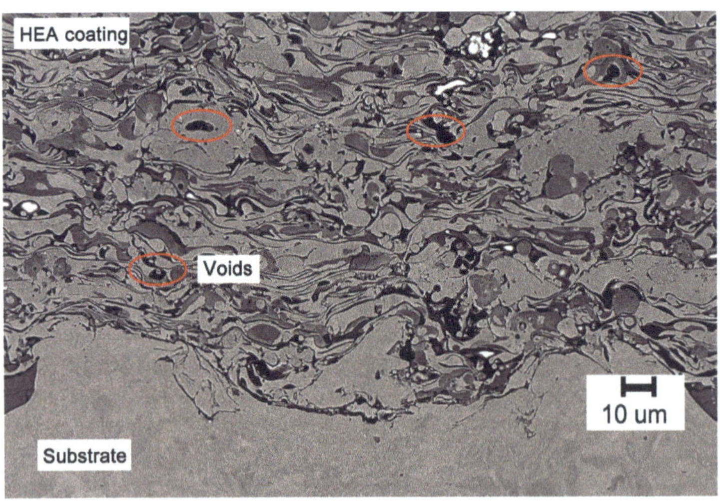

Figure 3.2 Plasma sprayed AlCoCrFeNi, a high-entropy alloy (HEA) showing entrained oxide bifilms and porosity (Aug et al., 2015).

(As reemphasized repeatedly in this book, and for which the patience of the reader is requested, porosity in cast metals does not occur as a result of a nucleation and growth process, as has been widely assumed, but is mainly entrained mechanically during liquid metal handling.)

Sprayed materials generally have properties intermediate between those of PM and cast metal. The material usually requires further mechanical consolidation but at best will be expected to achieve properties similar to those of PM products.

3.4.4 Additive manufacture

Some of the new 3D printing processes are classic examples of bifilm problems between added layers. Yuan (2019) introduces three reports in a series of investigations into defects in additive manufactured (AM) materials. Naturally, at this early date, the probable presence of bifilms between molten and semimolten additions has not been suspected in these studies.

However, reports can be found in which the presence of bifilms seems likely. Jelis et al. (2015) found evidence of cracking on the 4340 steel direct laser sintered component which they were unable to explain. "Inability to explain" is typical of bifilm problems because bifilms are usually invisible, or at least not easily seen by casual observation, and in any case not usually suspected. In laser additive manufactured Inconel 718, Zhang et al. (2013) found Al_2O_3 and Cr_2O_3 bifilms, clearly characterized by their "butterfly" symmetry on fracture pairs and further identified by the precipitated laves phases and carbides on the wetted exterior faces of the bifilms.

For the future of AM, full strength and toughness can only be realized if bifilms are not generated between additions. It is possible to envisage that bifilms might be ruptured or even entirely shifted out of a forming joint as part of a deliberate mechanical action during addition. Alternatively, efforts may have to be directed to additives which will reduce the melting point of the oxides.

In the meantime, it is a concern that AM is expanding its applications rapidly, while its vulnerability to the entrainment of bifilms seems not yet fully understood or controlled.

3.4.5 Welding

During the welding of metals, the weld itself requires to be made by a skilled operator who is usually very much aware of the problems of entraining the surface oxide which covers the weld pool. This aspect of avoiding entrainment defects is well understood, especially for the welding of "difficult" metals such as aluminum and stainless steels.

Welds are an interesting complicated case in which the multiple consolidations ensure that bifilms lie in wait to open as cracks in the heat-affected zone (HAZ). The HAZ cracks are not nucleated, they preexist, as is evident from the character of the second phases which have formed on them in the liquid state. They have been formed in the original casting process for the liquid steel and have clearly survived rolling into plate and sheet. The welding process forcefully encourages the opening of even small bifilms as a result of the extremely high stresses which can occur in this process. In the experience of this author, all weld cracks appear to be bifilm defects. An interesting prediction therefore follows: all HAZ and much other weld cracking phenomena should disappear when the parent metals are properly melted and cast to limit or eliminate their bifilm content. (It is worth emphasising that this is easily achieved at negligible additional cost.)

It is common for the HAZ to become so hot that it suffers a condition often called "incipient melting." This is a condition in which low melting point phases, usually in grain boundaries, melt. As we have noted in Part II, the melting of these phases opens up bifilm cracks on which the particles have deposited. However, resolidification of the phases does not result in the reclosing of the newly opened cavities: they remain open (Campbell, 2009).

The incipient melted cavities are usually relatively trivial compared to those cracks opened by the high stresses involved in the welding process. HAZ cracks are probably always bifilms resulting from the pouring of the original steel. They are identifiable with reasonable certainty from their appearance under the microscope. The shape of the crack is usually meandering, rather than exclusively transverse to the applied stress. Secondly, the phases either side of the crack are often dissimilar, as a result of different precipitates forming on the different sides of an asymmetric bifilm. The bifilm is often asymmetric as a result of having often formed on a thick old oxide from the surface of the melting furnace or from a charge material. Its composition will usually be a spinel such as $Al_2O_3.MgO$. This will contrast with the other half of the bifilm, folded in place during the pouring process, its time between creation and submergence being only a few milliseconds. This oxide film will be only nanometers thick and, with insufficient time for Mg or Cr or other solute to diffuse into the oxide, will therefore usually consist of pure Al_2O_3. The two sides of the bifilm could hardly be more different (Fig. 3.3).

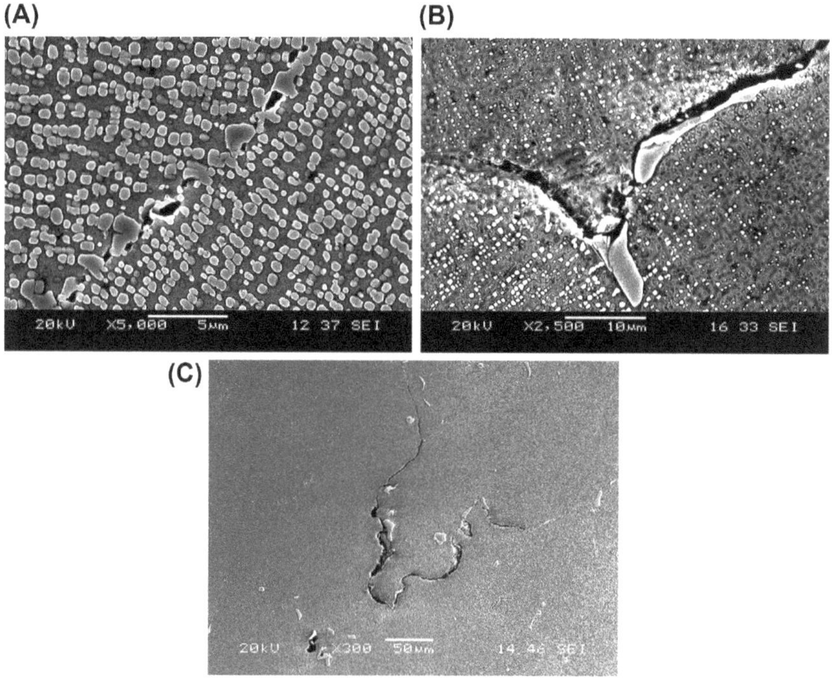

Figure 3.3 Grain boundary melting in the heat-affected zone (HAZ) of a Ni alloy 738LC. (A) Carbides cracked and decohered in the grain boundary; (B) asymmetric features of the crack suggesting an asymmetric bifilm; and (C) a convoluted crack reminiscent of a bifilm pushed into place by dendrites (Sidhu et al., 2005).

3.4.6 Melting and casting processes

The melting and casting route is the major route for the consolidation of most metals. However, how does the principle of consolidation apply in this case?

When liquid metals are stirred or poured, the surface becomes broken into droplets, jets, and breaking waves. When such bodies of liquid reassemble, consolidating, their surfaces necessarily come together by their dry oxide film surfaces mutually impinging. The impinged films create a dry, unbonded interface which becomes immersed in the bulk liquid. Very turbulent pouring can fill a liquid metal with cracks. Conversely, of course, the two outer interfaces of the double oxide film are the undersides of the original surface film on the liquid metal. They are therefore in perfect atomic contact with the bulk liquid. These are the distinguishing features of the bifilms, their double-film nature, with internal nonwetted, nonbonded crack, but perfect exterior wetting; its unique structure resulting from the necessary logic of the mechanism of entrainment.

Cast metals can suffer a bifilm content characterized by a randomly wide range of sizes, and randomly distributed, giving a wide scatter of properties from acceptable to disastrous. The higher temperatures for consolidation of liquid metals, compared to the consolidation of solid metals, necessarily lead to faster reactions and thicker oxides. In contrast, all the various particulate processes, such as PM, have fine and reproducible structures and might even benefit to some extent from their distribution of oxides. Even so, the necessary presence of oxide bifilms will be an ultimate limit to properties. In comparison, casting from the liquid state can have a wide spectrum of sizes of bifilm to give widely scattered and low properties, but the same liquid metal can, in principle, be perfectly cleaned by filtering, sedimentation, or flotation. It then has the potential to achieve a defect-free condition, with its expected uniquely excellent properties.

Looking further at the liquid metal route, the wide spectrum of defect sizes in solidified metals arises because of the widely different conditions experienced by the liquid metal at different stages of melting and casting. The large dispersion of bifilm sizes and number density corresponds to the significant scatter of mechanical properties. It reflects, regrettably, current widespread but unnecessarily sloppy and uncontrolled pouring practice. Recent casting developments (Section 1.6.4 and Campbell, 2018) can produce solidified metal with zero harmful bifilms.

Studies to date have indicated that the population of bifilms in Al alloys (Campbell, 2008) appears to be high, estimated to be often as high as 10^{12}–10^{18} m^{-3} (not per kg as stated in error in the original publication) based on the number of ductile dimples on fracture surfaces (Fig. 3.1). Steels and Ni-base alloys are expected to have similar populations.

3.4.7 Wrought metals

The casting defects are inherited by wrought materials, possibly with slightly less effect on scatter depending on the amount of plastic deformation involved. However, this is assuming that the working process does not crack and destroy the cast product on its first pass or other early passes through forges or mills. Ingots of Ni-base alloys poured badly, such as top-poured in a vacuum induction melting furnace, are well known for cracking on the first stroke of the forge.

Assuming the forging work piece survives the forging process, it seems that the deformation process merely pushes the cast bifilm defects around a little, but otherwise has only a minor effect on the defect population. Forgings therefore suffer nearly as much scatter in properties as castings. Even rolled sheet, extrusions, and drawn wire do not completely shed the disadvantages inherited from the original bifilm population. This is a consequence of the stability of the oxides and the residual air in the form of the 1% of insoluble argon which will always remain in association with the oxide inclusion. The residual argon bubbles are volume defects which allow the apparent nucleation of pores and cracks from the inclusion. The so-called "alumina stringers" seen on polished sections of wrought steels are almost certainly the residues of alumina bifilms (because the "stringer" on the 2D section has to be a plane in 3D), and porosity (the residual argon bubbles) is nearly always detectable in association with the fragments of oxide (Fig. 1.55).

The unbonded interface of the bifilm appears therefore to survive significant plastic deformation. This is perhaps to be expected from such stable ceramic interfaces and the remnants of air between the faces of the films, trapped in rucks and folds, and in the microscopic roughness of the interfaces. This reservoir of oxidizing or nitriding gas will continue to create new film, as a bifilm is extended by plastic working, thus preventing bonding. Only when the reservoir of air is consumed will new extensions of the crack start to bond. Finally, of course, the remaining 1% of argon in the air will remain as a permanent pore because argon appears to be completely insoluble in metals.

These thoughts match the observations of Toda et al. (2009) who studied hot and cold deformation of an Al–Mg alloy up to 22% strain which failed to permanently close pores. Even for the hot rolling of steel, Joo et al. (2014) observed the closure of some pores, but generally found that bonding was rare or not at all. Given that, on occasions, there will be a generous supply of residual argon in the bifilm, it is perhaps not surprising that the bifilm will have an extended life, if not an effectively infinite life, despite the punishment it may receive.

Al alloys retain their unbonded regions even after the severe extrusion required to produce millimeter-thick window frames, as is evident from filiform corrosion (FFC). This degradation of the surface is clearly seen by eye on unprotected extrusions, in which tens or hundreds of corrosion sites per square centimeter follow the elongated unbonded bifilms that tunnel through the metal, happening to intersect the surface from time to time to create long lines of corrosion sites. The survival of the bifilms during plastic working is probably the result of the reservoirs of air that remain trapped in the rucks and folds between the oxide surfaces. Even after the consumption of the oxygen and nitrogen, the residual argon in the air will, of course, assist to further delay the onset of bonding, as will any residual hydrogen in the alloy.

From much work carried out on the application of pressure during solidification and from compaction during such processes as hot isostatic pressing (HIP) of cast metals, the benefits to properties are almost certainly the result of the closing of defects such as pores and bifilms. There is evidence in many metals that pores and bifilms, especially if the oxides are relatively thick, are reluctant to bond or weld (Staley et al., 2007) or only bond in limited places (Raiszadeh and Griffiths, 2008), which is to be expected from the great stability of some of their surface oxides and nitrides (exceptions to this include those oxides that react at the temperatures of HIP (Lumley et al., 1999). Thus, although the defects will remain weak in tension between their surfaces, the increased strength properties almost certainly arise from the fact that the defects are now closed; their contacting surfaces can now at least resist shear as a result of friction and jogs as shown schematically in Fig. 1.36. Whether or not the defects will fully remain unbonded, or will bond, will now depend on the chemistry of the interfaces and the conditions of time and temperature.

3.5 The microstructures and properties of metals

3.5.1 Bifilms as cracks

At the time of writing, at the very beginning of the metal production process, cracks in steel ingots and continuous castings remain a major problem. This unsatisfactory situation has existed for years, with relatively little progress. For instance, Ge and workers (2019) describe the cracking problems in safety critical 33MnCrTiB, 35CrB and 40Cr continuously cast steels used for the forks of forklift trucks. Cui et al. (2019) report both microcracking of grain boundaries and macro longitudinal cracking of continuous cast slabs of 1.35%Al TRIP steel for automotive applications. Inclusions in low carbon pipeline steels have been studied by Sidorova and co-workers (2019) because of their association with the development of cracking in oil or natural gas containing H_2S and H_2O, leading to hydrogen-induced cracking (HIC) and sulfide stress cracking. In aluminum cast houses, semicontinuously cast slabs suffer cracks of length often measured in fractions of meters and sometimes in meters.

Cracking of metals is therefore horrendously frequent in metallurgy and engineering, right from the start of the production of metals, throughout the working processes, and into the final products. It is a massive loss of efficiency which costs the world dearly. Furthermore, we need to bear in mind that this massive loss continues despite the massive investment in research and development throughout the world for many decades. Clearly, progress continues to be made, but an unbiased observer would have to conclude that at the overall glacial rate of progress, the massive scale of the problem strongly suggests that the problem is not likely to be solved with present concepts and practices.

Having attempted to put the problem of cracks into a global perspective, we shall now proceed to consider the many details: the effect of the built-in population of cracks, usually from the casting process, often microscopic and unsuspected, on the many different properties and failure modes of metals as explored below.

3.5.1.1 Tensile properties

The yield (or proof) stress of metals tends to be insensitive to the presence of bifilms. This is because although the load bearing area of the bifilm is lost, the additional strain now acting on surrounding areas strengthens them by work hardening. This natural strengthening action tends to compensate for the loss of area.

Although yield is hardly affected, ultimate tensile strength is normally impaired by the reduction in area and by the reduced work hardening as a result of the reduced elongation to fracture. This is a straightforward but unspectacular reduction in properties.

The ductility is a completely different matter. The elongation to failure is highly sensitive to the presence of bifilms and can dramatically rise from brittle behavior, 0% elongation in the presence of a high concentration of bifilms, to 50% elongation or more as bifilm density is reduced.

A typical result is illustrated in Fig. 3.4. Tiryakioglu (2019) collate published elongation results from 18 foundries. The foundries were mainly aerospace operations in the United States which would claim to have all the very many manufacturing variables under strict control. However, the lamentable scatter of elongations illustrates perfectly the one variable outweighing all others, but unrecognized at the time was turbulence during the pouring of the castings. For some values of the proof stress, the elongation can vary between 1% and 20%. Knowing the industry, the 1% value was probably achieved at times by "rounding up"! These results date from some years back, so that probably better performance might be expected today (Fig. 3.4).

In terms of the microscopic mechanisms involved, a study from the Czech Republic (Ohnistova et al., 2019) seems typical. They used a tensile stage in a TEM to observe the tensile failure of a 7475 aluminum alloy in

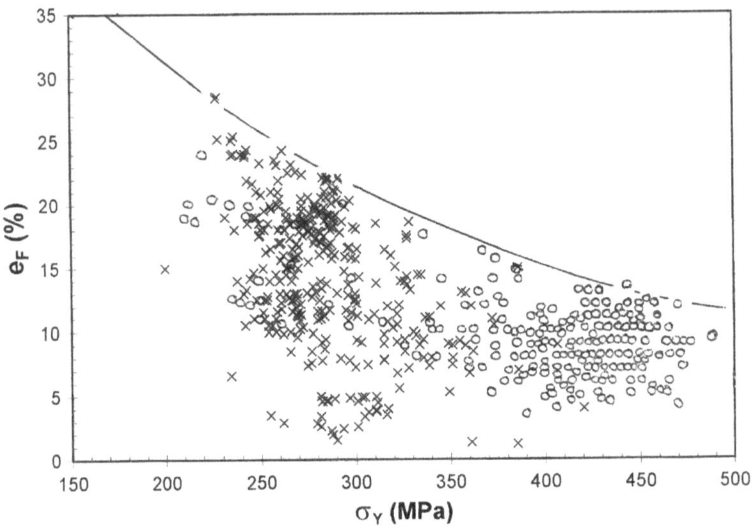

Figure 3.4 Scatter of tensile elongations of A206 (x) and A201 (o) alloys (Tiryakioglu et al., 2009).

detail. They recorded that "cracks nucleated at the interfaces between 'brittle particles' and the metal matrix. The cracks were observed to spread under the increasing tensile loading and the samples were broken before reaching the standard tensile strength value".

For carbon-manganese steels deoxidized with some aluminum, if entrained rapidly as a "young" oxide, the bifilm is often formed from a pair of aluminum oxide films. However, if air is also entrapped within the bifilm, the oxygen is quickly consumed to thicken the existing oxide films. However, the nitrogen in the air will subsequently be consumed to form aluminum nitride (extrapolating from Griffiths and Raiszadeh's 2007 work on aluminum alloys). At the time of writing, it is not known whether this is the aluminum nitride which is commonly seen on the fracture surfaces or whether, perhaps, the nitride is augmented by the precipitation of additional aluminum nitride during solidification. Whether augmented or not, the aluminum nitride would appear uppermost on the fracture surface; the remnants of the original alumina film which caused the fracture would probably be overlooked, being under the nitride, and thinner. A further complication can be envisaged if the nitride were to additionally form on the bifilm as a favored substrate as described in Section 3.4.3. The oxide film of the bifilm would then be coated with nitride on both sides. There is much careful electron microscopy to be carried out to clarify these possibilities. The accurately descriptive American term for aluminum nitride embrittled fracture surface is "rock candy fracture."

Ted Anderson, in his excellent tome "Fracture Mechanics" 2017, describes the significance of the ductile/brittle transformation temperature of steel. As the temperature is lowered, the failure mode of the steel changes from ductile to brittle. At high temperatures the fracture is ductile and fails by the linking of ductile dimples, whereas at low temperatures the fracture appears to be by cleavage. In the transition region, both mechanisms can contribute. This dual behavior is perhaps to be expected in steel as a result of its location in the transition region between ductile and brittle metals in Table 3.2. The role of bifilms in this behavior is not known at this time, but may be central. It is important to test whether steel, if cast with limited air entrainment and turbulence, will eliminate the descent into brittleness at low temperatures. A steel, ductile at all temperatures, would be welcomed.

The action of sulfur when seen as a type II sulfide may be similar, appearing as a precipitate to cover the original oxide bifilm, and so appearing to act as an embrittling feature. Although in this case the sulfide will form on the outside of the bifilm and so will lie under the oxide when seen on the fracture surface, yet the oxide will still probably not be detected; it will usually be expected to be so thin that the sulfide will appear clearly through it, and the microanalysis signal for oxygen will also be negligible in favor of the sulfur signal from immediately below.

At this point a digression concerning the measurement of the mechanical properties of castings, particularly tensile properties, should be raised. The practice of discarding tensile results from failures outside of the gauge length is a practice permitted by nearly all official testing specifications. The non-reporting of such an event turns a blind eye to the fact that a serious defect exists in the test bar. The practice was understandable when it was universally assumed that metals were essentially uniform and reliable; the unexplained failure of a test bar outside the gauge length could perhaps be dismissed as some kind of "mistake," perhaps on the part of the operator. The discarding of poor results is clearly seen to skew findings, yielding a potentially optimistic and perhaps dangerous picture.

Now that the concept of the bifilm has been introduced and while our casting techniques continue to be rather poor, variability of properties from point to point in the metals is, unfortunately, to be expected. All results should be required to be recorded, allowing a truer overall assessment of the reliability, or perhaps we should say, unreliability of our metals to be achieved. The evidently faulty result is valuable in the sense that it is a true measure of the strength of that particular sample of our metal and should therefore be recorded.

3.5.1.2 Impact toughness

Chinese researchers (Mao et al., 2019) studied the impact toughness of a magnesium alloy (AZ80 approximately Mg−8Al−0.5Zn−0.2Mn−0.2Ag). As is usual in much metallurgical reporting, the casting conditions for their material were not specified, neither the hot rolling temperature nor the rolling reduction. Nevertheless, it is safe to assume pouring conditions were turbulent, and bifilms would be prevalent throughout the structure. As a result, they found that in the as-rolled condition and when solution treated (T4) the elongation and toughness were good. However, in the solution-and-aged condition (T6) the elongation was approximately halved and the impact toughness was reduced to only a quarter of the solution treated condition. The authors attribute the superior toughness of the as-rolled and solution-treated material to the presence of profuse twinning during impact testing, while in the T6 sample the twinning activity is suppressed because of the presence of the β-$Mg_{17}Al_{12}$ phase.

More significantly, the lamellar-shaped β-$Mg_{17}Al_{12}$ particles are reported as “easy to fracture.” It seems probable therefore that the precipitation of β-$Mg_{17}Al_{12}$ has formed on bifilm substrates during the T6 treatment, transforming the closed bifilm cracks into open cracks. The arrival of a population of open cracks clearly impairs both elongation and toughness, but the toughness suffers far more.

3.5.1.3 Fatigue

By way of introduction to the topic of failure as a result of repeated loading, the author requests permission for the luxury of indulging in a reminiscence. He recalls setting up the countergravity casting process for Cosworth Engineering in 1980 in an effort to raise the quality of the castings by reducing turbulence during the filling of the mold. He was not aware of bifilms at that time, but the process clearly benefited from a seat-of-the-pants feeling for what was right. Looking back, the lengthy dwell time for the liquid metal in the holding furnace was correctly predicted to encourage cleaning of the melt by oxide sedimentation, and the uphill filling of the mold with the computer-controlled electromagnetic pump minimized the creation of new defects.

As an unlooked-for bonus for a new foundry with minimal experience of casting at that stage of its development, aerospace customers reported that the castings enjoyed fatigue results significantly better than any other European aerospace foundry at that time. In addition, while 50% of aluminum alloy gravity cast cylinder heads for Cosworth's Formula One and Indy racing engines commonly failed by thermal fatigue on the test bed and were scrapped, the failures fell to zero overnight when the new countergravity casting process was introduced. From that moment of history, no Cosworth casting was ever known to fail again by fatigue (Campbell, 2013).

Interestingly, while foundry text books warn of the dire consequences of tears or cracks from stress-raising sharp corners, as a result of the high ductility we appeared to enjoy with our metal, I never bothered from that time onward in the company of ensuring that stress-raising corners and notches were radiused; all sharp corners were cast as sharp as the designer wished to have them. We never suffered tears or cracks from such sources.

Following these early encouraging experiences in the practical elimination of failures, practical expertise has continuously improved, and bifilms have been gradually better understood, leading finally to the development of improved techniques for their control and elimination. The improved understanding has been particularly relevant for the understanding of fatigue.

Byczynski and Campbell (2005) draws attention to the shape of the bifilm as an ideal crack-initiating defect for the initiation of fatigue as a result of its very sharp tip radius. Birmingham UK researchers (Halliday et al., 1999) concluded that fatigue cracks *invariably* initiate at cracked intermetallic particles. It seems bifilms really may be necessary for fatigue.

To obtain some idea of the benefit to fatigue to be expected if bifilms could be eliminated from metals, the example of cast iron seems especially appropriate. Fig. 3.5 shows fatigue life data for gray iron and ductile iron. The gray iron is known to contain a dense but uniform population of graphite flakes because each flake (if the theory of this author is correct) is formed on a bifilm formed from an oxide of silicon (SiO_2). In contrast, the ductile (spheroidal graphite) iron has experienced its oxide bifilm population eliminated by an addition of magnesium in the nodularizing process. The magnesium addition effectively and immediately reduces silicon oxide to silicon metal which is taken into solution, eliminating bifilms. The remaining MgO particles form the basis for the nucleation sites for graphite nodules (Section 3.4.3 and Campbell, 2009).

$$SiO_2 + Mg = Si + MgO$$

The scale of benefits to be expected from the elimination of bifilms in a typical metal is seen to be an increase in fatigue life by a factor of 100,000.

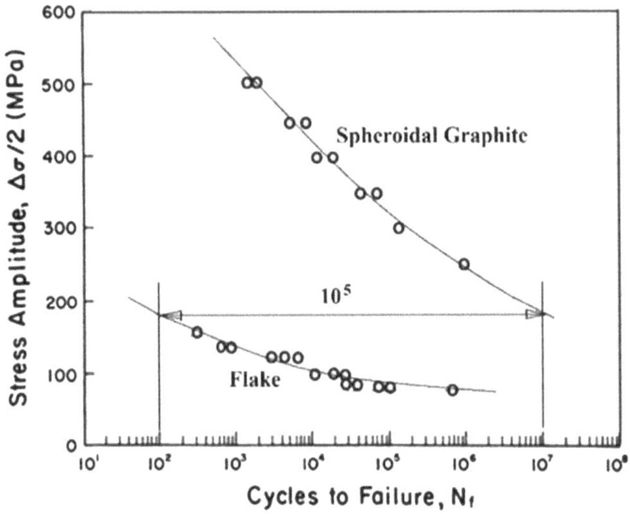

Figure 3.5 Servohydraulic fatigue life data for pearlitic gray and nodular irons. *Data from Socie, D.F., Fash, J., 1982.*

The huge scale of this factor makes sobering comparison with the efforts of traditional metallurgy, where alloying or heat treatment might give a gain of 20% in fatigue life, and would therefore be considered worth a published paper. The 100,000 factor reveals the overwhelming benefits to be gained from targeting defect elimination.

In fact, examining the 100,000 factor more closely, it is with certainty an *underestimate* of the potential benefits of defect elimination. This is because, unfortunately, following the nodularizing treatment, the ductile iron will have been cast turbulently, so that the alloy will have taken on a fresh delivery of unwanted defects, greatly reducing its fatigue resistance. Taking note of the central conclusion to this book, that bifilms may be the *only* initiation sites for crack formation, in the complete absence of biofilms, the real factor of increase may be infinity. This conclusion is hard to ignore. It reinforces the central conclusions of the book that metals without bifilms cannot fail.

In a similar demonstration of the effect of bifilms, but employing an exquisitely perverse negative study, Venugopalan and Alagarsamy (1992) have shown that a really bad turbulent filling system for their nodular iron castings resulted in fatigue lives well below the norm for reasonable castings. Significantly, their fracture surfaces displayed oxide defects and their polished microstructures contained defects which they called "stringers," which, of course, are clearly planar features seen in section, most probably bifilms.

Assuming the universal presence of bifilms explains the universal agreement that there appears to be no crack initiation stage in the fatigue of cast products. Tiryakioglu (2009) confirms this finding in a characteristically careful statistical study of data sets from the literature.

The neutral buoyancy of bifilms in Al alloys (with their slightly denser oxide encouraging sinking, but entrained air helping to float them), causing them to retain their bifilm populations during processing and freezing, contrasts with those in steels, in which bifilms detrain fairly rapidly by flotation, producing a much cleaner metal. This difference in defect density probably explains the observation that steels have a fatigue limit, below which stress fatigue does not occur, but Al alloys appear to have no lower limit, the SN curve (the stress vs. number of cycles to failure curve), but decreases progressively as fatigue cycles increase (I am grateful to Professor Tiryakioglu for pointing out this explanation for the absence of a fatigue limit in Al alloys).

For precisely the same reason that they are cleaner, steels have elongations to tensile failure in often in the region of 50%, whereas many cast aluminum alloys struggle to achieve single figures. Only with the modern careful techniques of handling and casting aluminum alloys is the elongation exceeding 20%, although 100% or more seems a reasonable target which may be achieved in the next few years now that the problem is understood.

The fact that entrainment defects (mainly bifilms, but occasionally foreign inclusions) appear to be the starting point of by far the majority, if not all, fatigue failures, are strongly suggested by numerous studies. For instance, work at Wright—Patterson (Szczepanski et al., 2008 and Sinha et al., 2018) observed that the fatigue initiation site exhibited facets on the basal of prism slip planes of their titanium alloy. These close-packed planes are those to be expected to organize bifilms by their directional encouragement among the growth of dendrites, promoting transgranular bifilm cracks on these favored planes as explained in Section 1.4.4. The development of facets as flattened bifilms is clearly shown by Tiryakioglu et al. (2011) in Fig. 1.42.

Importantly, a study by Ozdes and Tiryakioglu (2016 and 2017) demonstrates a 1-to-1 correlation between tensile ductility and fatigue life for Al alloy castings. In a personal note, Tiryakioglu admitted with wistful irony: "It took me only 25 years to answer the question that came to my mind when I first started (*my first job*) at Boeing; why do we care about (*specifying*) elongation if the stress in the part will never exceed the yield?" It seems probable that this hugely valuable correlation could also be valid for other metals. If so, it would be an immense saving of time and cost, quickly exploring the fatigue resistance of a new alloy simply by carrying out tensile tests (Tiryakioglu, 2017). The confirmation of fatigue properties could be checked later by just a limited number of fatigue tests.

In general, it seems reasonable to suppose that most fatigue failures occur by starting from the opening of a favorably oriented bifilm, followed by slow stepwise propagation of the crack across sound metal as revealed by beach marks, but on encountering a bifilm, jumping forward nearly instantaneously. A temporary arrest of the crack would be expected at the next misaligned bifilm, necessitating either some shear deformation to realign the crack path or a return to a stepwise fatigue progress across the next grain. A fatigue fracture surface therefore is expected to show a mix of planar featureless areas which are probably grain boundaries containing bifilms, plus fatigue striated regions (local regions devoid of bifilms, necessitating stepwise crack propagation across the matrix to reach the next preformed crack provided by a bifilm) on the same sample. The mix of striated grains and some kind of quasi-cleavage grains on the fracture surface is commonly seen and is to be expected. The ratio of featureless to striated areas will reflect the cleanness of the metal. However, it is understandable that researchers studying the fracture surface would hunt for grains showing striations, or beach marks, and record those images, ignoring the wealth of relatively smooth grains formed by bifilms.

Quasi-cleavage features are also commonly seen on tensile fracture surfaces. Planar grain boundaries and regions of ductile dimples occur on the same fracture surface, probably corresponding to relatively extensive grain boundary bifilms, and regions of matrix which fail from the coexistent population of minor bifilms. The minor bifilms are associated with their "cracked" inclusions.

The well-known team specializing in the study of fatigue at Worcester Polytechnic Institute led by Diana Lados (Gavras et al., 2013) has proposed an interesting fatigue map which combines metallurgical and engineering concepts to aid material design and selection. Among what appears to be the many benefits of this approach, the aspect of special interest here is the assumption that the fatigue crack initially proceeds intergranularly, only later becoming transgranular in its advance. In very general terms, this approach would be that to be expected from the initial growth using bifilms, until the stress at the crack tip became sufficient to initiate the progress of the crack across bifilm-free matrix. This would be the location at which "beachmarks" would be identified. As we have noted above, the simple delineation of the two forms of advance are expected in practice to be have a variable degree of randomness arising simply from the differences in bifilm population from batch to batch of metals.

Other differences of crack advancing processes can be noted at times. On occasions it is common to see facets generated by what appears to be crack propagation by a quite different mechanism. The mechanism seems to involve a ratcheting action between slip planes (Fig. 1.42). But in a recent study, Tiryakioglu et al. (2011) found that even when slip planes appear to make a clear contribution to a fatigue failure, the initiation of the slip process always seemed to be associated with the presence of a bifilm, as though the bifilm had initiated the fracture. Thus, it seems even when fatigue occurs by slip, bifilms may be inextricably linked in the initiation of this mechanism.

There remains potential for confusion for the origin of facets on fatigue fracture surfaces. While some facets appear likely to be the result of a ratcheting action of slip planes, having a crystalline perfection as seen in Fig. 1.42a, other facets (Fig. 1.42b) have minute bumps and dimples which suggest their origin is a bifilm crack straightened by dendrites, probably transgranularly. More work is required to clarify these two apparently different mechanisms of facet formation.

A widely appreciated fact about fatigue is the surface defects are more important in its initiation than subsurface defects. However, Tiryakioglu et al. (2011) have quantified this phenomenon, revealing that for Al—7Si-0.4 Mg alloy at 150 MPa, the average life of castings failed from subsurface defects is 14 times longer than that from surface defects and at 240 MPa is 6 times longer.

Another aspect of key interest in fatigue is the size range of typical bifilm defects involved in initiation and propagation. As we have seen earlier in the book, bifilms in liquid metals can occur in the size of newspapers. These tend to result from the encapsulation of the oxide films formed on large, chunky charge materials. The author is aware of such massive defects in his early years in the casting industry. Thankfully, nowadays, such vast defects do not normally find their way into metallurgical products. A team from Tuscaloosa, Alabama (Xue et al., 2007), set out to understand the large scatter in fatigue life in Al alloys, assuming the scatter to be the result of defect size. They found their simulations to give an upper bound to fatigue life at defects of 0.020 mm diameter, whereas defects of 1.2 mm diameter corresponded to a lower bound. Even so, it is significant that occasional large defects gave a fatigue life outside of their estimated bounds, indicating real but unseen extents of defects over 1.2 mm. Moser (1985) records sizes of defects up to 2 mm diameter. There seems little doubt that larger sizes are not uncommon (neglecting at this time the bifilms measuring 10 to 100 mm or more present at times in vacuum arc refined (VAR) steel ingot material as described in Section 1.6.3).

Turning briefly now to some other microstructural studies relating to fatigue, Wang et al. (2005) went to some lengths to compare the effects of pores versus bifilms in the initiation of fatigue in Al alloys. Naturally, they found pores to be the most effective. This is in line with our expectations that bifilms can withstand some stress, particularly shear stress in the plane of the bifilm, compared with pores which can clearly support nothing (Fig. 1.36). However, the major problem with this study is the determination of the size of the bifilms. This is almost certainly underestimated in every case, simply because the bifilm is often difficult to detect. The other aspects of this work which dominate the concepts involved are (i) the observed pores were almost certainly part of a much larger bifilm; the pore was simply a packet of gas which decorated part of the air gap in the bifilm; and (ii) both pores and bifilms are both entrainment defects forming within milliseconds during the turbulence of the casting process (as we have seen earlier, neither pores nor cracks is formed by nucleation and growth processes during solidification). Ultimately, because they are both entrainment defects, probably part of the same entrainment defect, differing only in the amount of air they have trapped, efforts to distinguish their roles seem superfluous.

Chinese workers (Horng et al., 2000) have observed many fatigue cracks in Al–Si alloys progress through the Si particles and, at higher Fe contents, occasionally through the beta-Fe particles. The passage of the crack through both the Si and beta-Fe is, of course, following the air gaps of straightened bifilms (Section 1.4.4). We may also surmise that, as a result of a crack being unable in principle to progress through the ductile Al alloy matrix, the 'propagation' through the matrix occurs mainly on bifilms. The linking of the progress of the crack from one bifilm to the next will often occur by plastic shearing of the matrix if the bifilms are at different levels. Where the distance to the nearest bifilm is large, it seems that cracks following planar grain boundaries adjoin regions of ductile dimples as observed by Jiang et al. (1991) as described earlier—an astonishing mix of something akin to pseudo-cleavage adjacent to regions of plasticity on the same sample. Traditional metallurgy would surely have a problem with this!

This heterogeneity of the medium through which the crack propagates seems to be reflected in the jumps in the rates of acoustic emission which accompany the progress of the fatigue process (Shanyavskiy and Banov 2019).

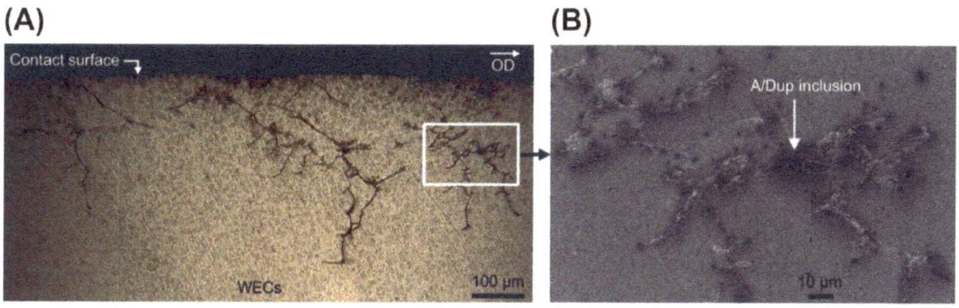

Figure 3.6 (A) The microstructure adjacent to the failure of the main roller bearing housing for a wind turbine, showing the extensive array of cast-in bifilm cracks, and (B) the minute localized patch of genuine fatigue in the form of butterfly wings. *Courtesy Martin Evans et al. 2013.*

In terms of engineering problems in which fatigue has come to the fore in recent years, the failure of the main bearings of wind turbines has been a great concern as reviewed by M-H Evans et al. in 2013. The turbines are designed for a 20 year life, but the main outer bearing ring of some are failing after only 5 years, undermining the economics of wind power. Fig. 3.6 shows a micrograph taken from a failed outer bearing ring. The steel is the typical roller bearing steel, 1%C, 1.5%Cr, and should be able to withstand the repeated 2000 MPa rolling contact stress experienced with the passage of each of the rollers. The researchers Ooi and colleagues (2017) report the presence of the branching cracks which lead to gross failure, with pieces of the bearing fragmenting and becoming detached, with great damage and rapid failure of the rest of the bearing.

The cracks have been widely assumed to be fatigue cracks. This is an error. The cracks are clearly bifilms which have originated with the casting of the steel. It is highly likely that steel for this special application would have been produced by top pouring into an ingot mold. The region of tangled bifilm cracks is surrounded by relatively sound metal. With each cycle of stress, the whole region will clearly shift with respect to the surrounding solid matrix and will experience shifts between individual blocks within the cracked region. Thus, it can be envisaged that "true" fatigue processes will be taking place at the rather randomly scattered junctions within and between blocks. These minute isolated foci of fatigue will eventually lead to local failure and the release of blocks of metal among the rollers of the bearing with disastrous results.

There has been much interest in their appearance as "white etching cracks" and the formation of minute areas of fatigue cracks "initiated" as "butterfly wings" from cracked inclusions (Fig. 3.6B). Ooi reports the conclusion common among investigators, "their origin is unknown." However, there seems no doubt that these regions are the isolated foci of fatigue in the junctions between blocks separated by cracks formed by the tangle of bifilms. The presence of the small cracked inclusion at the center of the wings may have been helpful in initiating this small local pocket of fatigue. It seems that the white etching phase is nanograined ferrite adjacent to the crack. We may speculate that ferrite usually forms on bifilms to reduce the strain energy of its formation as it cools from austenite; its fine grain size is probably the result of abrasive shearing between the bifilm halves; the copious production of vacancies during this severe plastic working might assist to diffuse out carbon to the crack surface where it can oxidize, reducing the dark etching carbide-containing pearlitic matrix to white etching ferrite. Nevertheless, these academic speculations are of minor importance. The main engineering issue is the presence of the extensive array of cracks in the form of bifilms entrained by a poor casting process.

The manufacturing technique for the main bearing ring is probably as follows. The ring is such a large piece of steel that it cannot be formed as part of a continuously cast product but requires to be separately cast as an ingot. Although many ingots are top-poured for maximum cheapness, the steel maker may go to more trouble, selecting what he believes to be the better process; the "uphill teeming" of the ingot. Nevertheless, the entrainment of air in the conical trumpet at the entrance to the filling system still creates masses of air bubbles through the steel, with the creation of bubble trails and bifilms. These severe oxide crack-type defects easily survive the forging process because the steel suffers relatively little deformation during the change of shape required for a large ingot to become a large ring; the bifilms are merely pushed around a little.

To create a steel work piece free from such large bifilm defects will require *contact pouring* to exclude air from the casting process. It may also require other sophistication such as traps for initial oxide-damaged metal in the filling system channels during the few seconds of the priming process before the complete filling of the channels. The provision of all these features is a trivial cost penalty compared with the immense cost of turbine failures and replacements.

Another deeply troubling example of bifilm-initiated fatigue is the tragic failure of helicopters. As mentioned previously (Section 3.4.4 "Pitting Corrosion"), the failure apparently initiated from a trivial etch pit baffles investigators because it is far too small to initiate the apparent fatigue failure in a component such as a rotor shaft which has been designed to resist failure by fatigue by a safety factor of five. Unfortunately, the failure investigators have not known that the etch pit signifies the emergence of a bifilm crack in an otherwise clean steel. The etch pit forms because of the enhanced corrosion opportunity provided by the bifilm, with its crack, permitting ingress of corrodents, and associated favored precipitates to provide corrosion couples. The so-called "fatigue crack" causing the shaft to fracture was in fact mainly already in position, as a bifilm, from day 1 of its service life.

Other occasional tragic failures in aviation can be cited. A Canadian single engine plane suffered a failure in one of its Ni-based alloy turbine blades which took out most of the remaining blades, causing the plane to crash, resulting in grievous injuries and loss of life. The failure was clearly the result of several cracks, occupying a large fraction of the cross-sectional area of the blade. The cracks were identifiable as bifilms from the casting process (Campbell and Tiryakioglu, 2012) from their characteristic spidery trajectory, plus a region which exhibited three planar parallel cracks across a grain, hardly to be expected for a crack resulting from stress (the first crack would relieve the local stress in the grain, making further parallel cracks across the grain improbable). However, parallel cracks are expected to be commonplace for bifilms because several could easily be ironed out by the progress of an array of dendrites, producing beautifully planar and parallel cracks across a single grain. In this instance, the oxides on the surface of the cracks were thick, so that the originating bifilms had been further oxidized in service by the ingress of the oxidizing environment, tending to open the cracks further and thereby probably encouraging the advance of the real fatigue propagation which led to the eventual failure.

Clearly, we need steels of exemplary reliability for components subject to conditions of high and repeated stress. In the opinion of this author, at the time of writing, it is extremely disappointing that totally reliable steels do not exist. VAR steels are ultraclean, but can contain an unknown number of very large bifilm cracks which successfully evade detection, but are sufficiently thin to have no observable effect on the total oxygen content of the steel (Section 1.6.3). ESR steels are our nearest approximation to reliable steels. In the near future, it is hoped that the secondary remelting industry will upgrade its electrode casting practice to make ESR steels totally reliable (Section 1.6.3). This is clearly a task of world importance.

3.5.1.4 Elastic modulus and damping capacity

In an engineering structure, the elastic modulus is the key parameter that determines the rigidity of the design. Thus, if a rigid structure is required, steel with an elastic modulus of 210 GPa is very much preferred to the light alloys, aluminum (70 GPa) and magnesium (45 GPa). However, of course, the specific modulus (the rigidity modulus divided by density) gives a somewhat more favorable comparison because they are all closely similar at about 26 $MPam^3kg^{-1}$.

The elastic properties of metals cannot normally be altered by normal metallurgical means such as alloying or work hardening because they arise directly as a result of interatomic forces.

(There is a lone exception to this general rule: the alloy Al—Li. The large volume fraction of Al—Li intermetallics practically fill the microstructure, not only raising the modulus by about 10% but also reducing the density by about 10%. The result is an increase in specific elastic modulus of 20%. This can hardly be claimed to be a massive increase in stiffness but is probably the highest metallurgical achievement so far in our engineering metals.)

It was all the greater surprise to the author, therefore, when he arranged for a computer simulation to be carried out on the largest bell in the United Kingdom. This was Great Paul, the 17,000 kg bell in St. Paul's Cathedral, London. The investigators, Hall and Shippen (1994), found that agreement with the fundamental note and the harmonics of the simulation could only be obtained by assuming that the elastic modulus was only 87 GPa, not 130 GPa as expected for alloys of copper. This alarming 33% reduction was almost certainly the consequence of Great Paul being full of defects, mainly porosity and bifilms. When standing inside Great Paul, the porosity and lap defects are all too evident! However, this poor structure is unfortunately true for most bells because they are top-poured via the crown of the bell. (The Liberty Bell in Philadelphia, USA, is a famous example of a crack opened from the massive bifilm constituted by the oxide flow tube that can be seen to have formed around the sinuous flow of the falling metal stream, starting high on the shoulder of the bell, and continuing down its side to its lip.) At the time of writing, the one bell foundry in the world known to the author (Nauen, Tonsberg, Norway) that uses bottom gating for its bells, placed mouth upward, claims that it has never suffered a cracked bell, and its bells sound for twice as long as conventional bells. The longevity of the sound is understandable in view of the reduced quantity of bifilms to be expected, so reducing the loss of energy by internal friction between the rubbing surfaces of the bifilms as the bell vibrates.

It is clear, therefore, that many castings will suffer from a lower elastic modulus than expected because of their high content of bifilms. Furthermore, because elastic modulus is very rarely checked, this fact is not widely known. For the future, the full stiffness of a metal will be gained only if the metal is relatively free from bifilms as a result of an entrainment-free melting and casting process.

The effect of a high density of cracks in castings reducing the elastic modulus of the cast material has been accepted in a well-known example for many years. The example is that of flake cast irons compared with steels. In their work to model the stress development in irons during solidification, Wiese and Dantzig (1990) assume the graphite flakes can withstand no tensile stress. This author goes further to suggest that the flakes act precisely as cracks as a result of having formed around bifilms (Section 1.4.4 "Furling and Unfurling"). The iron, behaving like a steel matrix laced with a multitude of cracks, finds its modulus reduced from 210 GPa to only 152 GPa, a reduction of 28%, a value comparable to the reduction found for the bronze bell, Great Paul. In addition, of course, flake graphite irons are renowned for their excellent damping capacity, its absorption of vibrations, and deadening of noise by friction across the "air gap" of the bifilms inside the flakes. It's high damping capacity, among other properties, made it for many years the ideal choice for the beds of machine tools, although nowadays resin-based composites are widely used.

Sand cast plates of Mg alloy ZE41A-T5 showed variations in elastic modulus in the as-cast condition that could be plotted on a Weibull distribution, with measurements generally in the 6%–37% range below the theoretical 44.7 GPa (Cao, 2009). The high damping capacity claimed for some aluminum alloys, particularly the Al–Zn alloys, seems most probably the accidental result of poor casting technology rather than any real intrinsic property of the alloys (Campbell, 2015).

Zildjian cymbals are a specially fascinating example. These brass instruments are much prized by percussionists and drummers because of their unique sound and mechanical reliability (lesser cymbals can turn inside out when struck!). They are first cast as a preform and tested by striking to see if the material has the correct ringing sound in its cast state. If it does not, it is remelted. If it passes it is then processed via forging and spinning to create the thin discs. Clearly, the effects of the presence of bifilm defects survive the intense working to influence the final product.

The writer reveals his age when he admits to recalling the wheel tapper, moving along the side of the rail track, tapping the wheels of the train with his long handled hammer. He was checking that the "ring" was correct. A cracked wheel would give a different sound. *How different* was a shock to me when I tried this for myself. The Manchester Science Museum in the United

Kingdom had an example of a good steel railroad wheel adjacent to a cracked wheel. The good wheel rang like a bell, the other with the dullness of lead. Less romantically, but perhaps more reliably, the wheel tapper has now been superseded by the automated ultrasonic testing station in which the train is rolled into position on a special piece of track, then remains stationary while the wheels are spun and tested.

Ultimately, therefore, although there are fundamental reasons based on the atomic structure of metals why the elastic modulus of many castings and their alloys cannot be improved, the reader may be able to raise the elastic modulus of a badly poured casting by up to 30% with negligible cost, exceeding the achievements of the famous Al—Li alloy whose development cost millions (although, of course, the Al—Li alloy itself might benefit from a similar uplift if cast well).

3.5.1.5 Ultrasound transmission

Although ultrasound is most commonly used to explore for defects in solid metals, their use in liquid metals has provided direct evidence for the existence of bifilms in metals. Bifilms floating about in a melt reflect sound waves with great efficiency as a result of their "air gap." The elastic wave cannot cross the air gap and so is 100% reflected. Mountford and Calvert (1959) used this technique in their wonderful pioneering work to observe bifilms in suspension in liquid aluminum, watching their tumbling and sedimentation behavior, and correlating the observed bifilm fog, or clarity, with bad or good quality, respectively, in their cast RR350 aluminum alloy.

Mountford has continued the development of the ultrasonic detection of bifilms together with Somerville and others (1992 and 1997), monitoring reflections of bifilms in liquid steels and liquid nickel-based alloys.

Similarly, for solid metals, ultrasonic testing is used to check for the presence of cracks and pores. Such testing has been considered not possible in thick sections of large-grained aluminum alloy castings because such grains were said to "scatter ultrasound." More probably, the fog of reflections was caused by bifilms sitting at the grain boundaries; once again, the air layer inside the bifilm generates a perfect specular reflection. The author agrees that backwall echoes in many thick Al alloy castings are usually not observable because of the fog, but on those occasions that the bifilm population is successfully reduced, the backwall echo becomes clear and sharp.

Similarly, in steels, stainless steels are said to pose difficulties for the detection of defects by ultrasound as a result of reflections from the austenite grain boundaries, in contrast to carbon steels which are easily inspected. The stainless steels involved are probably those for which bifilms are present in large amounts, partly as a result of the chromium-rich oxides and partly because of the lower melting temperatures of the steels in comparison with many carbon steels whose melting temperatures are high and whose oxides usually contain silicon; the silicate surface films melting and so forming liquid droplets instead of bifilms when entrained.

It seems therefore that there may be important information in the ultrasound reflections from the "fog"; the granularity of the "grassy" signals should give information on the size and number density of bifilms in metals. In addition, of course, such measurements are easily carried out on the cold, solid material. Such a key source of important information seems to have been overlooked so far. The knowledge of the bifilm density would yield much valuable data to researchers and manufacturers, helping to explain behavior which has previously been bafflingly inexplicable.

A possibly related effect to the scattering of ultrasound is the apparent scattering of X-rays sometimes observed in cast aluminum alloys. The scattering has usually been attributed to lattice reflections from large grains, giving a characteristic "X-ray mottle" on radiographs. The mottle contrasts with a radiograph favored by radiographers in which the image has a "wine glass" translucency. Once again, it is not clear whether the mottle is the result of reflections from lattice planes at random, glancing orientations, or whether it is because of reflections from random nanometer-thick bifilm cracks. It does not appear to be too difficult to clarify this situation. Clarification would be valuable. It is an excellent PhD topic for someone.

3.5.1.6 Conductivity

The presence of a high density of bifilms in a metal would be expected to increase the electrical and thermal conductivities because in both cases, the flux path would be tortuous, having to circumnavigate each bifilm particle as a result of the excellently insulating behavior of its air gap.

During early experimental work on Al–Si alloys, when modified by sodium, Jacob and Remy (1983) found the electrical conductivity of Al–7Si–Mg alloy increased by about 10% and Al–13Si alloy increased by 20%–30%. These results are as expected (Campbell and Tiryakiogly, 2012). In its unmodified state, the silicon phase is assumed to grow on bifilms, as suggested by their common longitudinal cracks. The simple cubic lattice of silicon promotes a roughly planar growth which, because of its attachment to its favored alumina bifilm substrate, effectively straightens and flattens the bifilm. Thus, the bifilm is changed from its original convoluted compact (furled) form into an extensive planar form, becoming a major impediment to conductivity. If sodium or strontium is added to "modify" the melt, the nucleation on bifilms is deactivated, causing the bifilms to remain compact or furled. Thus, the conductivity is increased by sodium additions.

Similar observations were made for thermal diffusivity measurements on Al−8Si alloy, finding a 25% increase on modification (Vijayan and Prabu, 2018).

The author has suggested that the large increase in hardenability observed when only 0.002−0.005 boron is added to steels (Campbell, 2016) is the result of the increase of thermal conductivity, allowing a quench action to penetrate deeper into the steel. The action is the result of boron forming low melting point liquid borides on the surface of the liquid steel, thereby preventing the formation of bifilms during the casting of the steel. The entrained liquid oxides are effectively converted from large planar barriers to heat flow into compact spherical inclusions. The boron steels are consequently renowned not only for their excellent hardenability but also for their toughness—they are difficult to crack, lacking a population of bifilms to provide crack initiation sites.

3.5.2 Bifilms as grain boundaries

3.5.2.1 Microstructure

From the liquid state, as we have seen, bifilms can be incorporated into the solid transgranularly, if dendrite growth happens to straighten them. However, very many bifilms, floating freely in the liquid, are effectively "pushed ahead" of growing grains, with the result that, on meeting other grains, they are automatically incorporated intergranularly; they form the grain boundary. Effectively, they *are* the grain boundaries.

In the solid state, recrystallization or grain growth occurs by the migration of grain boundaries, which will be unable to cross the "air layer" of the bifilm and so will become pinned in place. Thus, both liquid and solid states have mechanisms for favoring the sitting of bifilms at grain boundaries. The identification of those boundaries containing bifilms may be possible: those boundaries which display precipitates may result from the bifilm being a favored substrate.

The impressive persistence of prior austenite grain boundaries (PAGBs) in steels, which often survive a number of heat treatments and phase changes, has usually been attributed to segregation of alloying elements at the boundaries. It seems likely that bifilms could play a stronger role in many instances.

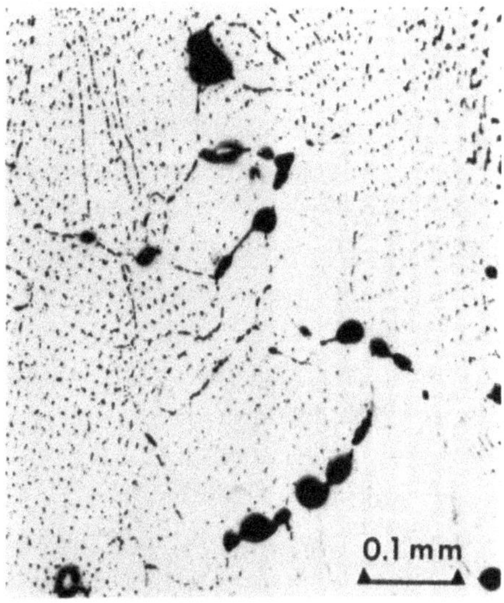

Figure 3.7 Pores decorating selected grain boundaries in a 316L weld subjected to creep (Le May, 1987).

The presence of pores on only some grain boundaries is seen in Fig. 3.7, although in this example from a creep study, there is the complication that the direction of stress will probably open up only suitably oriented boundary bifilms, it remains probable that boundaries without bifilms will not open pores.

It seems worth emphasizing that "clean" boundaries (free from bifilms) are probably not favored sites for either precipitates or pores. This issue is discussed in more detail in Section 3.4.3.

3.5.2.2 Creep

The slow extension of a metal sample under tensile stress at high temperature, when plotted on a time/extension plot, was historically interpreted in three stages. Stage I was an initial phase in which the rate of extension gradually slowed; stage II was the apparently linear "steady-state" regime; stage III was the terminal phase in which the rate of extension accelerated to failure. Over decades of study of creep rate, the emphasis has been centered on the "steady-state" rate. This has been unfortunate as recent work has revealed.

The great breakthrough in the interpretation of creep curves was pioneered by Brian Wilshire, the larger-than-life, booming voiced, head of metallurgy at Swansea University in the United Kingdom, known to his many friends as "George." Early papers appeared in 1986 (Brown et al., 1986) pointing out that a much more accurate description could be made assuming the curve was comprised of only two components: a decaying primary component and an accelerating final component (Fig. 3.8). The so-called steady-state regime was merely the overlap of the initial and terminal components. The early unhelpful interpretation of the creep curve had undermined our creep studies for years, resulting in creep standards worldwide unfortunately based on the three-stage concept. The elegant simplicity of the new interpretation has proved a breakthrough in creep studies, being fundament platform on which much subsequent sophistication has been possible (Evans, 2016, 2019). Long-term creep extrapolations of extraordinary accuracy, not previously believed possible, have now become possible.

Alloys which require long-term creep life in extraordinarily aggressive conditions include the single crystal Ni-base superalloy turbine blades. The famous result by Versnyder and Shank in 1970 showed the 10-fold benefit to creep life when using single crystals compared with polycrystalline castings for resistance to creep (Fig. 3.9). The long-held explanation of this amazing result is that grain boundaries are weak when subjected to creep, so that their elimination by the use of a single crystal naturally results in a major increase in creep resistance. However, it is now well understood that grain boundaries are strong, practically as strong as the matrix (Yamakov, 2011). The traditional explanation is therefore not correct. However, as we have seen, grain boundaries often contain bifilms, and the bifilms would be expected to decohere when subjected to creep. Thus, the confusion is understandable.

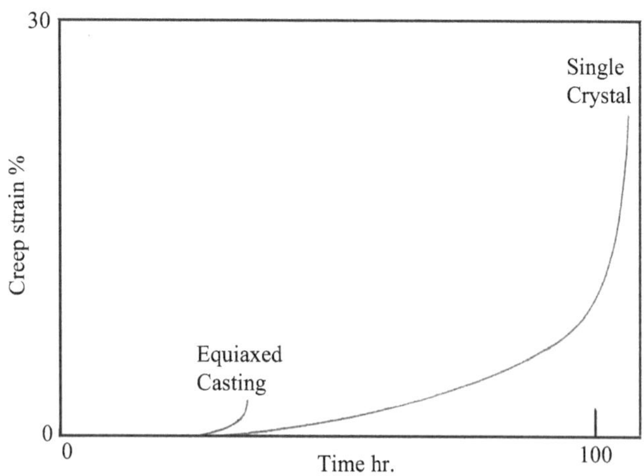

Figure 3.8 The creep to failure of a Ni superalloy polycrystal and single crystal. *After Versnyder, F.I., Shank, M.E., 1970. The development ... through directional solidification. Mater. Sci. Technol. 6 (4), 213–247.*

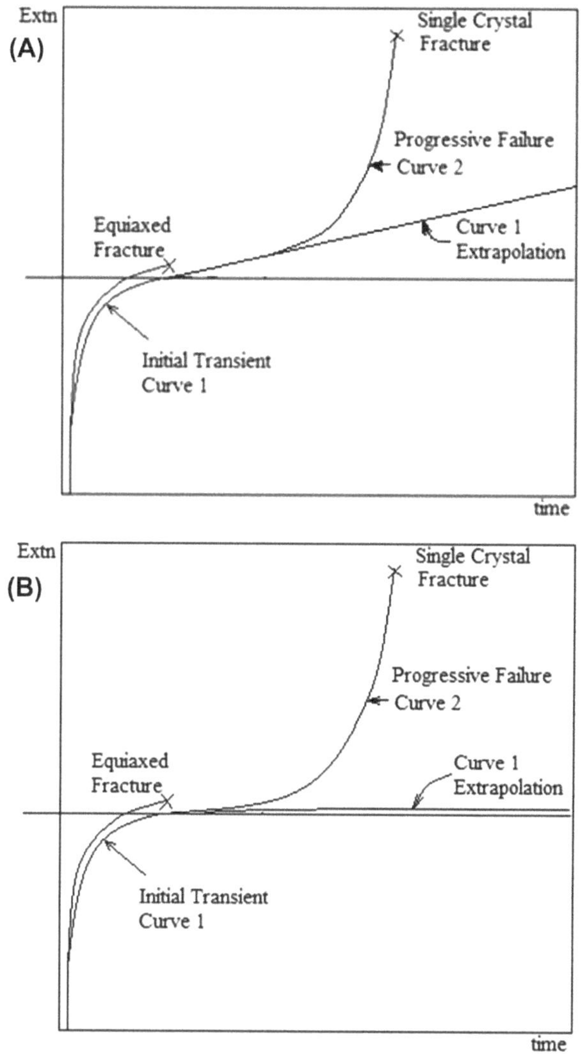

Figure 3.9 The two-part creep curve and the extrapolation of Curve 1; (A) showing the effect of a steady-state extension regime; (B) with no steady-state creep condition.

It seems, therefore, that the great benefit of the single crystal is its reduction in bifilm defects. This occurs because solidification starts at the base of the casting, and as the mold is slowly withdrawn from the hot zone, freezing gradually progresses upward on a macroscopically planar front. Bifilms are therefore mainly pushed ahead of the front during its unidirectional advance, and the long solidification time, perhaps an hour or more, aided by a tendency toward flotation, also in the upward direction. Thus, the single crystal becomes unusually free from bifilms.

Naturally, the polycrystal does not enjoy these advantages. It solidifies quickly, so there is little time for flotation of oxides. In addition, grains nucleate and grow throughout the casting, homing in on its bifilm population from all angles and trapping the bifilms between grains as grains impinge. Thus, the mode of solidification of the polycrystal is designed to maximize the damage potential of its bifilm population.

The Versnyder and Shank result is therefore a wonderful exemplar of the potential to be expected of bifilm-free material. We need to be aware that this result, and the behavior of single crystal blades in general, is an underestimate of the potential benefits. We know this because the single crystal finally fractures; the long upward-curving creep graph is an accelerating failure transient as expected by Wilshire; so we know that it does contain some bifilms; the upward slow solidification, although extremely useful, is not totally successful as a cleaning operation.

Pursuing further a close look at the result in Fig. 3.8, if now we replot these curves, interpreting them Wilshire style, as an initial transient and a final transient, the second curve represents the growth of defects and an acceleration toward failure. If, as we have asserted elsewhere in the book, bifilms are the *sole* initiators of failure in metals, it follows that the acceleration towards failure in the second curve is a demonstration of the effects of bifilms on creep. If bifilms were eliminated, the acceleration of the second curve would cease to exist. Creep would now continue by continuation of Curve 1. The decaying rate of Curve 1 effectively means that failure by creep has been eliminated.

If creep failure could be eliminated and perhaps creep rate also reduced for similar reasons, it would, of course, be of monumental importance. It would predict that in the absence of bifilms, creep would be expected to continue via the macroscopic processes of power law and diffusion-controlled deformation, but its rate would be expected to be low because the metal would enjoy the benefit of its full load bearing area. Its final failure would be massively delayed, and when it finally happens, it would take place by plastic yielding, not by fracture.

Whether the extrapolation of Curve 1 is as shown in (a) or (b) of Fig. 3.8 does not seem clear at this stage. However, the absence of the acceleration to failure curve (Curve 2) means that creep rate overall is at least reduced, and creep life is extended.

Let us examine the nature of Curve 2 in Fig. 3.8 in more detail. The acceleration to failure condition occurs by a number of mechanisms, principally grain boundary sliding and the opening up of cavities. The opening of cavities is definitely a process dependent on the presence of bifilms, recalling that cavities are not expected to be capable of nucleating in metals free from unbonded interfaces.

A study of cavity nucleation and growth during the creep of 12Cr—Mo—V steel by Wu and Sandstrom (1995) revealed the appearance of pores so early that the authors were driven to suggest that the pores were already present; the creep process merely opened them. Grain boundary sliding would also be expected to rely on bifilms. It seems reasonable to suppose that fully intact, defect-free boundaries would be reluctant to slide, whereas those making little contact across the air gap of a bifilm would be seriously weakened and expected to shear faster because of the loss of area. Related scanning electron microscope (SEM) studies on cavitation during superplastic forming (Kulas et al., 2006) have revealed features which they called "filaments" stretching across the interiors of the growing cavities, almost certainly indicating the presence of bifilms.

The accelerating failure curve may therefore be mainly controlled by bifilms; if bifilms were avoided, it seems possible that one of the major mechanisms of creep, grain boundary sliding and the opening of grain boundary cavities, would be eliminated. In addition to the evidence from the Ni-based single crystals, there is already a scattering of evidence from these and other alloy systems to support this prediction.

Ni-based superalloys have, of course, been intensively researched for improved creep resistance. Suave et al. (2018) found that their superalloy suffered because of cracked carbides at grain boundaries initiating cracks during creep—damage which can definitely be attributable to bifilms. Similarly, Hopgood and Martin (1986) found the creep behavior of their alloy to be inexplicably variable, with final fracture associated with the growth of cracks "which nucleate at voids in the casting"—more entrainment features! Liu et al. (2003) discovered that MC carbides formed during solidification were detrimental to creep because of associated micropores which provided crack initiation leading to final fracture.

Minute amounts of boron added to Ni superalloys are repeatedly claimed in the standard text on Ni superalloys (Reed, 2006) to strengthen grain boundaries, which we now know cannot be true because grain boundaries already have practically the full strength of the lattice; they cannot be strengthened further. In addition, such small additions in the region of

0.05%–0.005wt% clearly contribute negligibly to any alloying effect. Despite this, Antonov et al. (2018) found the addition of boron to their powder processed Ni superalloys increased grain boundary cohesion, increased ductility, and was beneficial to creep resistance, but the effect was not understood by the authors. They were unaware that these low B additions profoundly affect the melting point of the surface oxide film on the liquid alloy, suppressing the formation of bifilms as discussed in Section 1.4.9. As a natural consequence, the bifilm population was significantly reduced. A similar benefit to creep, as a result of fewer microcracks, was found by Yu et al. (2013) when they checked the effect of B addition to their superalloy.

Turning for a moment to related observations on steels. From their research on the $MnO-B_2O_3-SiO_2$ oxide system, Kim and Yung (2015) confirmed that as little as 5 ppm boron (0.005 wt%) can create a liquid surface layer on steels at temperatures even as low as 1000 C. However, there was competition between O and N for the boron, so that control over the boron level during steel casting could not be claimed to be easy.

Sahoo et al. (2018) reported that B suppressed creep cavitation in their 304 austenitic stainless steel at 700 C. He et al. (1998) had researched the creep rupture of 304 stainless steel and discovered film-like accretions on planar areas of their fracture surfaces. This seems likely to be an early report of bifilms. The benefits of B additions are cited by Mejia et al. (2013) in stress rupture experiments on an ultrahigh strength steel, only worked if the B addition was sufficient to exceed its loss by reaction with nitrogen to form BN.

Kang et al. (2013) investigated hot ductility, finding B to be effective in conferring freedom from transverse cracking in twinning-induced plasticity (TWIP) steels which they note as "difficult to cast." Carbon-manganese steels have been traditionally deoxidized with 0.1–0.05 Al, which has been known to give problems with reoxidation, creating alumina-rich inclusions. It is all the more understandable therefore to learn of difficulties for TWIP steels whose alloy content of aluminum is over 10 times higher (1.3Al). This would result in strong alumina bifilm problems when turbulently poured in their vacuum furnace. Better casting technology would probably revolutionize the performance of these steels, permitting the full potential of the Al alloy without the introduction of alumina bifilms.

Finally, a kind of direct result from an unlikely source, Srinivasan et al. (2006), compared the creep rate of magnesium alloy cast by a gravity die process (a pouring process, usually with very poorly designed filling channels) with the creep rate of material cast by low-pressure die casting (a basic, non-too sophisticated, but at least a countergravity process expected to significantly reduce bifilm production during casting). They found the alloy cast with reduced turbulence had double the creep performance (half of the creep rate).

It seems there is a real effect of bifilms on creep. We have not yet begun to gain what promises to be an overwhelmingly important benefit to the performance of metals at high temperatures.

3.5.2.3 Anisotropy and texture

There is a large literature relating to studies to understand the development of anisotropy and texture during the working of metals. However, at the time of writing, a relatively recent review (Hutchinson, 2015) of the possible causes of anisotropy in metals did not include any mention of bifilms. This is of course understandable. However, interestingly, and perhaps predictably, Hutchinson concludes that a better understanding would be a suitable area for further investigation. Naturally, the case presented here draws attention to the probable central role of bifilms in the development of anisotropy and texture of metals.

We need to recall thàt oxide bifilms are generally only nanometers thick, not easily seen, and can therefore be sitting at a grain boundary, or occasionally transgranularly, but in both cases, unsuspected. Furthermore, at this time, they are prolific and ubiquitous in most of our metals as a result of the near universality of turbulent casting processes. Finally, we need to take note that its central layer of air (Shafaei and Raiszadeh, 2014) is the perfect barrier to the migration of grain boundaries because of the clear impossibility of the boundary crossing the air layer.

During a plastic deformation process such as rolling, the bifilms will become progressively more aligned with the working direction, so that the grains will take on an elongated aspect ratio. The strength of the pinning action of the boundary on the bifilm means that the aspect ratio of the grains is usually capable of surviving treatments which would otherwise cause recrystallization and grain growth.

Furthermore, the concept of the migrating or recrystallizing boundary encountering a bifilm and being pinned in place is examined. In a sense, there is an obviousness about this effect, as typified by the numerous experiments to bond stacks of sheet metal by rolling or pressing. For instance, foil laminates of 7475 aluminum alloy press bonded to a reduction of between 10:1 and 15:1 are clearly observed to experience grain growth inhibition by the presence of the doubled oxides separating the surfaces to be bonded (Iwasaki et al., 1998). Similar behavior is observed by Sureth et al. (2013) who worked with copper, repeatedly folding and rerolling to create a structure with 256 layers subjected to a total strain of 4.8. Even in this case, with the very weak oxides associated with copper, the doubled oxides are noted to strongly influence the grain size and shape.

An interesting indirect influence of bifilms on texture is seen in the research by Wen et al. (2016). These researchers investigated the iron content on the recrystallization texture and earing behavior of Al−2.4 Mg rolled sheet with Fe contents 0.12% and 0.47%. After annealing they found the 45/135 degree ears were raised by the higher iron. The behavior seems the result of the iron precipitating on bifilms as the monoclinic beta form (Al5FeSi) which straightens and rigidizes the bifilms. They are then highly effective in affecting the movement of grain boundaries and would therefore be expected to achieve a strong influence in the development of texture.

In the cases of severe plastic deformation of metals, such as the laboratory technique of equal channel angular pressing (Bazarnik et al., 2013), or the industrial practice of groove pressing (Jain and Chakkingal, 2013), it seems probable that the simultaneous pressing and severe shearing of bifilms lead to some fragmentation and bonding of the bifilm population. These effects would be expected to contribute to the raised mechanical properties, in addition to the predicted benefits from reduced grain size and increased dislocation density. Zhou et al. (2013) record a further strengthening mechanism as a result of the formation of precipitates during the deformation. Interestingly, they observe that the zones in which precipitates form are coincident with the zones of fine grains. It seems possible that these zones are those regions containing bifilms, or clusters of bifilms, in which strain would be likely to be concentrated.

The phenomenon of the fibrous "grain" of metals is now easily understood. Traditionally, forging was always carried out to ensure that the direction of the grain followed the shape of a product. A crane hook is a good example. Analogous to the "grain" of wood, the forging benefits from significantly enhanced toughness and resistance to failure because cracks would tend not to propagate transversely, but would turn to follow the aligned bifilms: the "grain" of the metal. It is useful to recall that the concept of the "grain" of forgings dates from the early days of steelmaking in which bifilms would be present in copious quantities, so the alignment of the bifilms was essential. An identical modern forging made from an extremely clean steel would be expected to show no "grain" structure and no asymmetry of mechanical properties. This prediction would be valuable to forges and not difficult to confirm when improved steels become available.

The preoccupation of metallurgists with longitudinal properties (aligned with a rolling or wire drawing direction) versus transverse properties in steels, especially "through-thickness" properties, shows that the effects of aligned bifilm cracks remain with us today. In addition to the many theories of crystallographic contributions to texture development, it seems certain that bifilms have a major role, possibly *the* major role.

3.5.3 Bifilms as favored substrates

We may note some earlier discussion of this concept in Section 1.4.4 as "Number 5" in the long list of mechanisms for the conversion of furled bifilms to unfurled bifilms. It is perhaps a surprising fact that in the case of *every* precipitate in a cast product, the author has found that the precipitate has formed on a bifilm. No exceptions have been observed at this time of writing, apart from, naturally, nanoscale precipitates formed in the lattice such as GP zones.

In the case of many precipitates formed in the liquid state, it is probable that nucleation and/or growth on the bifilm reduces the energy of formation. The theory that second phases and intermetallics form on bifilms in the liquid state by favored nucleation is not easily proven. In general, the various interfacial energies are not usually known, and in any case subject to the vagaries of changes as a result of the formation of adsorption layers, possibly only a monolayer, of an impurity element in the alloy. The apparent precipitation of all intermetallics and second phases on bifilms in the liquid (no exceptions have been found to date) appears to be a fact, but whose cause remains a mystery that will clearly be resistant to solution.

Some tendency to favor nucleation may result from the huge surface area of bifilms in most liquid metals. Any other substrate, as a result of its more compact form and/or smaller volume fraction, would have a much reduced chance of successful nucleation.

Nucleation on bifilms in the solid state, in contrast to nucleation in the liquid, is easily explicable. Here, precipitates almost always favor formation on a bifilm simply because the "air gap" allows the newly forming precipitate to reduce its strain energy associated with its shape change and volume change by growing partly into the "air gap." As is seen at many points throughout this book, this is a major mechanism whereby bifilms are forced open, becoming significant paths for the spread of fluids throughout the metal, such as corrodents and hydrogen.

There are many examples of bifilms acting as favored substrates. Several are mentioned below.

3.5.3.1 Si particles in Al–Si alloys

In hypoeutectic Al–Si alloys, generally containing close to 7%Si, the sequence of reactions for the initiation and growth of Si seems to be (i) the favored nucleation of AlP particles on the alumina bifilm; (ii) the nucleation of Si on the AlP particles; and (iii) the growth of Si, spreading out from the AlP initiation site to cover and to some extent flatten the Al_2O_3 bifilm as a result of the simple cubic structure of Si which favors its roughly lamellar growth form (Campbell and Tiryakioglu, 2010). The straightening of the bifilms leads to their increased effectiveness as equivalent to engineering-type cracks. In this way the coarse (unmodified) Al–Si alloy is weakened by the bifilm cracks alongside or within the coarse Si particles as straightening agents.

If Na or Sr is added the AlP particles are deactivated so that, the Si no longer nucleates and grows on bifilms, so the bifilms are now not straightened but remain convoluted and compact. Having its easy nucleation site now unavailable to it, the Si now has to precipitate on less favorable substrates which require higher driving undercoolings. Once nucleated at the lower temperature, the Al–Si alloy now grows as a coupled eutectic on a substantially planar front; the Al and Si phases growing together, their spacing now dictated by diffusion in the liquid, and so controlled to be regular and fine. As a result of the nonstraightened bifilms and the finer eutectic structure, Al–Si alloys modified with Na or Sr display improved properties.

In the hypereutectic Al–15Si alloy, Al-Helal and fellow researchers (2013) investigate the effect of different levels of Ca. This alloy has a rather more complicated behavior: at all levels of Ca which they study, some Si appears always to form as chunky, facetted crystals. Perhaps this is to be expected as a consequence of the high Si content of the hypereutectic Al–Si alloys, and its early arrival (because of its high liquidus temperature) during the freezing process. The eutectic Si is, however, significantly altered, following a similar behavior as for the hypoeutectic alloy described above.

With low values of typical impurities, 20 ppm Ca and 20 ppm P, the alloy solidified with a coarse unmodified eutectic as a result of sufficient AlP nucleating on oxide bifilms and itself in turn acting as a nucleant for Si, whose growth as a tabular morphology while attached to the clinging bifilm unfurls the bifilm in pace with the extending Si crystal.

If Ca is raised to 200 ppm, typical of a commercial purity alloy, the preferential formation of CaP effectively eliminates AlP as a nucleant for Si. In this case, Si is prevented from forming easily on bifilms at a relatively high freezing temperature and therefore is now forced to form on less favorable nuclei at lower temperatures. A coupled eutectic growth typical of a modified eutectic then forms.

Their results for a 5000 ppm addition seem to be complicated by the arrival of large quantities of entrained oxide films, the arrival of a new phase Al_2CaSi_2 together with extremely fine eutectic Si, and the continued presence of large chunky facetted crystals of Si as might be expected from its hypoeutectic composition. However, overall, these results seem to this author are not easily understood and would benefit from additional clarifying research.

3.5.3.2 Carbon in cast iron

As has been mentioned before, one of the metal alloys for which bifilms are probably most clear is cast iron. This is because the "A-type" graphite flakes have each formed on an oxide bifilm consisting of silica (SiO_2). Graphite flakes are relatively straight because they grow as graphite crystals with negligible restraint while forming on bifilms suspended freely in the liquid. The originally convoluted bifilms are themselves straightened by this process, reducing the strength and ductility from that of a steel to that of a relatively weak and brittle material as is well known.

The silica bifilm substrate for graphite flakes was a proposal put forward by the author in 2009 but was only confirmed experimentally by Boisvert et al. in 2016. The impressive density of graphite flakes in cast irons reflects a typical, population density of bifilm cracks in metals. The density is similar to that shown for alumina bifilms in aluminum alloys in Fig. 1.14.

The bifilm theory appears to be the only theory to explain the action of Mg in the instant, and almost "magical," conversion of flake iron to spheroidal graphite iron, as described in Section 3.4.1 under the subject heading "Fatigue."

In passing, we may note that cast iron founders tend to reject contact pouring because they are unable to carry out a late inoculation—a technique giving excellent nucleating response for graphite formation. By contacting the pouring ladle and mold, no room is now available to introduce the stream of finely divided inoculant into the metal stream as it enters the mold. However, it seems likely that the late inoculation may be unnecessary if secondary oxidation during pouring is eliminated by contact pour—the late inoculation may have only served the purpose of making up for the huge consumption of inoculant oxidized by entrained air when using a conical trumpet intake.

The formation of graphite on silica bifilms in cast irons is described earlier in Section 1.4.4 as an unfurling mechanism. Similarly, Si in Al—Si alloys has been mentioned above. Additional examples are described below, including topologically close-packed (TCP) phase in Ni superalloys and cracked carbides in steels.

3.5.3.3 Carbides in steels

Fig. 3.10 is worth a closer study. The bifilms on which the carbides have formed will be raveled randomly, with parts of the bifilm at all orientations. However, only those perpendicular to the stress will have opened. Those sections of the bifilms which are parallel to the carbides will remain closed and invisible (although one central particle has a small decohesion on its lower right-hand side). The multiple cracks of some particles may indicate the thicket of tangles which is typical of bifilms, interacting with the carbides multiple times. Alternatively, the vertical direction of stress may not have opened the main longitudinal bifilms on which the particles have grown on one side only as a result of some difference in the two films; furthermore, the favored side was also the largest area, creating the transverse folds which led to the formation of the subsidiary cracks, transverse to the stress, which have opened. The structure of the main crack (unopened in this case) and transverse subsidiary cracks is very clearly shown in the Al—Si alloy (Fig. 1.47) and the Ni superalloy (Fig. 1.48).

During the cooling of steels in the solid state, the well-known formation of ferrite grains on prior austenite grain boundaries (PAGBs) is to be

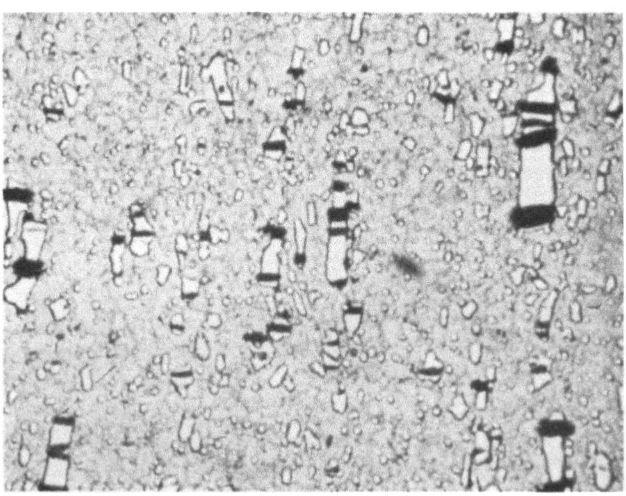

Figure 3.10 Cracked carbides in martensitic matrix in a D2 tool steel bar (Kattus, 1987).

expected as a result of the presence of bifilms having been trapped in a high proportion of the boundaries during freezing. The volume and shape change of the ferrite would be more easily accommodated by the air gap of the bifilm. It seems doubtful whether ferrite would form on a simple, bifilm-free, austenite grain boundary.

Other embrittling phenomena in steels are likely to be the result of the presence of bifilms. The "rock candy fractures" seen in some steel castings, in which the fracture surface exhibits aluminum nitride, seem to be simply a bifilm. The bifilm might originally be an aluminum oxide when first entrained, the aluminum being the residue of the aluminum addition used to deoxidize the steel. The quantity of entrained air in the "air gap" will react to create aluminum nitride, as a result of air being 80% nitrogen, and with the possible additional contribution of the nitrogen already in solution in the steel.

Similarly, it seems probable that type II sulfides in steels will also be a bifilm phenomenon. Type II sulfides are quite different from Type I and Type III and have never been satisfactorily explained. However, they fit all the features to be expected of bifilms. Sulfides will be expected to precipitate on oxide bifilms, taking on the appearance of sulfide films because the original oxide bifilms will usually remain difficult to detect. However, of course, the deleterious effect on properties will be from the bifilms themselves. As usual, the precipitated sulfur-rich nature of the inclusion is probably irrelevant to the failure mode.

3.5.3.4 The fish eye

One of the most intriguing and relatively common features on steel fracture surfaces is the "fish eye" (Fig. 3.11A). This extraordinary symmetrical feature has proven difficult to understand. For instance, what is the explanation of its circular shape on a fracture surface? Why is the inclusion nearly always central? Why is this circular symmetry not influenced by the direction of propagation of the crack? What is the nature of its mirror-like surface? There may be answers to all of these questions in terms of bifilm theory as follows.

As we have indicated at several points within the book, the closed bifilm exhibits some ability to withstand a limited amount of stress, particularly shear stress. The precipitation of an inclusion such as an oxide or carbide on the bifilm is favored as a result of the new phase being able to accommodate its volume and shape change in its "air gap." If the inclusion is large, or if it is accompanied by a large enough volume or shape change, then the air gap will be forced open as the inclusion grows. The degree of opening

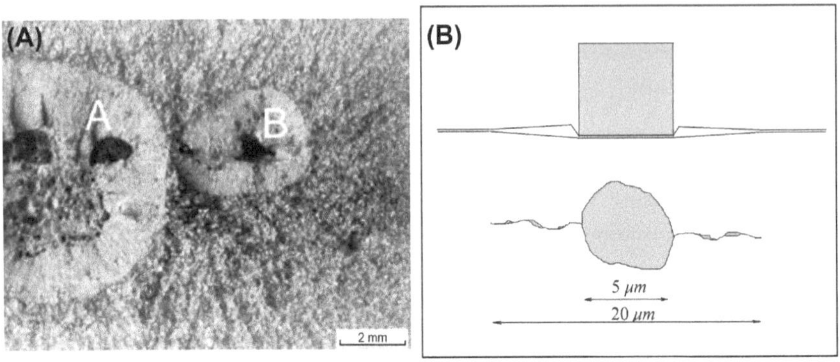

Figure 3.11 (A) A "fish eye" on a bend test fracture surface (Moser et al., 1985) and (B) a suggested model section of a fish eye, and a sketch of section of a fish eye produced by hydrogen charging observed by Todoshchenko et al. (2014).

will taper off in all directions, eventually reaching some limiting distance where the bifilm remains undisturbed and closed. The oxide film will be expected to be stretched taught by the levering open action of the inclusion. At this point, the bifilm resembles a shallow angle bell tent, with the inclusion being the central supporting pole (Fig. 3.11B). The passage of a crack will favor its direction through the now-weakened, opened patch with its tightly stretched bifilm and central inclusion, revealing the "fish eye" formation.

The smoothness of the facet surrounding the central inclusion indicates, I think, that the original unopened bifilm was straightened somehow, perhaps by dendrite growth, so that it would achieve its mirror-like appearance. Radial stretch marks are evident in Fig. 3.11A which are an interesting puzzle. These could surely only be generated if the formation of the fish eye occurred in the liquid state. Some epitaxial registration between the inclusion and film is to be expected and might be sufficient to separate the bifilm when immersed in the liquid, forming a fish eye.

However, the main reduction of strain energy driving force will be applicable only in the solid state, but now there is a problem to explain the radial folds or creases resembling stretch marks because the bifilm will be firmly attached to the solid matrix. The radial stretch marks could arise in the solid state if the bifilm were not a single bifilm, but was a multiple, a heap of bifilms piled up on the grain boundary. Such multiple thick bifilms have been frequently observed. The upper layers of the bifilms would become free to stretch, being out of contact with the matrix. Multiple layers of bifilms in grain boundaries are probably to be expected, and the stretch mark observations might be the initial evidence for them.

The two types of inclusions present in the "eye" of Fig. 3.11A suggest inclusions precipitated at different times, so that liquid and solid state contributions to the fish eye may be present in this example.

Alternatively, fish eye "B" appears to have been generated from the black inclusion, but fish eye "A," which contains similar black inclusions, appears to have formed from the rather larger area of granular inclusion. Perhaps the granular inclusion was active in the liquid state, but the black inclusions are solid state precipitations.

Although, therefore, some contribution to some fish eye development may occur in the liquid state, there is no doubt that many of its features can be explained by its development in the solid, as a result of the precipitation of inclusions by tempering reactions. It will be interesting to check whether they are encouraged by the formation of large inclusions with large volume and shape changes. Only careful study of fish eyes will resolve these speculations.

Murakami in his famous book on Fatigue (2002) discusses fish eyes at length but reaches no conclusion about a formation mechanism. He gives a number of examples, most of which appear to have been formed in the solid state, associated with the precipitation of inclusions from tempering reactions. They come in widely varying diameters from 0.2 mm (Murakami, 2002) to 2.0 mm (Socie and Fash, 1982), although Hornik and co-workers (2019) record a fish eye in a superalloy 4 mm diameter.

Murakami and others report fish eyes to occur commonly on fatigue fracture surfaces. For instance, Hornik et al. (2019) report a large fish eye on one of their MAR-M superalloy samples. However, they also seem to occur for other fracture processes, such as the simple bend test shown in Fig. 3.11. They are also reported during hydrogen charging when studying hydrogen embrittlement (HE) (Todoshchenko et al., 2014).

It may be that fish eyes are the standard precursors of the fracture processes. Only comparatively rarely will they be sufficiently aligned and sufficiently alone to take on the fish eye appearance. In most situations they will be expected to be present throughout the metal, at many grain boundaries, and in many orientations. A metallographic cross section shown in Fig. 3.11 illustrates how difficult it would be to identify a fish eye on a metallurgical section.

At low stresses such as are used for fatigue testing, fish eyes may be observed more commonly as Murakami seems to find. However, at the higher stresses of tensile testing, with densely arrayed inclusions at grain boundaries, the overlap of opened eyes may obscure any pattern. The common condition for the tensile fracture surface may be, therefore, that there are too many eyes to be seen!

3.5.3.5 Topologically close-packed phase in Ni superalloys

The appearance of the TCP phase in Ni superalloys is traditionally dreaded as a crack-initiating particle. This, of course, is a misunderstanding of its formation on bifilms, as elegantly demonstrated in Fig. 1.48, which contains all the classical features exhibited only by bifilms. These include the long fairly straight edge, straightened during the growth of the phase; the transverse cracks from only one side formed by the folds in the larger of the two films forming the bifilm; the apparent porosity on the opposite side of the particle formed by the separation of that side of the bifilm not attached to the phase. The transverse "cracks" further confirm their bifilm origin because of the virtual absence of any significant stress capable of forming a stress crack during the solidification and cooling of the alloy. As always with particles which form on bifilms, it is probable that if the bifilm were not present, the particle could not form.

3.5.4 Bifilms as invasive pathways

If the bifilm contacts the outer surface of the metal, there is the danger that any environmental corrodent will find its way deep into the interior of the metal. Often, this will be following grain boundaries as a result of the boundaries being occupied by bifilms. In the case of highly diffusible elements such as hydrogen, the bifilm does not need to be in contact with the surface. Invasive possibilities leading to corrosion and embrittling reactions are explored in this section. The various contributors to such mechanisms include the following phenomena:

1. When first formed, the bifilm will be expected to have a layer of air trapped in the roughness and corrugations of the two films. However, the air will be expected to react with the films, causing the bifilm to tend to close (except for the 1% argon in the air) but not necessarily creating bonding between the films.
2. Second phases and intermetallics will precipitate onto the outer surface of the bifilms to reduce surface energy and (in the solid state) reduce their strain energy of formation. As a result, the closed bifilm will tend to be forcibly reopened.
3. The impingement of bifilms with the outer surfaces of a product, particularly in the case of machined surfaces which will have cut into bifilms, would reveal the bifilm as a route into the interior of the metal. Because the corrodent, if liquid, is likely to wet the interior surface of the bifilm, the ingress of liquid corrodents will probably be enhanced by capillary attraction into the reopened crack.

4. Second phases and intermetallics of various kinds will have different electrochemical potentials but will be separated from the incoming corrodent by only a few nanometers of oxide. This oxide barrier will in any case almost certainly be defective, containing crack damage from the cooling or working of the metal. Thus, the formation of one or more galvanic corrosion couples will be created, leading to enhanced dissolution.

The phenomena associated with bifilms, the easy ingress deep into the matrix; the presence of second phases to form corrosion couples and separate the two halves of the bifilm, must feature strongly in corrosion mechanisms. It seems reasonable to suppose, therefore, that several different types of invasive corrosion might be expected to be the result of the presence of bifilms. Some likely examples are discussed below.

3.5.4.1 Pitting Corrosion

The simple etch pit is an important example. The outside surface of a metal may be tolerably free from defects and largely resistant to corrosion, but where a bifilm emerges, breaking the surface, local corrosion at this point is likely to facilitate the formation of an etch pit (Fig. 3.12). Thus, the presence of an etch pit most probably marks the point at which a bifilm, of unknown area and unknown extent of penetration, contacts the surface. Fatigue failures apparently from such insignificantly small features as etch pits can therefore be understood.

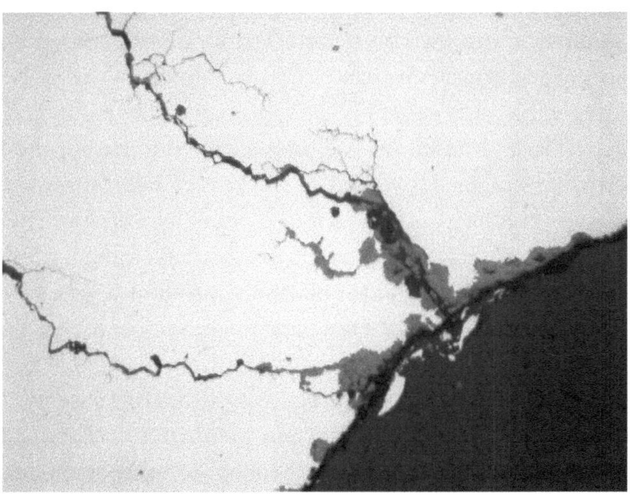

Figure 3.12 Where bifilms met the surface, an etch pit formed in a steel blade in a steam turbine (Metallurgical Associates, 2014).

In the Al alloy automotive wheel casting business, the formation of etch pits in Al alloys associated with the presence of oxide inclusions is well known (although the identification of the oxides as bifilms has not generally been realized). Wheels pass or fail the salt spray tests depending on the population of oxides in suspension in the melt. These "oxides" sometimes cause the wheel to leak, deflating the tire. Good quality melts (i.e., liquid metals free from oxides in suspension) make wheels which are not subject to etch pitting. This gritty shop floor experience is supported by more fundamental and elegant reviews (Szklarska-Smialowska, 1999).

Almaraz et al. (2014) find that a corrosive environment of their Al alloy led to cracks which appeared to form most frequently from two or more corrosion pits, resulting in an inexplicably severe loss of toughness. Lu et al. (2014) studied an austenitic stainless steel which seemed to initiate SCC at the bottom of a corrosion pit, the crack propagating along the boundary between austenite and ferrite. In both cases the cracks would be expected to be preexisting bifilms in contact with the surface, which would lead to the formation of etch pits. As we have seen at several points in this book, the austenite/ferrite boundary is predicted to be a location favored by bifilms. These authors could not understand why the reduction in properties was so severe and why their results were inconsistent. However, of course, the size of the bifilm crack would be highly variable and decorated with variable second phases to encourage galvanic reactions, and no-one will have bothered to record whether the metal was poured excessively turbulently on the day that metal was cast.

Fig. 3.12 illustrates an etch pit in a steel turbine blade for a power generating turbine (Metallurgical Associates, 2014). Although the etch pit itself is clearly not negligible, it is generally far too small to have initiated failure of the blade by cracking. Somehow, the pit is assumed to trigger the formation of stress corrosion cracks. However, as the reader will be now aware, the bifilm cracks are predicted to preexist as a result of entrainment of oxides during the casting process. At the point at which such defects intersect the surface, their "open mouths" automatically provide a gateway through any protective passive layer. The etch pit is the mouth of the crack, enlarged by corrosion, and the visual evidence that a bifilm emerges at this point.

The extensive bifilm cracks leading away from the base of the pit are usually interpreted as fatigue cracks when seen on a fracture surface. The two separate ditchings in the North Sea by two separate Super Puma helicopters within months of each other in 2012 and the loss of a Norwegian helicopter with the loss of all lives in 2017 were, at the time, difficult to understand. All resulted from the fatigue failure of a VAR steel shaft, from what appeared in each case to be a harmlessly small etch pit (Campbell, 2016; AAIB, 2012; AIBN, 2018).

3.5.4.2 Filiform corrosion

Those forms of FFC at which a self-sustaining filament of corrosion wanders under a thin surface paint or plastic film do not appear likely to be associated with bifilms.

However, for rolled metal products protected by their own surface oxide film, this type of FFC appears to be the result of a bifilm mechanism. A common test is the scribing of a scratch across the surface of a rolled product at right angles to its deformation direction. When subjected briefly to a corrosion environment, and subsequently held in air at a controlled temperature, characteristic corrosion markings, generally called "filaments," are seen to grow away from the scratch. When examined in detail, the damage is seen to be a local attack, but which can repeatedly tunnel under the surface and repeatedly emerge, following the deformation direction. Such filamentary lines on Al alloy rolled sheet are commonly 20 mm long, but can, on occasions, be much shorter or longer. Speculation has been made on the filament length being a function of the distance to which the corrodent can penetrate. Furthermore, experiments to quantify the problem are bafflingly scattered from batch to batch of alloy. In their review of the research on FFC of Al alloys, Leth-Olsen and Nisancioglu (1998) record their exasperation at the lack of repeatability of experiments. The source of the scatter has remained a mystery. Afseth et al. (2000) has demonstrated that heat treating the alloy to produce a solid solution reduces FFC, suggesting that the precipitation of electrochemically noble second phase particles enhances the phenomenon.

Careful examination of the filaments clearly reveals them to be bifilms. Their length is almost certainly the "length" of the bifilms as extended by plastic deformation of the matrix in the rolling direction. The scatter in FFC results reflects the hugely different conditions for the source of the metal to be melted, such as, for instance, whether the charge might contain a large proportion of machining "chips." Additional scatter is expected from melting conditions, pouring conditions, and hydrogen content during freezing which may have tended to open up bifilms, making them more effective as cracks. Every batch will be expected to be different because the bifilm content of melts is currently either not controlled at all or at best poorly controlled.

3.5.4.3 Exfoliation corrosion

The 2000 and 7000 series Al alloys are particularly prone to exfoliation after severe rolling or extrusion followed by exposure to corrosive conditions such as weathering. The damage often appears initially as pitting but leads to flaking and sometimes blisters. A well-developed exfoliated region of the surface of a metal sheet appears to rise in flaky layers like puff pastry, pointing to the presence of multiple layers of bifilms, probably indistinguishable from a dross inclusion. Lin et al. (2012) noted that exfoliation corrosion (EXCO) is a significant source of life-limiting degradation of air frames. The phenomenon appears to be associated with an unusually elongated grain structure—a structure to be expected of material containing a high quantity of oxide bifilms as discussed in Section 3.4.3. All these observations are compatible with masses of bifilms, aligned by plastic working, and separating easily across their central unbonded interface as corrosion of associated intermetallics and matrix gradually release the bifilms in turn. Their separation, probably one at a time, may be driven partly by the accumulation of corrosion products as suggested by Lin, but perhaps also by the relief of high residual stresses in the flakes of intervening matrix which are successively released.

3.5.4.4 Stress corrosion cracking

SCC is a term used to describe one of the most puzzling of failure modes in which slow, time-dependent crack propagation, resulting in major cracks, occurs deep into the interiors of metals. The phenomenon occurs with the cooperation of only low tensile stress, which may be externally applied or may be an internal residual stress. A compressive stress can halt the process.

The review by Russel Jones (2017) emphasizes the baffling and confusing aspects of the phenomenon, typical of a process in which bifilms are not known or suspected, but are in control. Once again, the action of bifilms offers a simple explanation and appears consistent with the monumental volume of data in the review by Jones and previous reviews (Raja and Shoji, 2011; Gibala and Hehemann, 1984).

The proposed bifilm action is simply the ingress of a corrodent along the "air gap" of bifilms, accelerated by the actions of precipitates on the bifilms: (i) opening of the "air gap" and (ii) forming corrosion couples.

As in other insidious corrosion failure modes, the loss of material associated with the growth of a seriously damaging crack is minimal. SCC is sometimes noted to be predominantly transgranular or predominantly intergranular depending on conditions. It is noteworthy that bifilms occur mainly transgranularly in cast material near to the mold surface, where columnar dendrites actively straighten and flatten bifilms, coercing them to become mainly transgranular and parallel to dendrite growth directions. Conversely, in situations in which the grains are more equiaxed, such as the centers of ingots, bifilms would be expected to develop no preferred direction, but would allow themselves to be pushed into random intergranular regions. An investigation into such potential structural links between bifilm types and SCC has yet to be carried out.

It is worth dwelling briefly on one of the most famous examples of SCC. Austenitic stainless steels generally enjoy excellent resistance to corrosion. However, they can be "sensitized" by slow cooling from a high temperature to precipitate chromium carbides, $Cr_{23}C_6$, on grain boundaries. The local loss of Cr from the vicinity of the boundary reduces the corrosion resistance of this zone, leading, it is assumed, to intergranular corrosion particularly by chloride solutions. There may, of course, be some truth in this long-held traditional view.

However, it also seems possible that an important contributor to SCC in this instance may be the precipitation of the carbides, leading to the prizing open of the "air gap" of the bifilm on which they have grown, as the growing carbide forces its way into the gap, accommodating its volume and shape changes. There may be some evidence for this in the observations by Moss and Was (2017) who studied SCC in Ni–Cr alloys 600 and 690 in hydrogenated supercritical water. They found that cracks propagated along the grain boundaries but not in chromium-depleted regions.

Much other work lends support to the involvement of bifilms in SCC. Some examples are listed below.

SCC behavior of an austenitic stainless steel in high temperature water was concluded to initiate SCC at the base of corrosion pits, which then propagated along the austenite/ferrite phase boundaries (Lu et al., 2014). However, the complaint was once again the lack of consistency in the results, typical of a bifilm mechanism. As before, the bifilms would preexist from the steel casting process, and the pits would form where they emerged at the surface. When forming from austenite, the ferrite would preferentially form on bifilms to reduce the strain energy of formation, so the final appearance would be that the apparent SCC appeared to follow the austenite/ferrite phase interfaces.

Recent studies on Ni alloy 690 (60Ni, 30Cr, 10Fe, 0.03C) by Yonezawa et al., (2015) testing different crack growth rates in simulated PWR primary water were reported to be not explicable—the typical bifilm scenario!

However, the bifilm role is confirmed once more by the precipitates of TiN and $M_{23}C_6$, as "beads on a string" typical of the evidence for the presence of bifilms at grain boundaries. In addition, the carbide "beads" were cracked transversely to the stress (any longitudinal cracks parallel to the stress were not visible); importantly, the stress would have been too low to explain the fracture of the precipitates, but would have assisted to open those bifilm substrates at right angles.

Research by Raman and Muddle (2003) in ductile cast iron and Ye et al. (2018) in a generally corrosion-free Al alloy clearly show concentrations of tangled cracks which resemble nothing but bifilms. The SCC fracture surfaces of a 7xxxx series Al alloy by Peng et al. (2018) are strongly suggestive of rucks and folds of an oxide bifilm, possibly elongated by plastic work. Lu et al. (2014) investigated an austenitic stainless steel, noting that the SCC crack preferred to initiate at the bottom of a corrosion pit and propagate along the phase boundary between austenite and ferrite. Raman and Muddle found the SCC crack path to follow a number of different boundaries in their ductile iron; these included (i) ferrite grain boundaries; (ii) ferrite/graphite interphase boundaries; and (iii) ferrite colony boundaries.

A study of a lightweight steel containing 5.8%Al undertaken by Sohn et al. (2016) found that if cold rolled and deep drawn, an environment of 0.1M HCl solution resulted in a time-dependent fracture. In this case, their results cannot be claimed to be exclusively explained by bifilms but are simply consistent with the bifilm concept and are in any case interesting. The steel would typically fracture within a day. If the steel was annealed at 750 C for 1 h, fracture was delayed for 18 days. The researchers concluded this showed the importance of internal stress. This seems reasonable. However, the fact that the failure appears to be not prevented, but is only delayed, seems an overriding concern.

Finally, it is sobering to note the failure of austempered ductile iron (ADI) when placed in water (Boeri and Martinex 2013). The tensile elongation falls from close to 15% down to 1%. Even more dramatically, if water is applied to a tensile bar already under stress, the failure is practically immediate. There is an urgent need to test whether ADI can be made resistant to embrittlement by water if cast without bifilms.

3.5.4.5 Hydrogen embrittlement

The research on HE is dauntingly extensive, detailed, and complex, making it inappropriate and impractical to summarize the many theories here.

However, the general features of HE are simply described. Steel pressure vessels containing hydrogen at high pressure gradually absorb hydrogen and become embrittled and can fail catastrophically. Similarly, natural gas pipelines suffering chemical corrosion from "sour" gases (such as natural gas containing some sulfur dioxide) gradually embrittle and may fracture. Although the gas pressure in the pipeline is not necessarily high, the action of generating atomic hydrogen in contact with the metal creates a flux of hydrogen equivalent to a high pressure of hydrogen gas. A number of other chemical environments have been found to result in HE. However, intriguingly, if the embrittled metal is taken out of its hydrogen-rich environment it can gradually recover its ductility.

Nevertheless, HE is serious, as the example from Zhonqiu (1990) shows for his heavy section 13Cr 5Ni steel which he subjects to electrolytic charging with hydrogen. He found the ductility, measured as reduction in area of a tensile failure, was reduced to near zero as hydrogen in solution was increased (Fig. 3.13).

The reactions and the crack growth all take time as shown in a second example (Fig. 3.14) by Gibala and Hehemann (1984). They follow the initiation and growth of a crack by electrical resistive measurements on a high strength steel of that date electrochemically charged with hydrogen. The sample was sharply notched and loaded with a dead weight. On receiving the applied tensile stress, the bifilms would open somewhat, causing the sudden increase in resistance seen in the figure. Subsequently, the bifilm cracks would now slowly accumulate hydrogen, the pressure increasing until

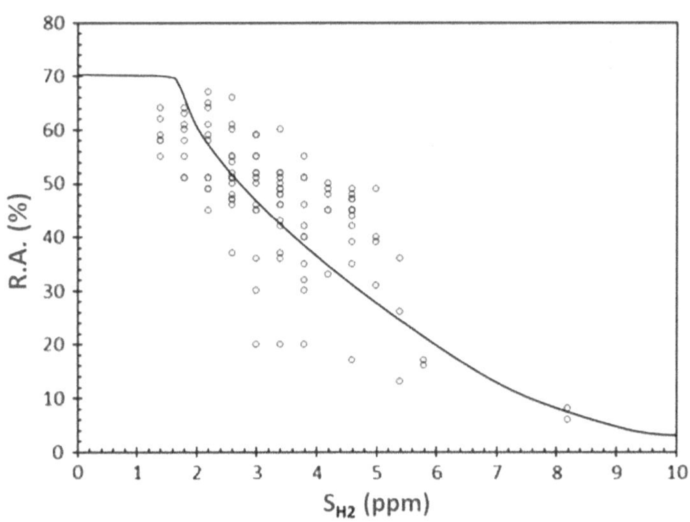

Figure 3.13 Typical result showing the disastrous loss of ductility with increasing hydrogen content in a steel. *Data from Zhongqiu, 1990 replotted by Tiryakioglu, 2019.*

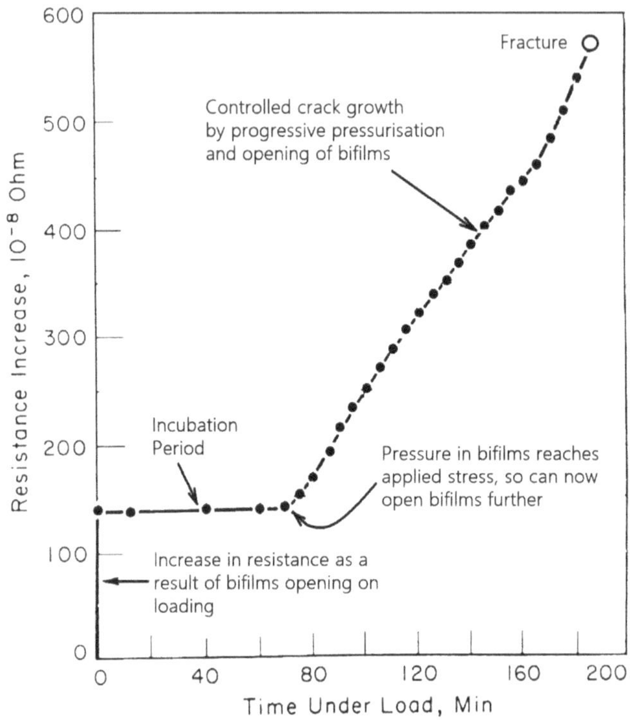

Figure 3.14 Typical result showing gradual crack propagation with time for a steel under constant load (Gibala and Hehemann, 1984 reinterpreted JC, 2019).

it reached the applied stress, at which stage the bifilms could now begin to open and extend further. The effective crack area is seen to grow progressively until final fracture.

The embrittlement of metals by the absorption of hydrogen is one of metallurgy's significant problems. Metals absorb hydrogen in many ways. Right at the start, before casting, liquid steels can be "sensitized" to the pick-up of hydrogen by the deoxidation practice (Svensson, 1980). An excess of Al or Ca leads to reaction with moisture in the environment, releasing H into the steel. Thus, the weather during steelmaking might be more important than we think. However, many hydrogen problems arise in the service environment of the solid metal, particularly steels. The presence of H_2O, H_2S, and chloride ions in an aqueous environment can all seriously degrade properties in some steels and can lead to failure by fracture.

Despite the decades of research, admissions in the research literature that the problem is not understood remain as frequent as ever. There seem good reasons why bifilms should be involved, so it is hoped that some contribution to understanding at least a part of this morass of mystery can be achieved. Because no-one has been checking for the role of bifilms, the current evidence must be gleaned from experiments not targeted to elucidate their role. Even so, the evidence, such as it is, seems generally supportive; targeted research would, of course, be so much better.

The theories proposed to explain hydrogen induced cracking (HIC) are many; but, as already admitted, a review is beyond the scope of this short text. There is no doubt that at times chemical and electrochemical processes are at work, especially in hydrogen environment assisted cracking and hydrogen-enhanced local plasticity. The reduction in the surface energy of the crack, accelerating its rate of growth, also features strongly in some theories.

Researchers have speculated that hydrogen "traps" in metals are helpful to reduce HE because of the reduction in diffusible hydrogen remaining in solution in the lattice. Although grain boundaries and dislocations have been suggested as traps (Depover and Verbeken 2018), more emphasis has generally been placed on the role of inclusions whose stress fields in the lattice have generally been supposed to provide traps of greater volume.

With this mechanism in mind, Van den Eeckhout (2018) found that carbides in quenched and tempered martensitic steels varied in order of diminishing effect: V, Ti, Cr, and W. For welded steels, Beidokhti et al. (2017) found TiC and MnS were especially strong traps. A team led by Bhadeshia (Ramjaun et al., 2018) set out to design a special steel to resist HE. They targeted V and Mo carbides precipitated in the martensitic matrix by tempering because these carbides maintained their coherency with the matrix, thus maximizing the associated stress fields and thereby reducing the danger of diffusible hydrogen.

Although hydrogen is likely to be trapped in lattice defects, there seems no doubt that in many cases a large proportion of, if not most, hydrogen will be trapped in bifilms, simply because of the profusion of such entrainment defects to be expected in many current steels. Bifilms are therefore *expected* to be present and would be expected to contribute toward a hydrogen interaction not only by provision of traps, but by providing short-circuiting routes for hydrogen through their unbonded, open pathways. The matrix would, in contrast of course, have a much lower permeability for hydrogen.

The speeding up of hydrogen permeation through the cracks provided by bifilm-filled metals has an interesting analogy in the work of Gulsoy and Was (2015). In their Ni alloy IN617, they found that oxygen can travel along "incoherent alumina-metal interfaces," effectively five orders of magnitude faster than the rate of oxygen diffusion through the matrix. The "incoherent alumina-metal interfaces" seems an accurate description of bifilms, suggesting the mechanism whereby gases can travel rapidly along their central cracks. Solano-Alvarez et al. (2015) found the rate of diffusivity of hydrogen in their bearing steel increased by a factor of 3.3 if the sample was cracked. This apparently small increase is probably the result of them unknowingly assuming their base steel to be uncracked, which, of course, we would expect to be already densely populated with bifilms too fine to be resolved in their experiment.

Other fascinating evidence for the short-circuiting of gas flows through liquid metals is provided by the amazing experiments of a young Scandinavian researcher, Tiberg, in 1960. He observed the growth of CO bubbles in liquid steel by looking directly into the expanding bubble. He recorded this exciting miracle on cine film through the walls of a silica tube, watching the nucleation and growth of CO bubbles on the tube walls. The speed of growth was constant, but slowed and became different for different bubbles after deoxidation with Al or Si. However, most significantly, when a small bright spot appeared in the bubble, growth suddenly speeded up. It seems to me that the only cogent explanation of Tiberg's observations is that the bright spot is the point of contact with a bifilm created during the deoxidation procedure. Its brightness results from the observer looking directly into the "air gap" between the films; the air gap will act as an effective black body cavity in the bulk liquid, radiating the full spectrum corresponding to the temperature of the steel. This would be in contrast to the bubble surface which would appear to be cooler because of its mirror reflection of the cool surroundings outside the silica tube. The sudden speeding up of the growth of the bubble is the result of the large area of the bifilm and its "air gap," its large area gathering gas and funneling it into the bubble. (I happened to meet Tiberg by accident in the 1970s and congratulated him on his research as the most exciting I had ever read. He immediately agreed that it was the most exciting research he had ever carried out!)

Tiberg's observations seem to be supported by Kita (1979) who used the SEM to study the interiors of pores in solidified steels. He found the inner surfaces of pores exhibited curious markings, radiating like petals of a flower, suggesting that this location in the pore was the site of the ingress of gas via some unknown route. The details of the markings were different for pores created by hydrogen, nitrogen, and carbon monoxide. Some markings also appeared to consist of oxides.

It seems certain that bifilms, in addition to providing "short-circuiting" routes for hydrogen, would provide areas in the matrix which could accumulate high loads by only modest pressures of hydrogen acting over the cumulative area of the population of bifilms. These accumulated expansive loads will, of course, add to any externally applied tensile load on the specimen. In addition, the resulting internal stresses would be concentrated at the sharp tips of the bifilm cracks.

The diffusion of hydrogen to internal cracks, where it collects as hydrogen molecules and pressurizes the crack, is a long-established theory for HE. It is all the more curious that this theory has been dismissed by some researchers in recent times (Bal et al., 2016). A dismissal of this appealingly simple theory seems counterintuitive, particularly because there is the most obvious evidence for it which can be observed by anyone with their unaided eye as described below.

During the charging of metal samples with hydrogen, a precursor to many HE studies, blisters are observed to grow on the surface of the sample. Ren et al. (2008) study blister formation in steels, noting that the many researches have so far been unable to explain the nucleation of hydrogen-induced blisters. For instance, Gibala and Hehemann (1984) report that during the early days of electrochemical charging of hydrogen, the samples often cracked or created blisters during the charging period. This was, they said, unsatisfactory as it made controlled experiments impossible. (The reader will by now suspect that this lack of reproducibility is a standard result of the presence of bifilms of sundry sizes and locations.)

If we recall the central thesis in this book, that cracks and fissures can *only* be formed by bifilms, and that bifilms are likely to be present throughout the metal, then it is clear that bifilms close to the surface of the sample will be pressurized by hydrogen. Because of their proximity to the surface, the pressure results in the plastic expansion of the fissure as a blister. Clearly in the interior of the sample, the same pressurization by hydrogen will be expected to occur, but the expansion of the bifilm will be resisted by the strength of the surrounding metal. However, it is important to realize that the blister is the *obvious* demonstration that the pressurization of bifilms is a factor in HE. *It could not be more obvious.* In a morass of tentative theories and half facts, this simple observation leading to such a definite conclusion is a providential gift.

Recent work by Handrigan et al. (2018) corroborates this suggestion. Samples of 13%Cr steel were imaged using computer tomography before and after charging with hydrogen. They measured the void fraction to increase 18 times, supporting the theory that the role of hydrogen in promoting failure involves void expansion by the recombination of hydrogen into diatomic molecules.

We shall pursue the blister phenomenon a little further. While blister formation is found by researchers who deliberately chemically or electrochemically charge the metal with hydrogen, hydrogen blisters are also found in other circumstances. For instance, during the heat treatment of Al alloys, blisters are often formed, especially if the atmosphere of the heat treatment oven contains water vapor. I recall that when managing the Cosworth Casting foundry, my countergravity castings never showed blisters during heat treatment (the forbearance of the reader is requested to overlook such awfully proud boasts—my excuse is that the countergravity castings were truly excellent!). However, forgings from conventional DC cast material invariably grew blisters up to 20 mm diameter which were sufficient to scrap the part. (We should be sympathetic toward forgings, they suffer terribly from the defects which the foundry cast in!)

Celik and Bennett (1979) record that filtering of an Al alloy melt is an effective way to control blister formation during heat treatment of the rolled product, confirming that defects from the liquid state, which are capable of being filtered out, are necessary, as well as confirming that rolling is not especially effective in bonding or welding these defects.

A completely different approach to the blister problem is related by Dorward (2001). He finds that the introduction of a little $NaBF_4$ in the atmosphere of the heat treatment furnace completely solves the problem of blister formation (Fig. 3.15). Similarly, Zurecki (1996) found that blister

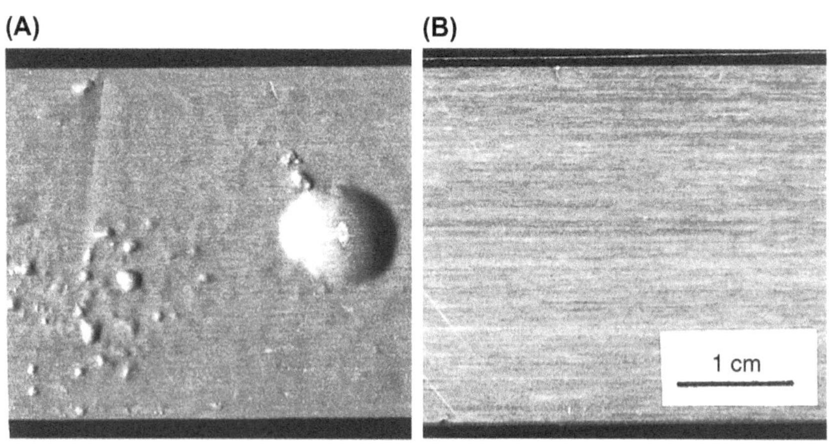

Figure 3.15 Surfaces of stretched sheet of 7475 (Al–Zn–Mg–Cu) alloy heated to 515 C (A) in air and (B) in air containing some NaBF4 (Dorward, 2001).

formation during the heat treatment of Al—Mg alloys could be suppressed by low levels of sulfur dioxide in the furnace atmosphere.

Clearly, the diffusion of hydrogen into metals and the consequent pressurizing of defects, which, in these light alloys, will be bifilms, are well-documented phenomena. While many theories of HE require hydrogen diffusion along dislocations and grain boundaries, or reductions in surface energy at the crack tip, or enhanced plasticity, and, of course, all such phenomena may contribute, they are clearly not needed to support the explanation of HE as the pressurization of bifilms by hydrogen. The pressurization of bifilms by hydrogen gas is the simplest explanation which appears to fit most, if not all, of the observations.

We can now turn to assess the pressures which may be involved. If we assume that in the extreme, the blister is so close to the surface that minimal resistance to the gas is required to swell the blister, then the pressure in the blister will be close to 1 atm. If we further assume the interior of the blister is of the order of a millimeter thick, compared to the thickness of an "average" bifilm in the interior, of perhaps a micrometer thick, the pressure in an interior bifilm might be as high as 1000 atm (100 MPa). Allowing for the fact that the pressure in the blister will certainly be higher than 1 atm, and perhaps as high as 10 atm or more, this rough estimate illustrates that an interior pressure might easily approach or exceed the yield stress of some metals but, in any case, would certainly contribute to an apparent reduction in strength.

Ren et al. (2008) estimate that if the metal is saturated with H at 1000K (727C), then on cooling to room temperature, the pressure in pores will be of the order of 1 GPa. If this is true, the pressurization and opening of bifilms in the interior of metals, because of elastic yielding, and perhaps even some plastic yielding, will favor the path of a crack. This seems probable because, for want of a better term, the fracture surface is called "quasi-cleavage" because it seems to have so far defied metallurgical explanation (McMahon et al., 2008). It does seem consistent with the type of surface expected from arrays of opened bifilms, linked by regions of plastic shear.

Stress in the plastic domain was found by Kim et al. (2017) to open up interfaces in second phase particles in ferritic steels under load and form blister cracks. These additional hydrogen traps helped to slow down the rate of permeation of hydrogen. Traps for hydrogen, by reducing the concentration of hydrogen in solution, have been viewed as a benefit. However, this appears to counter to the suggestion that bifilms are pressurized and contribute to effective loss of strength. This potential conflict is examined further below.

A recent study (Ramjuan et al., 2018) designed a 2.5Cr 1.5Mo steel to possess excellent hydrogen trapping capacity by precipitating V and Mo carbides in the martensitic matrix. The carbides maintained their coherency, with the result that their associated strain fields were assumed to act as trapping sites, reducing the danger of diffusible hydrogen. At first sight, as we have noted above, this interesting concept appears to clash with the bifilm concept. The precipitation of carbides seems often to occur exclusively on bifilms, with the result that the formation strain of the particle is accommodated by the forcible opening of the bifilm, thus providing a trapping site, but one which we have assumed acts to promote HE rather than reducing HE.

The resolution of this apparent conflict may be that the bifilm opening acts to reduce diffusible H at low levels of H, but beyond a critical H threshold, the opening of bifilms degrades properties. Interestingly, this effect is seen in the experiment of Fig. 3.14. The two-stage condition is probably to be expected. The critical threshold probably corresponds to the H content at which H pressure equals the yield strength of the alloy, at which point, of course, the bifilm crack will start to expand by plastic, irreversible deformation of the matrix.

Naturally, this behavior will not only be different in different steels, but different in the same steels subjected to different strengthening heat treatments, but even the same grades of steels will be substantially differently depending on how they were cast, and each batch of poured ingots or continuously cast material will contain different populations of bifilms. This subject will require much care to disentangle the numbers of important variables.

It is clear that the action of precipitated inclusions such as carbides may be acting not so much because of their strain fields, but by precipitating on and opening up bifilms. In fact, of course, those inclusions especially selected by Ramjaun et al. (2018) and others referred to above, as retaining their coherency with the matrix, will most benefit from the reduction in strain energy by precipitating on and opening bifilms. It may be therefore that the "trap" is constituted mainly by the bifilm rather than the lattice strain.

The role of bifilms (alias microvoids) as traps for hydrogen seems to be confirmed for aluminum in a fascinating recent experiment. Toda et al., (2014) describes the tensile test of an Al alloy in an evacuated enclosure. When the sample fractured, hydrogen was found to be suddenly released from traps. The immediacy of the release of hydrogen strongly suggests that hydrogen was trapped as hydrogen gas in voids, most probably, of course, bifilms. If the hydrogen had been trapped in lattice defects such as dislocations or lattice strain fields, the rate of release of hydrogen would have been limited by the rate of diffusion as atoms to the outside surface and the recombination as a diatomic gas, all of which might have been so slow as to make H release practically undetectable. The repeat of this experiment for steels would be an important critical test.

The action of bifilms is seen in the work by Harris et al. (2016) in Cu–Ni alloy Monel K-500. These researchers found that hydrogen appeared to cause cracking of its carbides, which were therefore discovered as initiators of tensile failure at the bottom of ductile dimples.

Other results suggestive of bifilm action were observed by Shibata et al., (2017) who noted that for martensitic steels, H cracks in higher strength steels followed PAGBs, but which exhibited facets of {110}, whereas lower strength steels the cracks tended to be quasi-cleavage on {110} planes. These crystallographic relationships are typical of bifilms which have been straightened (actually "flattened") by the growth of arrays of dendrites during solidification.

Peet and Hojo (2016) observed the dramatic drop of elongation from about 22% down to about 0.2% when their nanostructured bainitic steel was saturated with H. Their fracture surfaces were a mix of quasi-cleavage and plasticity, but, perhaps significantly, they found a strong correlation between total H content and the total area of alpha-Fe/gamma-Fe boundaries. As we have seen elsewhere, ferrite seems to form preferentially on austenite grain boundaries in which bifilms are expected to be concentrated.

A telling report of HIC in X65 pipeline steel by Du et al. (2016) reveals cracks were only 4 mm long when deoxidized with Zr + Ti whereas they were 10 mm long when deoxidized with Al. The authors also report the formation of MnS inclusions on ZrO_2 and Al_2O_3. Although the MnS precipitation is interesting it is probably of little further consequence to the damage process by the oxides (which would probably have been in the form of bifilms), in contrast to the concern by the authors.

Evidence for the mechanical effects of bifilms on the fracture process is strongly suggested by the results of a study by Takakuwa et al. (2017). These workers found that H appeared to cause the most extraordinary shapes of the macroscopic fracture surface of hydrogenated steels, some of the shapes being highly asymmetrical and irregular, somewhat resembling craggy mountain tops. It is not easy to think of any mechanism for such surprising lack of the expected conical symmetry other than the presence of "precracks," in the form of bifilms, lying at random angles throughout the test pieces.

Mention must be made of the fascinating experiment employed by Japanese researchers (Matsunaga et al., 2014). They used a photographic technique based on an emulsion containing silver bromide which could precipitate silver, marking the location of hydrogen emitted from different regions of the microstructure of ductile cast iron. The approach is so informative it deserves to be much more widely used; it would greatly help to elucidate the complexities of this complex subject.

Japanese researches (Nie et al., 2012) studied a tempered martensitic structure 1500 MPa vacuum induction melting (VIM) steel (0.6C, 2Si, 1Cr) by slow strain rate testing. In common with many other studies, including some of those described here, they found that cracking occurred along the boundaries of its prior austenite grains, where, as we know, bifilms will have been concentrated during solidification.

It is important and chilling to find that researchers have found HE in steels which have been subjected to VIM and casting. If our above hypothesis for the mechanism for HE is correct, this steel, usually selected for its "cleanness," must therefore have similar concentrations of bifilms to conventional steel as a result of its turbulent top pouring technique. Its apparent cleanness results from its bifilms merely being thinner. However, they still seem to act as effective cracks.

Other Japanese researchers (Koyama et al., 2016) studied 20%Mn steel also prepared by VIM, finding it suffered severe loss of ductility when electrochemically charged with hydrogen during a tensile test. Higher Mn steels, up to 40% Mn, were less affected, particularly an optimum 34% Mn. This may be the result of the presence of MnO_2 as a liquid oxide on the surface of the molten steel during steelmaking and casting, inhibiting the formation of bifilms for the higher Mn steels. The authors report the H cracking to coincide with the gamma Fe to epsilon martensite boundaries, which, once again, seem likely sites for bifilms.

Whatever the detailed mechanism, the HE damage to the 20Mn steel is likely to be the result of oxide entrainment, producing bifilms, which were then pressurized by hydrogen. The fact that this has happened after melting and casting in vacuum shows the limitations and danger of the VIM process; it is ineffective in preventing oxidation and it uses disappointingly turbulent gravity casting practice which create bifilms.

In summary regarding HE, it seems there is good evidence for hydrogen pressurizing bifilms. The time-dependent build-up of pressure depends of course on the effective hydrogen pressure in the environment and on its rate of diffusion into the metal. It results in bifilms near the surface swelling to form blisters and those in the interior contributing to an effective loss of strength and ductility. The precipitation of inclusions seems effective in reducing HE only for low levels of hydrogen, but at higher hydrogen levels develops into the main mechanism for failure.

3.6 Conclusions

Naturally, the conclusions of this book will have to remain tentative at this time. We need to await critical, targeted experimental corroboration for many of the proposals which are contained.

Nevertheless, if there is some truth in the concepts presented in this book (and the author has to admit to perhaps more than a hint of prejudice) we should lament over our mediocre and unreliable metals with their multiple failure modes, which are not the fault of our metals, but of our inappropriate manufacturing techniques.

For the present it is postulated that mesoscopic or macroscopic unbonded defects entrained during manufacture dominate the failure of metals and that these features constitute the main, if not the only, source of cracks in both cast and wrought metals. Therefore,

1. Powder metallurgy and other particulate consolidation processes such as spray forming, etc., necessarily result in a dense population of uniform, fine, bifilm-type cracks. Intensively sheared liquid/solid mixtures may be analogous, although the processing may cause the films to evolve into fine particles. The fine dispersion of oxides from particulate processing may benefit strength, but will limit ultimately high strength attainments because of the associated entrapped argon which will provide unbonded initiation sites for failure.
2. Unnecessarily poor melting and casting techniques cause casting defects of a wide spectrum of sizes, accounting for the currently wide scatter of properties of most of our engineering metals. In contrast, good melting and casting techniques appear to have the potential to create metals with reliable and immensely high theoretical strengths and toughness.
3. Solidification is merely a phase change and cannot lead by in itself to the formation of damaging defects such as pores and cracks. It is only *entrainment processes,* especially casting by pouring, which can create the serious populations of cracks suffered by most of our metals at this time.
4. Pores and cracks which seem to occur naturally in metals are therefore *extrinsic* defects, usually introduced into metals by the turbulence of poor casting processes. They are not necessarily present in metals.
5. Probably most if not all types of failure only occur from entrained unbonded interfaces, mainly bifilms. An immense and dense population of bifilms is to be expected in all our current engineering metals; the various alternative metallurgical failure theories which have been suggested from time to time, including atomic and lattice mechanisms, seem to be not viable. Failure by cracking is not a phenomenon to be expected in metals.

6. Failures that appear to have initiated by decoherence or fracture of foreign inclusions will in fact have originated on its associated bifilm acquired during prior entrainment, usually in the liquid state.
7. Entrained defects, mainly bifilms, introduced during the manufacture of metals appear to survive significant plastic working processes, tending to be relatively permanent, becoming usually elongated and possibly fragmented along the working direction.
8. Bifilms which have undergone significant plastic deformation or high temperatures, which encourage their transformation into arrays of separate, compact, solid inclusions, are to be expected never to lose the argon from their residual entrapped air. The associated argon bubble or film will remain to enable apparent decoherence, cracking, or pore initiation. We conclude that damage introduced during casting can never be completely eliminated by subsequent processing.
9. The opening of bifilms by inclusions such as carbides precipitating onto the bifilms as favored substrates appears to be a key operating mechanism for SCC and HE.
10. Currently, the majority of the world's metallurgical research is carried out in vacuum induction melting and casting furnaces. Unfortunately, this is neither a good process for producing good steel nor a good process for carrying out research. It seems probable that the number and density of the bifilm population is not reduced by pouring in vacuum. Logic indicates that the geometry encouraging entrainment is not altered. Only the thickness of bifilms is expected to be reduced, giving the (misleading) impression of a cleaner steel. It is therefore very regrettable that nearly all our metallurgical research up to the present date has suffered the complication of a dense population of bifilms which has often controlled and confused results. Improved air melting and casting methods could provide significantly better steels, particularly better steels for research. Filling systems for ingots and shaped castings are now available for avoiding most entrainment problems.
11. It is very much to be regretted that, at the time of writing, reliable steels for safety critical applications do not exist. Currently, the ESR process appears to produce our most reliable steel. However, its upgrading to achieve 100% reliability is strongly recommended as an urgent requirement.

In an appropriately cast and solidified metal, within the scope of current technology, pores and cracks need not exist. Thus, metals without such defects would be expected to exhibit the following microstructural behavior and property response:

1. Intermetallics should never crack because they are in general extremely strong (furthermore, most would probably never form as primary or secondary phases in the absence of the favorable entrained bifilm substrate. The elements which would normally form such precipitates will remain in solution in the matrix.).
2. Inclusions or second phases formed in situ in the melt or matrix should never decohere from the matrix.
3. Grain boundaries should never decohere, for instance, in creep or superplastic forming.
4. Failure of metals by creep, fatigue, SCC, and HE appears to be preventable. These metallurgical problems may now be solvable.
5. Failure of metals should therefore occur either by (i) ductile failure at extremely high elongations and 100% reduction of area or (ii) brittle fracture only at the extremely high theoretical ultimate stress, of the order of tens of GPa.
6. The brittle failure of metals stressed to their theoretical limit is, of course, predictably explosive with an energy similar to dynamite. The use of metals at these ultimate stresses is not therefore recommended. The benefits of defect-free metals will be found in more modest but valuable increases in strength, but most valuably in their reliability—their resistance to failure by any of the many fracture modes.

Appendices

APPENDIX I

It is not widely known that my well-known Canadian friend and colleague, Robert Puhakka, used his training as a metallurgist to take an early job as a Failure Investigator. The job entailed receiving failures of all kinds of components and metal alloys, studying the fractures in the scanning electron microscope and carrying out what additional tests appeared to be necessary such as the determination of metallurgical structure and hardness. A report then had to be written and sent back to the purchaser of the investigation. The work could hardly have been less rewarding; it seemed none of the investigations made any metallurgical sense.

By chance, Bob came across a copy of my book "Castings," which dealt at length with bifilms. Bob devoured this overnight, returning to the lab the following morning to go through the failed samples again, exclaiming to the effect, "Yes! They are all bifilms!" Interpretation of the failures as initiated by bifilms appeared to explain them all. The reasons for the failures could be understood for the first time.

As many in the casting industry will know, Bob went on to apply the principles of bifilm reduction to cast metals, achieving unprecedented success in cast steel and later, in his own foundry, unprecedented success in the manufacture of aluminum alloys. His demonstration of the routine production of engineering castings of unparalleled perfection continues to be unequaled.

APPENDIX II

In a recent meeting with an industrialist, he described to me the testing of a gear wheel from ingot-cast steel, in which the typical fracture surface of the component revealed an inclusion in the center of a fatigue fracture. Naturally, it was concluded that the inclusion initiated the failure. However, with better quality steel, continuously cast, the fracture surface contained an identical fatigue fracture but, inexplicably, no initiating inclusion.

The mystery is easily explained if the fracture surface is not principally a fatigue crack, but a bifilm, and therefore exists in the liquid state, before the formation of inclusions. The inclusion, whether successful to precipitate on the bifilms or not, is largely irrelevant, merely serving to witness that a bifilm exists on which it could form (otherwise, interestingly, without the prior existence of the bifilm, it probably would not have formed).

APPENDIX III

I have reason to be grateful to Professor John F Knott, our Head of the School of Metallurgy and Materials at the University of Birmingham, UK. It was in 2006, during his time as editor of the respected Materials Science and Technology journal, that he asked me to write for him a "Materials Perspective" paper on bifilms, which was intended to present a viewpoint, possibly controversial, held by the author, rather than a conventional experimental reporting paper or review of prior work. I greatly appreciated this generous offer, more particularly because John was clearly taking a risk; bifilms were even more controversial in those days than they are now!

JFK's position as a world leader in fracture mechanics was, of course, known and respected by us all in the Department. One day in the lab a curious incident occurred. A student was struggling at his microscope, trying to explain the cracks in the microstructure of a weld. John was leaning over him, dominating the scene, his powerful analytical mind in top gear, firing off in quick succession one suggested explanation after another, but seemingly without success to explain the observations. Mike Wise, a senior lecturer, and I were silent bystanders. In an effort to be helpful, I ventured that the cracks might be bifilms (this was in the very early days, before such thoughts were common). Mike Wise immediately chimed in "Oh! So the cracks were there first!" John immediately fell silent and froze motionless. He finally stood upright and turned and strode out of the door.

Whether it was from this time that John no longer spoke to me again I cannot recall. It is true that during our time at Birmingham he would neither talk to me on technical matters nor ever answer emails, which is why the invitation to write the Perspective was such a surprise. In my early days, cautiously feeling my way in the world of fracture, John's help would have been welcome.

In trying to piece together these events, it may be that our great Fracture King somehow knew that the bifilm concept contained some truth but, in some way, could not bring himself to admit it. This is despite his standard phrase to explain the importance of Griffith's theorem, in his clear didactic voice, "A crack initiates from a *pre-existing* crack." He always emphasized "*pre-existing*." In his tragic and untimely death, metallurgy has lost one of its great leaders in the field of fracture. We shall never know what he really thought.

APPENDIX IV

The development of knowledge in metallurgy is suggested, in broad terms, in Table 3.3, showing the logical progression of the adoption of defects of higher dimensions which, if the concepts in this book are correct, contribute to the behavior of metals.

The list could, in principle, be extended by the addition of bubbles as 3D defects, but this trivial academic detail is ignored here. Naturally, the bubble is relatively well-known and understood, and compared to the other defects in the table; bubbles in themselves (in contrast to their bubble trails) contribute in only a minor way to the behavior of metals.

Table 3.3 A tentative history of metallurgical development.

Approximate date	Dimension	Discovery	Solid property
1900	0D	Atoms	Strength and rigidity
1900	0D Point defects	Vacancies	Diffusion
1950	1D Line defects	Dislocations	Plasticity
2000	2D Planar defects	Bifilms	Fracture

APPENDIX V

A Summary of Practical Recommendations for Metal Quality

The essential concepts to be kept in mind are the following:

(i) On melting, be aware that your solid starting materials will probably be crammed with entrained defects from previous poor melting and casting procedures. Furthermore, during melting, the surface oxide on the pieces of charge material will thicken and can be entrained as complete oxide skins into the melt, forming major bifilm defects. Thus, pieces of charge for the melt will contribute both their external and internal oxide films.

(ii) On casting, avoid air entrainment into the liquid, forming bubbles (bubble trails, which can be meters long, are the most damaging entrainment defects).
(iii) Also, avoid entrainment of the melt surface to form bifilms.
(iv) Modify the deoxidation and/or alloy composition to liquefy the surface oxide film so that bifilm formation becomes impossible.

The achievement of these requirements varies somewhat from metal to metal.

Aluminum

For the melting and casting of aluminum, it is regrettable that the quality of much charge material, if recirculated from the foundry, can be appalling. Furthermore, the primary skins from the charge materials are necessarily entrained if, as usual, melting is carried out in a crucible or bath. In the great cast houses, their ownership of melting baths containing 20 tons or more liquid metal is often proudly pointed out. However, such furnaces are a mistake. This sea of liquid metal cannot be treated successfully to remove all the primary skins; even if 80% or 90% of the primary skins could be removed, the potential damage from the remaining skins would be dire.

Attempts to eliminate the primary skins by rotary "degassing" is, of course, never guaranteed to eliminate 100% of skins. In addition, the breaking of argon or nitrogen bubbles at the surface of the liquid, as the bubble bursts open to admit air, appears to generate large populations of small alumina bifilms, whose effect on the melt is at this time unknown.

The only method I recommend for the melting of aluminum is the dry hearth tower furnace. This furnace has the benefit of countercurrent heat exchange with waste gases ascending the stack as charge gradually descends by gravity, as melting takes place on the hearth. The melted metal runs out of the oxide skin, down the slope of the hearth, and into the holding furnace. The primary oxide skin remains on the hearth. The skins gradually build up and are removed from time to time via a side door. Thus, from the first moment of melting, the oxide skins and metal are separated.

The furnace has other interesting advantages for continuous supply of metal for constant level distribution and for the benefit of reducing the control difficulties of countergravity casting units, avoiding the complications of having to adjust pressures to take account of falling levels.

The continuous delivery of metal also facilitates such features as furnaces designed for sedimentation of bifilms, possibly aided by heavy metal additions such as Ti or Fe.

The criticism of continuous melting comes from traditional metallurgical thinking in which batch production of alloys was said to be required to ensure the homogeneity of alloys. This is true, but ignores the fact that the oxide content of aluminum alloys is often of far greater importance than small variations in alloy content; our slavish devotion to accuracy of alloy content is an unhelpful anachronism for many alloys. If small additions of alloying adjustments and grain refining additions, etc., are required downstream, they should be made enclosed in a bag of glass fiber filter cloth with pore size in the region of 1.5–2.0 mm. This will trap the primary skins from the additions but, of course, permit the melted liquid to flow freely into the main melt.

If gravity pouring of the metal is carried out, this should use preferably contact pouring, or less preferably an offset step basin with the correct use of a stopper, to prevent air entrainment into the filling system. The filling system itself requires to be designed to avoid further entrainment and requires to prevent priming bubbles of air to enter the mold. Techniques to achieve this are described in the author's Mini Casting Handbook Second Edition.

Preferably, casting should be carried out by countergravity, using a pump which does not introduce severe turbulence into the melt. A pneumatic pump can work well in this respect, whereas electromagnetic and centrifugal pumps are both highly turbulent in their working volumes. The speed of filling should be set to ensure that surface turbulence never occurs so that bifilms are not introduced during the filling process. A superb metallurgical quality is possible via this procedure.

Steel

Although liquid steel enjoys the significant benefit of rapid detrainment of oxides as a result of the large density difference, our horrendous handling of this vulnerable liquid seems designed to destroy every vestige of this valuable benefit. Nevertheless, criticism and complaint is easy, whereas improvements to practice are a nontrivial challenge.

The oxygen steelmaking methods are naturally prolific oxide generators, even nonoxygen melting furnaces require the furnace to be tilted and poured into a ladle, seated in a pit meters below. These universal pouring processes for steel introduce massive damage to the steel quality, which is probably never completely reversed despite all the subsequent processing.

The modern steelmaking approach, separating the melting in the furnace from the treatment required for the metallurgy of the steel, carrying out these subsequent operations in a heated ladle, may be an opportunity to repair some damage by bifilm-reducing treatments. Two key aspects are (i) extended time to encourage detrainment and (ii) the design of a liquid phase as a component of the inclusion or a component of the slag layer. This second feature assists the inclusion to detrain more rapidly as a spherical or compact inclusion rather than a film. Even if the oxide on the liquid steel surface is only partially liquefied, the agglomeration of partially liquid, sticky inclusions will aid the speed of detrainment. A sticky or liquid slag layer will capture inclusions which happen to make contact, gradually reducing the inclusion content in the metal.

Deoxidation with calcium is a major benefit in this respect providing the steel is sufficiently hot to ensure the oxide eutectic of calcia with alumina or chromia is above its melting point at around 1400 C. This temperature is easily achieved in low carbon steels, but less easily in some lower melting point stainless steels, and not easily achieved at all in many Ni base air-melting alloys. Thus, the great benefit of a liquid oxide inclusion, which cannot generate a damaging bifilm crack and will retrain rapidly, is probably not achievable in many stainless steels and is definitely impossible in many Ni alloys.

This raises the potential importance of boron. This element reacts with oxides in steels and Ni alloys to form a low melting point borate liquid on the liquid metal. The melting points of most of the borates are around or even below 1000 C, making them highly effective in liquefying inclusions and encouraging rapid detrainment. It seems to me that B would be valuable to use more frequently as an alloy in steels and Ni alloys; only very small additions in the region of 0.002—0.005 wt% seem to be effective.

The casting of ingot steels should never use a conical trumpet but should employ a nonentraining pouring technique such as contact pouring. Contact pouring is essentially zero cost and would make a massive difference to the inclusion contents of steels. The running systems should be changed, abandoning the drain pipe variety of hollowware, to use channels molded in sand which can fit the profile of the falling metal stream. These systems are described in detail in *Complete* and *Mini Casting Handbooks* (second editions).

Continuously cast steel is currently superior to ingot cast steel, but both require to be hugely increased in quality.

For continuously cast steels, the current use of parallel bore nozzles and parallel bore submerged entry nozzles in metal transfer is a perverse technique which requires to be changed to designs which follow the profile of the falling stream. This would prevent much unnecessary air entrainment and blocked nozzles. Studies to ascertain *detrainment* from launder systems might be abandoned immediately and the effort put in to eliminate *entrainment*. Similarly, the doubtful benefits of entraining argon with the metal stream might also be critically evaluated with a view to eliminating yet another probably unhelpful practice. The initial fall of steel into the empty mold also has to be eliminated in some way; a possible idea is floated in the text.

It is with regret to conclude that the quest for improved cleanness of steels and Ni alloys by vacuum casting seems to be a snare and delusion. The oxygen and nitrogen present in the vacuum environment is practically always sufficient to ensure some kind of oxide or nitride film on the liquid metal. Thus, surface turbulence will ensure the production of bifilms exactly as when casting in air—although, of course, the bifilms will be much thinner and therefore more difficult to detect. Vacuum casting would be beneficial if the precautions for bifilm avoidance, exactly as carried out in air, were also carried out in vacuum. This is a challenge for vacuum processing of liquid metal but does hold out a prospect for, finally, achieving reliable products from vacuum processing.

Unfortunately, reliability cannot be envisaged for vacuum arc remelting at this time. The slow formation of layers of metal creates excellent conditions for the growth of thick oxide films, guaranteeing major cracks in this material. This is a refining process to be avoided unless these fundamental problems of major crack formation can be resolved.

Ultimately, steel very low in oxide bifilms, and therefore of great reliability, can be generated by surrounding the whole steel ingot production process with a liquid oxide, thereby excluding air, as in the electroslag remelting process. Even this excellent process could be improved if its electrode were cast using a good contact pour and properly molded filling system. This may represent a solution to the achievement of bifilm-free steels and Ni alloys for the first time. We will have finally achieved what has become for me a personal maxim designed to appeal and be appreciated by all: "Good metal; metal you can trust."

Postscript

It is proposed that many, if not most, metal failures originate from entrained defects, particularly bifilms, introduced during their production. But the presence of entrained defects is not necessary. Microscopic cracks to initiate metal failure by fracture need not exist in metals processed from the liquid state (they remain unavoidable for particulate processing of course).

In a new era of engineering design, adopting the benefit of improved metals, the designer would only design for strength, simply to counter the one failure mode: yielding, causing distortion (or perhaps rarely ductile failure). There would be no need to specify ductility or any other mechanical property; there would be no requirement to design against failure by cracking, fatigue, creep, stress corrosion cracking, or hydrogen embrittlement.

For the metallurgist, the design of new alloys would now be easy; strength is the easy property to achieve in alloy design and would be the only requirement (resistance to all other failure modes being automatically met). The potential for improved performance for both cast and wrought metals by improved casting techniques, giving control over fracture, promises benefits to transform both metallurgy and engineering.

The scope for simplifying and reducing the environmental and economic costs of metal production is significant. It would not be necessary to forge to "close up" defects. In many cases, the high cost of processing of metals in vacuum would become unnecessary (only those reacting strongly with air, such as Ti, would require to be melted and cast in vacuum). In addition, superalloy turbine blades would probably require vacuum to achieve thinned oxide or nitride film on advancing liquid fronts, which need to fill extremely narrow sections, etc.

Amid the overwhelming mass of valuable benefits from cleaner metals, a final caution may be relevant. If the target of defect-free metals proves elusive, the status of relatively few remaining bifilm defects in metals may, perversely, enhance danger in such processes as hydrogen embrittlement: the hydrogen would tend to concentrate in the fewer defects. In the progress, therefore, from our current poor metals to defect-free metals, the danger of failure may get worse before it gets better. However, it seems necessary to emphasize the view that at least we shall know what the problem is.

Furthermore, such a disadvantage has not been encountered so far; all experience to date records improved properties from improved casting technology. However, of course, our duty of care will always demand continuing vigilance.

References

AIBN Accident Investigation Board Norway 2018 July SL2018/04. (see also 'Air Accidents')

Afseth, A., Nordlien, J.H., Scamans, G.M., Nisancioglu, K., 2002. Effect of thermo-mechanical processing on filiform corrosion of aluminium alloy AA3005. Corrosion Sci. 44, 2491–2506 and "Filiform corrosion of binary aluminium model alloys" 2529-2542.

Aguado, E., Baquedano, A., Uribe, U., Fernandez-Calvo, A.I., Niklas, A., 2013. Comparative study of different interfaces of steel inserts in aluminium castings. Mater. Sci. Forum 765, 711–715, 2013.

Ahmed, A.K.M.A., Kato, H., 2008. Int. J. Cast Metals Res. 21 (1–4), 162–167.

AAIB Air Accidents Investigation Branch Report AAR 2/2014 Eurocopter EC225 LP Super Puma, GREDW and Eurocopter EC225 LP Super Puma, October 22, 2012. GCHCN.

Al-Helal, K., Wang, Y., Stone, I., Fan, Z., 2013. Effect of Ca level on the formation of silicon phases during solidification of hypereutectic Al-Si alloys. Mater. Sci. Forum 765, 117–122, 2013.

Almaraz, G.M.D., Aburto, A.D., Gomez, E.C., 2014. Ultrasonic fatigue endurance of Al alloy AlSi 6061-T6 on pre- and non-corroded specimens. Metall. Mater. Trans. A 45A (Jan), 280–286.

Anderson, T.L., 2017. Fracture Mechanics, fourth ed. CRC Press.

Antonov, S., Hu, J., Feng, Q., Isheim, D., Saidman, D.N., Sun, E., Tin, S., 2018. Boron additions in powder processed Ni-base superalloys containing Nb. Metall. Mater. Trans. A 49A (March), 729–739.

Atkinson, H.V., Favier, V., 2016. Does shear thickening occur in semi solid metals? Metall. Mater. Trans. A 47A (April), 1740–1750.

Aug, A.S.M., Berndt, C.C., Sesso, M.L., et al., 2015. Plasma-sprayed high entropy alloys: AlCoCrFeNi and MnCoCrFeNi. Metall. Mater. Trans. A 46A (Feb), 791–800.

Badia, F.A., 1971. Trans. Amer. Found. Soc. 79, 347–350.

Bahreinian, F., Bourtorabi, S.M.A., Campbell, J., 2005. In: Tiryakioglu, M., Crepeau, P.N. (Eds.), Shape Casting: The John Campbell Symposium TMS, pp. 463–472.

Bai, Q.L., Li, Y., Li, H.X., Du, Q., Zhang, J.S., Zhuang, L.Z., 2016. Metall. Mater. Trans. A 47A (Aug), 4080–4091.

Bal, B., Sahin, I., Uzun, A., Canadine, D., 2016. A new view toward prediting the role of hydrogen embrittlement in metallic materials. Metall. Mater. Trans. A 47A (Nov), 5409–5422.

Barenblatt, G.I., 1959. The formation of equilibrium cracks during brittle fracture. PMM 23 (3), 434–444.

Barenblatt, G.I., 1962. The mathematical theory of equilibrium cracks in brittle fracture. Adv. Appl. Mech. 7, 55–129.

Barton, R., February 28, 1985. Found. Trade J. 117–126.

Bazarnik, P., Romelczyk, B., Kulczyk, Lewandowka, M., 2013. The strength and ductility of 5483 aluminium alloy processed by various SPD methods. Mater. Sci. Forum 765, 423–428.

Beck, A., Schmidt, W., Schreiber, O., 1928. US Patent 1788185.

Beidokhti, B., He, P., Kokabi, A.H., Dolati, A., 2017. Control of H cracking in welded steel using microstructural traps. Mater. Sci. Technol. 33 (4), 408–414.

Birch, J., 2000. Diecast. World 174 (March), 18–19, 35; (September) 28.

Boeri, R.E., Martinez, R.A., 2013. Embrittlement of ADI by liquids. Mater. Sci. Technol. 29 (6), 665–671.

Boisvert, M., Christopherson, D., L'Esperance, G., 2016. Metall. Mater. Trans. A 47B (Oct), 2971–2978.
Bombac, D., Tercelj, M., Kugler, G., Perus, I., 2018. Surface cracking during hot rolling of AISI D2 tool steel. Mater. Sci. Technol. 34 (14), 1723–1736.
Brown, S.G.R., Evans, R.W., Wilshire, B., 1986. Exponential descriptions of normal creep curves. Scripta 20, 855–860.
Byczynski, G.E., Campbell, J., 2005. Crack initiation sites in high cycle fatigue of 319 Al alloy castings. In: Tiryakioglu, M., Crepeau, P.N. (Eds.), Shape Casting: The JC Symposium, pp. 235–244.
Campbell, J., 1967. The solidification of a metal sphere. Trans. Met. Soc. AIME 239, 138–142.
Campbell, J., 1968. Iron & steel inst. Special Report. Solid. Metals 110, 18–26.
Campbell, J., 1969. The non-equilibrium freezing range.... Trans AIME 245, 2325–2334.
Campbell, J., 1988. Confluence welds. Mater. Sci. Technol. 4, 194–204.
Campbell, J., 1991. Castings. Elsevier, Oxford, UK.
Campbell, J., 2000. Critical velocity for Cast Iron. Ingenia 1 (4), 35–39.
Campbell, J., 2004. Casting Practice; the 10 Rules for Casting. Elsevier.
Campbell, J., 2006. Modeling of entrainment defects during casting". In: Tiryakioglu, M., et al. (Eds.), Shape Casting Symp, TMS Annual Congress San Antonio.
Campbell J, 2006. Entrainment defects: a materials perspective. Mater. Sci. Technol. 22 (2), 127–145 and (8) 999-1008.
Campbell, J., 2008. Fatigue fracture of ductile cast iron. AFS Internat. J. Metalcasting 2 (no 2), 43–46.
Campbell, J., 2009a. The modification of Al-Si casting alloys" (Including estimates of bifilm density per kg) AFS Internat. J. Metalcasting 3 (no 1), 65–77.
Campbell, J., 2009b. Discussion "Fatigue crack initiation for cast 319-T7 Al alloy". Metall. Mater. Trans. A 2 (2), 43–46.
Campbell, J., 2009c. Incipient grain boundary melting. Material Sci. Technol. 25 (1), 125–126.
Campbell, J., 2009d. Metall. Mater. Trans. 40A, 1009–1010.
Campbell, J., 2009e. Discussion: on the role of beta particles in stress corrosion cracking of Mg-Al alloys. Metall. Mater. Trans. A 40A (Feb), 273–274 (July) 1510-1511.
Campbell, J., 2009. A hypothesis for cast iron microstructures. Metall. Mater. Trans. A 40B, 786–801.
Campbell, J., 2009. Incipient grain boundary melting. Discussion Mat. Sci. Technol. 25 (1), 125–126.
Campbell, J., 2011a. HAZ cracking in Rene 80 superalloy. Mater. Sci. Technol. 27, 1610–1612.
Campbell, J., 2011b. Cavitation during superplastic forming. Cavitation during superplastic forming. Materials 4 (7), 1271–1286.
Campbell, J., 2011c. Mater Sci Technol "On Preventing HAZ Cracking in Laser Welded DS Rene 80 Superalloy.
Campbell, J., 2011d. The origin of griffith cracks. Metall. Mater. Trans. A 42B, 1091–1097.
Campbell, J., 2012. Stop Pouring; Start Casting. Trans AFS Paper 12-093. American Foundry Soc, Schaumburg, IL, USA, pp. 1–11.
Campbell, J., 2015a. The casting of steels and Ni-base alloys. Amer. Found. Soc. Trans 1–15. Paper 15-012.
Campbell, J., 2015b. Complete Casting Handbook, 2nd Revised Edition. Elsevier, Oxford, UK.
Campbell, J., 2013. Quality Castings: A Personal Account of the Development of the Cosworth Casting Process". Aspect Design 2013 Malvern, UK, and American Foundry Soc., Schaumburg, Illinois, USA.

Campbell, J., 2016a. The Consolidation of Metals: the origin of bifilms. J. Mater. Sci. 51, 96–106.

Campbell, J., 2016b. Melting, remelting and casting for clean steel. Steel Res. Int. 87, 1–13.

Campbell, J., 2018. Mini Casting Handbook, second ed. Aspect Design, Malvern, UK.

Campbell, J., Tiryakioglu, M., 2010. Review of effect of P and Sr on modification and porosity development in Al–Si alloys. Mater. Sci. Technol. 26 (3), 262–268.

Campbell, J., Tiryakioglu, M., 2012. Bifilm defects in Ni-base alloy castings. Metall. Mater. Trans. A 43B (August), 902–914.

Cao, X., 2009. Personal Communication. National Research Council Canada, Ottawa.

Cao, X., Campbell, J., 2004. Effect of precipitation and sedimentation of primary a-Fe phase on liquid metal quality of cast Al–11.1Si–0.4Mg alloy. Int. J. Cast Metals Res. 17 (1), 1–11.

Chen, X.-G., Fortier, M., 2010. J. Mater. Progress Technol. vol. 210, 1780–1786.

Clyne, I.W., Davies, G.J., 1977. In: Solidification and Casting of Metals. Metals Society Conf Sheffield, pp. n275–278. Published 1979.

Cook, R., Kearns, M.A., Cooper, P.S., 1997. Effects of residual transition metal impurities on EC grade aluminum. In: Huglen, R. (Ed.), Light Metals (TMS), pp. 809–814.

Cottrell, A.H., 1958. Brittle fracture in steels and other materials. Trans Amer Inst Mech Engrs. 212 (April), 192–205.

Cottrell, A.H., 1963. Proc. Roy. Soc. A 279, 1–10, 1963.

Cottrell, A.H., 1993. The Griffith Centenary Meeting" 23–25 March. Inst of Materials, London, pp. 4–15.

Cox, M., Wickins, M., Kuang, J.P., Harding, R.A., Campbell, J., 2000. Mater. Sci. Technol., 16, 1445–1452.

Cui, H., Zhang, K., Wang, Z., Chen, B., Liu, B., Qing, J., Li, Z., 2019. Formation of surface depression during continuous casting of high-Al TRIP steel. In: Li, Z., Davis, C. (Eds.), Ironmaking and Steelmaking. Published by MDPI, Switzerland, pp. 320–328.

Cunliffe, E.L., 1996. The Minimal Gating of Aluminium Alloy Castings (PhD Thesis) Metallurgy and Materials Dept. University of Birmingham, UK.

Dantzig, J.A., Rappaz, M., 2009. Solidification. CRC Press, Taylor & Francis Group, p. 500.

Das, A.A., Chatterjee, S., 1981. Metal. Mater. Technol. 13 (3), 137–142.

Das, P., Samanta, S.K., Bera, S., Dutta, P., 2016. Microstructural evolution and rheological behaviour of cooling slope processed Al-Si-Cu-Fe alloy slurry. Metall. Mater. Trans. A 47A (May), 2243–2256.

Decost, B.L., Holm, E.A., 2017. Abnormal grain growth in systems with nonuniform grain boundary mobility. Metall. Mater. Trans. A 48A (June), 2771–2780.

Depover, T., Verbeken, K., 2018. The detrimental effect of hydrogen at dislocations on the hydrogen embrittlement susceptibility of Fe-C- X alloys: an experimental proof of the HELP mechanism. Int. J. Hydrogen Energy 43, 3050–3061.

D'Errico, F., Rivolta, B., Gerosa, R., Perricone, G., 2008. J. Met. 70–75.

Din, T., Kendrick, R., Campbell, J., 2003. Trans Amer Found Soc vol. 111, 91–100 (paper 03-017).

Dispinar, D., 2005. Personal Communication at University of Birmingham.

Divandari, M., Campbell, J., 2000. Alum. Trans. 2 (2), 233–238.

Dorward, R.C., 2001. Oxid. Metals 55 (1/2), 69–74.

Doutre, D.A., Hay, G., Wales, P., 2000. European Patent 20000951137 Filed 26 July.

Du, Y., Wu, K.M., Heng, L., Li, Y., Isayev, O., Hress, O., 2016. Effect oX65 pipeline steel. Mater. Sci. Technol. 32 (7), 728–735.

Dybalska, A., Eskin, D., Patel, J.B., 2017. Evaluation of shearing time sufficient for effective liquid metal processing. JON 69 (4), 720–724.

Emamy, M., Campbell, J., 1995a. The flowability of particulate TiB_2-containing metal matrix composite. Cast Metals 8 (1), 13–20.
Emamy, M., Campbell, J., 1995b. Solidification shrinkage in titanium boride/aluminium MMCs. Cast Metals 8 (2), 115–122.
Emamy, M., Abbasi, R., Kaboli, S., Campbell, J., 2009. Internat. J. Cast Metals Research 22 (6), 430–437.
Emami, S., Sohn, H.Y., 2015. Protective layer over Mg melt. Metall Mater Trans A 46B (June), 226–234.
Erzi, E., Tiryakioglu, M., 2019. On the fracture pressure of liquid metals. Material Sci. Technol. 35 (13), 1656–1659.
Evans, J., Runyoro, J., Campbell, J., 7-10 July 1997. The flow of liquid aluminium in thin sections. In: Solidification Processing SP 97 4th Decennial International Conference on Solidification, pp. 74–78. Sheffield.
Evans, M.-H., 2012. Literature Review: white structure flaking (WSF) in wind turbine gearbox bearings: effects of 'butterflies' and white etching cracks (WECs). Mater. Sci. Technol. 28 (1), 3–22.
Evans, M., 2016. A statistical test for identifying the number of creep regimes when using the Wilshire equations for creep property predictions. Metall. Mater. Trans. A 47A (Dec), 6593–6607.
Evans, M., 2019. Semi-parametric estimation of the Wilshire creep life prediction model: an application to 2.25Cr-1Mo steel. Mater. Sci. Technol. 35 (16), 1977–1987.
Fan, Z.I.S., Fang, X., Liu, G., Patel, J., Das, A., 2007. In: Crepeau, P.N., Tiryakioglu, M., Campbell, J. (Eds.), Shape Casting, the 2nd Internat Symp TMS, pp. 299–306.
Fisher, J.C., 1948. J. Appl. Phys. 19, 1062–1067.
Fletcher, N.H., 1993. Van der Waals' equation and nucleation theory. Eur. J. Phys. 14, 29–35.
Fox, S., Campbell, J., 2000. The visualisation of oxide film defects during solidification of aluminium alloys. Scripta Mater. 43, 881–886.
Fuoco, R., Correa, E.R., Goldstein, H., 1998. Part I: interdendritic feeding evaluation. Trans Amer Found Soc 104, 1151–1155.
Gall, K., Horstemeyer, M.F., Schilfgaarde, M.V., Baskes, M.I., 2000. J. Mech. Phys. Solid. 48, 2183–2212.
Garcia-Garcia, O., Sanches-Araiza, M., Castro-Roman, M., Escobedo, B.J.C., 2007. In: Crepeau, P.N., Tiryakioglu, M., Campbell, J. (Eds.), Shape Casting 2nd Internat Symp TMS Annual Congress.
Gauthier, J., Samuel, F.H., 1995. Trans. Amer. Found. Soc. 103, 849–857.
Gavras, A., Spangenberger, A., Lados, D., 2013. Fatigue crack growth in cast and wrought aluminium alloys. Mater. Sci. Forum 765, 574–579.
Ge, Y., Zhao, S., Ma, L., Yan, T., Li, Z., Yang, B., 2019. Inclusions control and refining slag optimization for fork flat steel. In: Li, Z., Davis, C. (Eds.), Ironmaking and Steelmaking. Published by MDPI, Switzerland, pp. 272–283.
Gernez, M., 1867. Phil. Mag. 33 (4), 79.
Ghomashchi, M.R., Chadwick, G.A., 1986. Met. Mater. 477–482.
Gibala, R., Hehemann, R.F. (Editors), (1984) Embrittlement and Stress Corrosion Cracking. Amer. Soc. Metals. USA ISBN 0-87170-185-5
Giuranno, D., Ricci, E., Arato, E., Costa, P., 2006. Dynamic surface tension measurements of an Al-O system. Acta Materiialia 54, 2625–2630.
Goklu, S.M., Lange, K.W., 1986. In: Proc Conf Process Technology, vol. 6. Iron & Steel Soc, Washington, USA, pp. 1135–1146 (April).
Gordon, J.E., 1988. The Science and Structures and Materials. Scientific American Books Inc, ISBN 0-7167-5022-8.
Granath, O., Wessen, M., Cao, H., 2008. Int. J. Cast Metals Res. 21 (5), 349–356.

Gray, R.J., Perkins, A., Walker, B., 1977. Quality of continuously cast slabs. Sheffield conference. In: Solidification and Casting of Metals. Metal Soc, pp. 300–305, 1979 book 192.

Green, N.R., 1995. Univ Birmingham, UK unpublished work.

Green, N.R., Campbell, J., 1994. Trans. Amer. Found. Soc. 102, 341–347.

Griffith, A.A., 1920. The phenomenon of rupture and flow in solids. Phil. Trans. Roy. Soc. A 221, 163–198.

Griffith, J., Oliver, E.C., Fitzpatrick, M.E., Finlayson, T.R., Viano, D., Wang, Q., 2007. Shape casting. In: Crepeau, P.N., Tiryakioglu, M., Campbell, J. (Eds.), The 2nd Internat. Symp. TMS (The Minerals, Metals and Materials Soc), pp. 127–134.

Griffiths, W.D., Lai, N.-W., 2007. Metall. Mater.Trans. A 38A, 190–196.

Griffiths, W.D., Raiszadeh, R., Omotunde, A.O., 2007. The effect of holding time on double oxide film defects in aluminium alloy castings. In: Crepeau, P.N., Tiryakioğlu, M., Campbell, J. (Eds.), Shape Casting: 2nd International Symposium. TMS (The Minerals, Metals & Materials Society).

Grill, A., Brimacombe, J.K., 1976. Ironmak. Steelmak. 3 (2), 76–86.

Gulsoy, G., Was, G.S., 2015. Mechanism of internal oxidation of alloy 617 in He-CO-CO_2 environments at 1123K (850 C). Metall. Mater. Trans. A 46A (Jan), 525–535.

Gupta, N., Satyanarayana, K.G., 2006. The solidification processing of metal-matric composites: the Rohatgi symposium. J. Occup. Med. 1047–4838 58 (11), 92–94.

Guthrie, R.I.L., 1989. Engineering in Process Metallurgy. Clarendon Press, Oxford, UK.

Hadian, R., Emamy, M., Campbell, J., 2009. The modification of cast Al-Mg2Si MMC by Li. Metall. Mater. Trans. A 40B (Dec), 822–832.

Hall, F.R., Shippen, J., 1994. Eng. Fail. Anal. 1 (3), 215–229.

Halliday, M.D., PooleP, Bowen, P., 1999. Slip band decohesion as a unifying fracture event during fatigue crack growth. Mater. Sci. Technol. 15 (4), 382–390.

Halvae, A., Campbell, J., 1997. Trans. Amer. Found. Soc. 105, 35–46.

Hammond, D.E., 1989. Modern Castings, pp. 29–33 (August).

Handrigan, S.M., Nakhia, S., Morrissey, L.S., 2018. Quantifying void formation and changes to microstructure during H charging; a precursor to embrittlement and blistering. Metall. Mater. Trans. A 50 (March), 1460–1467, 3.

Harris, Z.D., Dolph, J.D., Pioszak, G.L., Troconis, B.C.R., Scully, J.R., Burns, J.T., 2016. Hydrogen environment-assisted cracking of monetl K-500. Metall. Mater. Trans. A 47A (July), 3488–3510.

He, J., han, G., Fukuyama, S., Yokogawa, K., 1998. Creep rupture of 304 stainless steel. Mater. Sci. Technol. 14 (12), 1249–1256.

Hellio, C., de Novion, C.H., Marraud, A., Boulanger, L., 1987. Materials sci. Forum 15–18, 937–942.

Hoover, W.R., 1991. In: Hansen, N., et al. (Eds.), Proc 12th Riso Internat Symp Materials Science: Metal Matrix Composites – Processing, Microstructure and Properties, pp. 387–392.

Hopgood, A.A., Martin, J.W., 1986. Creep behaviour of a Ni-base single crystal superalloy. Mater. Sci. Eng. 82, 27–36.

Hornik, V., Fintova, S., Smid, M., Hutar, P., Hrbacek, K., KunzL, 2019. Interaction of fatigue and creep in MAR-M 247 Superalloy. In: 9th Internat Conf Mater Structure & Micromechanics of Fracture (MSMF9) Brno, Czech Republic.

Horng, J.H., Jiang, D.S., Liu, T.S., Chen, L.H., 2000. Fracture of A356 + Fe under resonant vibration. Int. J. Cast Metals Res. 13 (4), 215–222.

Hoyle, G., 1983. Electroslag Processes : Principles and Practice. Applied Science and Distributed by Elsevier, London UK and New York, NY, USA.

Hutchinson, B., 2015. Critical Assessment 16: Anisotropy in metals. MST 31 (12), 1393–1401.

IBF, 1960. Technical subcommittee TS 35. British Foundryman 53, 15–20.
Iwasaki, H., Mori, T., Higashi, K., 1998. Grain size refinement and mechanical properties of foil-pressed 5475. Mater. Sci. Technol. 14 (6), 597-584.
Jacob, S., Remy, A., 1983. Conductivite electrique et morphologie du silicium dans l'A-S13 et l'A-S7G. Fonderie-Fondeur 22, 33–41.
Jain, E., Chakkingal, U., 2013. Influence of pressing process on the drawability (R Value) of aluminium alloy AA 5052 sheets. Mater. Sci. Forum 765, 363–367.
Jelis, E., Clemente, M., Kerwien, S., Ravindra, N.M., Hespos, M.R., 2015. Steel 4340 produced by direct laser sintering. J. Met. 67 (3), 582–589.
Jiang, D.M., Hong, B.D., Lei, T.C., Downham, D.A., Lorimer, G.W., 1991. Tensile and fatigue fracture of Al alloy 6063. Mater. Sci. Technol. 7 (11), 1010–1014.
Joo, S.-H., Jung, J., Chun, M.S., Moon, C.H., Lee, S., Kim, H.S., 2014. Closure and bonding of pores during hot rolling of steel. Metall. Mater. Trans. A 45A (Aug), 4002–4011.
Jones, R.H., 2017. Stress Corrosion Cracking – Materials Performance and Evaluation, second ed. ASM International, Metals Park, Ohio, USA.
Karsay, S.I., 1980. Ductile Iron: The State of the Art 1980. QIT Fer et Titane, Canada.
Kattus, J.R., 1987. ASM Handbook "Fractography" Figure 756. Association Metallurgical Consultants Inc.
Keehyun, K., Withey, P., 2017. Response to discussion of 'Formation of an intermediate layer between grains in nickel-based superalloy turbine blades. Metall. Mater. Trans. A 48A (10), 5160–5163.
Kim, S.J., Yun, D.W., Jung, H.G., Kim, K.Y., 2017. Hydrogen permeation of ferritic steel under load. Mater. Sci. Technol. 33 (2), 149–161.
Kim, Y.-M., Jung, I.-H., 2015. $MnO-B_2O_3-SiO_2$ system in steels. Metall Mater Trans A 46A (June), 2736–2747.
Kita, K., 1979. Amer. Found. Soc. Internat Cast Metals J. 4 (4), 35–40.
Kang, S.E., Banerjee, J.R., Maina, E.M., Mintz, B., 2013. Influence of B and T on hot ductility of high Al containing TWIP steels. Mater. Sci. Technol. 29 (10), 1225–1232.
Knott, J.F., Mahidhara, R.K., et al., 1997. In: Recent Advances in Fracture. TMS Warrendale, PA, USA.
Kong, G., Zhang, D., Brown, P.D., McCartney, D.G., Harris, S.J., 2003. Thermally sprayed stellite 6. Mater. Sci. Technol. 19 (Aug), 1003–1011.
Koyama, M., Okazaki, S., Sawaguchi, T., Tsuzaki, K., 2016. Metall. Mater. Trans. A 47A (June), 2656–2673.
Kuhlmann-Wilsdorf, D., 1965. In: Lattice Defects in Quenched Metals. Academic Press, New York.
Kulas, M.-A., Green, W.P., Taleff, E.M., Krajewski, P.E., McNelly, T.R., 2006. Metall Mater Trans A 37A, 645–655.
Lai, N.-W., Griffiths, W.D., Campbell, J., 2003. Modeling of casting. In: Stefanescu, D.M., Warren, J.A., Jolly, M.R., Krane, M.J.M. (Eds.), Welding and Solidification Processes X.
Lashkari, O., Ghomashchi, R., 2007. Mater. Sci. Eng. 454–455, 30–36.
LaVelle, D.L., 1962. Trans. Amer. Found Soc. 70, 641–647.
Lee, J.-K., Kim, S.K., 2007. Mater. Sci. Eng. 449–451, 680-683.
Le May, I., 1987. ASM Handbook "Fractography" Figure 691. Metallurgical Consultancy Services Ltd.
Le May, I., 1987. ASM Metals Handbook, ninth ed. Fractography.
Leth-Olsen, H., Nisancioglu, K., 1998. Corrosion Sci. 40, 1179–1194 and 1194-1214.
Lett, R.L., Felicelli, S.D., Berry, J.T., Cuesta, R., Losua, D., 2009. Trans. Amer. Found. Soc. 09–043, 8 paper.
Li, D.Z., Campbell, J., Li, Y.Y., 2004. Filling system for investment cast Ni-base turbine blade. J. Processing Technol. 148 (3), 310–316.

Lin, L., Liu, Z., Li, Y., Han, X., Chen, X., 2012. Effects of severe cold rolling on exfoliation corrosion of Al-Zn-Mg-Cu-Cr alloy. J. Mater. Eng. Perform. 21 (6), 1070–1075.

Liu, J., Wang, Q., Qi, Y., 2019. In: Tiryakioglu, M., Griffiths, W.D., Jolly, M. (Eds.), Atomistic Simulation of Bifilms. Shape Casting Symposium TMS, pp. 45–51.

Liu, L.R., Jin, T., Zhao, N.R., Sun, X.F., Guan, H.R., Hu, Z.Q., 2003. Formation of carbides and their effects on stress rupture of a Ni-base single crystal superalloy. Mater. Sci. Eng. 361 (1–2), 191–197.

Livingston, H.K., Swingley, C.S., 1971. Surf. Sci. 24, 625–634.

Lu, Y.H., Chen, Z.R., Zhu, X.F., Shoji, T., 2014. Stress corrosion cracking behavior of autstenitic stainless 23CN20-09M in high temperature water. Material Sci. Technol. 30 (15), 1944–1950.

Lumley, R.N., Sercombe, T.B., Schaffer, G.B., 1999. Metall. Mater. Trans. 30A, 457–463.

Majidi, S.H., Griffin, J., Beckermann, C., 2018. Effect of pouring conditions and gating system design on air entrainment during mold filling. Metall. Mater. Trans. A 49B (5), 2599–2610.

Mandal, A., Karmakar, A., Chakrabarti, D., Davis, C., 2018. Effect of alloying and coiling temperature on the microstructure and bending performance of ultra-high-strength-strip steel. Metall. Mater. Trans. A 49A (Dec), 6359–6374.

Mao, L., Liu, C., Wan, Y., Wei, J., Gao, Y., Jiang, S., Chen, Z., 2019. Influence of heat treatment on microstructures and impact toughness of Mg-al-Zn alloy. J. Metals 71 (8), 2874–2883.

Matsunaga, H., Usada, T., Yanase, K., Endo, M., 2014. Ductility loss in ductile cast iron with internal hydrogen. Metall Mater Trans A 45A, 1315–1326.

Mayer, A.E., Mayer, P.N., Krasnikov, V.S., 2014. Dynamic fracture of metal in solid and liquid states under ultra-short intensive electron or laser irradiation. In: 20th European Conference on Fracture; Procedia Materials Science, vol. 3, pp. 1890–1895.

Mayer, A.E., Mayer, P.N., Krasnikov, V.S., Voronin, D.S., 2015. Strengths of many liquid metals. J. Phys. Conf. 653, 012093.

Mejia, I., Altamirano, G., Bedolla-Jacuinde, A., Cabrera, J.M., 2013. Effect of boron on the hot ductility of a low carbon advanced ultra-high strength steel (A-UHSS). Metall. Mater. Trans. A 44A (Nov), 5165–5176.

Mendelson, S., 1962. J. Appl. Phys. 33 (7), 2182–2186.

Metallurgical Associates Inc, Waukesha, Milwaukee, Wisconsin, USA. info@metassoc.com.

Meyers, M.A., Traiviratana, S., Lubarda, V.A., Benson, D.J., Bringa, E.M., 2009. J. Metals 61 (No 2), 35–41.

Mi, J., Harding, R.A., Campbell, J., 2002. The tilt casting process. Int. J. Cast Metals Res. 14 (6), 325–334.

Mi, J., Harding, R.A., Wickins, M., Campbell, J., 2003. Entrained oxide films in TiAl alloys. Intermetallics 11, 377–385.

Midson, S.P., September 2008. Semi solid casting of aluminium alloys – an update. Die. Cast. Eng. (9), 100–110.

Miguelucci, E.W., 1985. Amer. Found. Soc. Trans. 93, 913–916.

Milstein, F., Zhao, J., Maroudas, D., 2004. Phys. Rev. B 70 (184102), 1–6.

Miranda, A., Alba-Baena, N., McKay, B.J., Skin, D.G., Ko, S.H., Shin, J.S., 2013. Study of mechanical properties of an LM24 composite alloy reinforced with Cu-CNT Nano-fillers, processed using ultrasonic cavitation. Mater. Sci. Forum 765, 245–249, 2013.

Molina, J.M., Voytovych, R., Louis, E., Eustathopoulos, N., 2007. The surface tension of liquid Al in high vacuum. Int. J. Adhesion Adhes. 27, 394–401 (s).

Moser, M., 1985. Hydrogen-induced cold cracking - cracks adjacent to welds (Bruch durch Wasserstoff-Kaltrissigkeit - nebennahtrisse). Schweiss Tehnik (Berlin) 34/35 (3), pp140–143.

Moser, P., Corbel, C., Lucasson, P., Hautojarvi, P., 1987. Materials sci. Forum 15–18, 925–930.
Moss, T., Was, G.S., 2017. Accelerated SCC initiation of alloys 600 and 690 in hydrogenated supercritical water. Metall Mater Trans A 48A (April), 1613–1628.
Mountford, N.D.G., Calvert, R., 1959-1960. J. Inst. Met. 88, 121–127.
Mountford, N.D.G., Sommerville, I.D., 1993a. Steel Technology International, London, UK, pp. 155–169.
Mountford, N.D.G., Sommerville, I.D., From, L.E., Lee, S., Sun, C., 1993b. A Measuring Device for Quality Control in Liquid Metals. Scaninject Conf, Lulea, Sweden.
Mountford, N.D.G., Sommerville, I.D., Simionescu, A., Bai, C., 1997. Trans Amer Found Soc 105, 939–946.
Murakami, Y., 2002. Metal Fatigue – Effects of Small Defects and Non-metallic Inclusions. Elsevier, Oxford, UK.
Nadell, R., Eskin, D., Katgerman, L., 2007. Mater. Sci. Technol. 23 (11), 1327–1335.
Narayanasamy, R., Ramesh, T., Pandey, K.S., 2005. Cold forging of pure Al powder met. Mater. Sci. Technol. 21 (8), 912–916.
Nie, Y., Kimura, Y., Inoue, T., Yin, F., Akiyama, E., Tsuzaki, K., 2012. Hydrogen embrittlement of a 1500 MPa steel. Metall. Mater. Trans. A 43A, 1670–1687.
Nieswaag, H., Deen, H.J.J., 1990. 57th World Foundry Congress (Osaka, Japan).
Nili-Ahmadabadi, M., Feghahati, S., 2007. Int. J. Cast Metals Res. 20 (2), 53–57.
Nyahumwa, J., Green, N.R., Campbell, J., 1998. Trans. Amer. Found. Soc. 106, 215–223.
Nyahumwa, J., Green, N.R., Campbell, J., 2000. Metall. Mater. Trans. A 31A, 1–10.
Oki, S., 1970. Imono. J. Japn Found. Soc. 42 (11), 9–19. English translation in bulletin of the faculty of engineering, Yokohama National University, 1969 18 (March), 127–136.
Ohnistova, P., Piska, M., Dluhos, J., Horikova, J., Sandera, P., Petrenec, M., 2019. A study of crack initiation mechanism in the aluminium alloy 7475-T7351 when tensile loading. In: 9th Internat Conf Mater Structure & Micromechanics of Fracture (MSMF9) Brno, Czech Republic.
Ooi, S.W., Gola, A., Vegter, R.H., Yan, P., Stadler, K., 2017. Evolution of white-etching cracks and associated microstructural alterations during bearing tests. Mater. Sci. Technol. 33 (14), 1657–1666.
Owen, M., 1966. British Foundryman 59, 415–421.
Ozdes, H., Tiryakioglu, M., 2016. On the relationship between structure quality index and fatigue life distributions in Al aerospace castings. Metals 6, 81–90.
Ozdes, H., Tiryakioglu, M., 2017. The effect of structure quality on fatigue life in 319 Al alloy castings. J. Mater. Eng. Perform. 26 (2), 736–743. February.
Panchanathan, V., Seshadri, M.R., Ramachandran, A., 1965. British Foundryman 58, 380–384.
Peet, M.J., Fielding, L.C.D., Hamedany, A.A., Rawson, M., Hill, P., Bhadeshia, H.K.D.H., 2017. Strength and toughness of clean nanostructured bainite. Mater. Sci. Technol. 33 (10), 1171–1179.
Peet, M.J., Hojo, T., 2016. Hydrogen susceptibility of nanstructured bainitic steels. Metall. Mater. Trans. A 47A (February), 718–725.
Peng, L., Zeng, G., Su, T.C., Yasuda, H., Nogita, K., Gourlay, C.M., 2019. Al8Mn5 particle settling and interactions with oxide films in liquid AZ91 Mg alloy. J. Met. 71 (7), 2235–2244.
Peng, X., Li, Y., Guo, Q., Xu, G., 2018. Stress corrosion cracking behaviour of 7xxx series (Al-Zn-Mg-Cu) alloy. J. Occup. Med. 70 (11), 2692–2697.
Popel, G.I., Esin, O.A., 1956. Zh. Fiz. Khim. 30, 1193.
Pumphrey, W.I., Moore, D.C., 1949. J. Inst. Met. 75, 236–727.
Qu, J., Yang, S., Chen, Z., Du, J., Li, J., Wang, D., 2019. Effect of turning amount on metallurgical qualities of GH4169 superalloy. Materials 12 (11), 1852–1862.

Raiszadeh, R., Griffiths, W.D., 2008. Duration of atmosphere in a double oxide film defect in pure Al. Metall. Mater. Trans. A 39B, 298–303.
Raja, V.S., Shoji, T., 2011. Stress Corrosion Cracking – Theory and Practice. Woodhead Publishing Limited, Cambridge, UK.
Raman, R.K.S., Muddle, B.C., 2003. Caustic stress corrosion cracking of a spheroidal graphite cast iron. Mater. Sci. Technol. 19 (12), 1746–1754.
Ramjaun, T.I., Ooi, S.W., Morana, R., Bhadeshia, H.K.D.H., 2018. Designing steel to resist hydrogen embrittlement: Part I – trapping capacity. Mater. Sci. Technol. 34 (14), 1737–1746.
Rao, G.A., Kumar, M., Srinivas, M., Sarma, D.S., 2001. In: Loria, E.A. (Ed.), Powder Met Superalloy 718; Superalloys 718, 625, 706 and Various Derivatives. TMS, pp. 605–616.
Rashid, A.K.M.B., Campbell, J., 2004. Metall. Mater. Trans. A 35A (7), 2063–2071.
Reed, R., 2006. The Superalloys. Cambridge University Press.
Ren, X.C., Zhou, Q.J., Shan, G.B., Chu, W.Y., Li, J.X., Su, Y.J., Qiao, L.J., 2008. A nucleation mechanism of hydrogen blister in metals and alloys. Metall. Mater. Trans. A 39A, 87–97.
Rohatgi, P., 1990. Adv. Mater. Process. 2, 39–44.
Rosenberg, R.A., Flemings, M.C., Taylor, H.F., 1960. Trans. Amer. Found. Soc. 68, 518–528.
Runyoro, J., Boutorabi, S.M.A., Campbell, J., 1992. Trans. Amer. Found. Soc. 100, 225–234.
Sabochick, M.J., Yip, S., Lan, N.Q., 1987. Materials Sci. Forum 15–18, 857–862.
Sahoo, K.C., Goyal, S., Parameswaran, P., Ravi, S., Laha, K., 2018. Creep deformation, damage and rupture life of 304HCu austenitic stainless steel under multiaxial stress. Metall. Mater. Trans. A 49A (March), 881–898.
Salamci, E., Cochrane, R.F., 2003. Spray deposited and extruded 7000 series Al alloys. Mater. Sci. Technol. 19 (August), 1130–1136.
Salehi, M., Shabestari, S.G., Boutorabi, S.M.A., 2014. Hot consolidation of amorphous melt-spun flakes. Metall. Mater. Trans. A 45 (December), 6344–6350.
Schaffer, G.B., 2004. In: Nie, J.F., Morton, A.J., Muddle, B.C. (Eds.), Powder Processed Aluminium Alloys; Materials Forum, vol. 28. University of Queensland, Australia, pp. 65–74.
Sen Gupta, A., Moser, P., Bourret, A., Corbel, C., Naidu, S.V., Sen, P., Hautojarvi, P., 1987. Materials Sci. Forum 15–18, 931–936.
Shafaei, A., Raiszadeh, R., 2014. Metall. Mater. Trans. B 45B (December), 2486–2494.
Shanyavskiy, A.A., Banov, M.D., 2019. Durability diagnostics of turbine blades based on non-destructive acoustic emission testing. In: 9th Internat Conf Mater Structure & Micromechanics of Fracture (MSMF9) Brno, Czech Republic.
Shi, W., Yang, S., Dong, A., Li, J., 2018. Corrosion of spring steel induced by MnS inclusions. J. Occup. Med. 70 (11), 2513–2522.
Shibata, A., Momotani, Y., Murata, T., Matsuoka, T., Tsuboi, M., Tsuji, N., 2017. Hydrogen-related fracture in lath martensitic steels. Mater. Sci. Technol. 33 (13), 1524–1532.
Shivkumar, S., Ricci Jr., S., 1990. D Apelian. AFS Trans. 913–922.
Sidhu, R.K., Richards, N.L., Chaturvedi, M.C., 2005. Mater. Sci. Technol. 21 (10), 1119–1131.
Sidorova, E., Karasev, A.V., Kuznetsov, D., Jonsson, P.G., 2019. Modification of non-metallic inclusions in oil-pipeline steels by Ca-treatment. In: Li, Z., Davis, C. (Eds.), Ironmaking and Steelmaking. Published by MDPI, Switzerland, pp. 284–295.
Sinha, V., Pilchak, A.L., Jha, S.K., Porter III, W.J., John, R., Larsen, J.M., 2018. Correlating scatter in fatigue life with fracture mechanisms in forged Ti-6242Si alloy. Metall. Mater. Trans. A 49A (April), 1061–1078.

Smallman, R.E., Bishop, R.J., 1999. Modern Physical Metallurgy and Materials Engineering, sixth ed. Butterworth-Heinemann, Oxford, UK, pp. 284–291.
Socie, D.F., Fash, J., 1982. Fatigue behaviour and crack development in cast irons. Trans. Amer. Found. Soc. 90, 385–392.
Sohn, S.S., Song, H., Kim, J.G., Kwak, J.-H., Kim, H.S., Lee, S., 2016. Delayed fracture properties in ferrite-austenite duplex lightweight steels. Metall. Mater. Trans. A 47A (Feb), 706–717.
Solano-Alvarez, W., Song, E.J., Han, D.K., Suh, D.-W., Bhadeshia, H.K.D.H., 2015. Cracks in martensite plates as hydrogen traps in a bearing steel. Metall. Mater. Trans. A 46A (Feb), 665–673.
Spenser, D., Mehrabian, R., Flemings, M.C., 1972. Met. Trans. 3 1925.
Spittle, J.A., Cushway, A.A., 1983. Met. Technol. 10, 6–13.
Srinivasan, A., Pillai, U.T.S., Swaminathan, J., Pai, B.C., 2006. Counter-gravity Mg alloy casting creep properties. Int. J. Cast Metals Res. 19 (5), 265–268.
Staley, J.T., Tiryakioglu, M., Campbell, J., 2007. Effect of various HIP conditions on bifilms and mechanical properties in aluminum castings. In: Crepeau, P.N., Tiryakioğlu, M., Campbell, J. (Eds.), Shape Casting: 2nd International Symposium. TMS (The Minerals, Metals & Materials Society), pp. 159–166.
Steinbach, S., Ratke, L., Zimmermann, G., Sturz, L., Roosz, A., Kovacs, J., Fautrelle, Y., Budenkova, O., Lacaze, J., Dost, S., Grun, G.-U., Warnken, N., Wu, M., Sillekens, W., 2017. Investigation of the effect of fluid flow on microstructure evolution in Al-Si-Fe alloys: the MICAST Project. In: Solidification Processing 2017 (SP17). BCAST, Brunel University, UK.
Storaska, G.A., Howe, J.M., 2004. Mater. Sci. Eng. Met. 368, 183–190.
Suave, L.M., Munoz, A.S., Gaubert, A., Benoit, G., et al., 2018. Thin-wall debit in creep of CS200+Hf alloy. Metall. Mater. Trans. A 49A (September), 4012–4028.
Sureth, K.S., Rollett, A.D., Suwas, S., 2013. Evolution of microstructure and texture during deformation and recrystallisation of heavily rolled Cu-Cu multilayer. Metall. Mater. Trans. A 44A (Aug), 3866–3881.
Suzuki, S., 1989. Modern Castings (October) 38–40.
Svensson, I., 1980. Influence of deoxidation practice on hydrogen pick-up in steel melts. Solidification Technology in the Foundry and Cast House. Univ Warwick Conference, pp. 253–262 (Published by The Metals Soc. 1983).
Szczepanski, C.J., Jha, S.K., Larsen, J.M., Jones, J.W., 2008. Microstructural influences on very-high-cycle fatigue crack initiation in Ti-6246. Metall. Mater. Trans. A 39A (Dec), 2841–2851.
Szklarska-Smialowska, Z., 1999. Pitting corrosion of aluminum. Corrosion Sci. 41, 1743–1767.
Takakuwa, O., Yamabe, J., Matsunaga, H., Furuya, Y., Matsuoka, S., 2017. Comprehensive understanding of ductility loss mechanisms in various steels with external and internal hydrogen. Metall. Mater. Trans. A 48A, 5717–5732.
Thomason, P.F., 1990. Ductile Fracture of Metals. Pergamon Press, Oxford and NY.
Tiberg, L., 1960. Jerkont. Ann. 144 (10\0), 771–793.
Tiryakioglu, M., 2009. Relationship between defect size and fatigue life distributions in Al-Si-Mg alloy castings. Metall. Mater. Trans. A 40A (7), 1623–1630.
Tiryakioglu, M., 2017. On estimating high-cycle fatigue life of cast Al-Si-Mg-(Cu) alloys from tensile results. Mater. Sci. Eng. A688, 9–15.
Tiryakioglu, M., 2018. On the intrinsic fracture pressure of liquid and solid aluminum around its melting temperature. Metall. Mater. Trans. A 49 (12), 5953–5955.
Tiryakioglu, M., 2019a. The Myth of Hydrogen Pores in Aluminum Castings. Shape Casting Symposium TMS San Antonio, Texas.
Tiryakioglu, M., 2019b. North Florida Univ Personal Communication.

Tiryakioglu, M., Campbell, J., Alexopoulos, N.D., 2009a. On the ductility of cast Al-Si-Mg alloys. Metall. Mater. Trans. A 40A (April), 1000–1007.
Tiryakioglu, M., Campbell, J., Alexopoulos, N.D., 2009b. On the ductility potential of cast Al-Cu-Mg (206) alloys. Mater. Sci. Eng. 506, 23–26.
Tiryakioglu, M., Campbell, J., Nyahumwa, C., 2011. Fracture surface facets and fatigue life potential of castings. Metall. Mater. Trans. B 42, 1098–1103.
Toda, H., Minami, K., Koyama, K., et al., 2009. Healing behavior of preexisting hydrogen micropores in aluminum alloys during plastic deformation. Acta Mater. 57, 4391–4403.
Toda, H., Oogo, H., Horikawa, K., Uesugi, K., Takeuchi, A., Suzuki, Y., nakazawa, M., Aoki, Y., Kobayashi, M., 2014. The true origin of ductile fracture in Al alloys. Metall. Mater. Trans. A 45A (February), 765–776.
Todoshchenko, O.M.I., Yagodzinskyy, Y., Saukkonen, T., Hanninen, H., 2014. Role of nonmetallic inclusions in hydrogen embrittlement of high strength carbon steels with microalloying. Metall. Mater. Trans. A 45A (October), 4742–4747.
Traiviratana, S., Bringa, G.M., Benson, D.J., Meyers, M.A., 2008. Acta Mater. https://doi.org/10.1016/j.actamat.2008.03.047.
Turner, G.L., 1965. British Foundryman 58, 504–505.
Uberuaga, B.P., Hoagland, R.G., Valone, S.M., Voter, A.F., 2007. Phys. Rev. Lett. 99, 135501–135503.
Ucsnik, S., Gradinger, R., Becirovic, A., Waldhor, A., 2013. Enhanced performance of steel-aluminium cast nodes through cold metal transfer. Mater. Sci. Forum 765, 736–740.
Van den Eeckhout, E., Depover, T., Verbeken, K., 2018. Effect of microstructure on hydrogen permeation. Metals 8 (10), 779–789.
Venugopalan, D., Alagarsamy, A., 1992. Trans Amer Found Soc 100, 337–342.
Vernede, S., Jarry, P., Rappaz, M., 2006. Acta Met. 54, 4023–4034.
Versnyder, F.I., Shank, M.E., 1970. The development ... through directional solidification. Mater. Sci. Technol. 6 (4), 213–247.
Vijayan, V., Prabu, K.N., 2018. Effect of Sr modification on thermal diffusivity of Al-8Si alloy. Int. J. Cast Metals Res. 31 (2), 80–86.
Wang, L., Rhee, H., Felicelli, S.D., Sabau, A.S., Berry, J.T., 2009. Oxide film and porosity defects in magnesium alloy AZ91. Shape Casting Symposium. In: 3rd International Symposium TMS USA Annual Meeting February 2009.
Wang, Q.G., Crepeau, P.N., Griffiths, J.R., Davidson, C.J., 2005. In: Tiryakioglu, M., Crepeau, P.N. (Eds.), Oxide Films and Porosity on Fatigue. Shape Casting: The JC Symposium, TMS, pp. 205–214.
Wang, Y., Li, H.-T., Fan, Z., 2017. Grain Refinement of Al- and Mg-alloys by Native Oxide Particles. Solidification Processing 2017. BCAST, Brunel University, UK (SP17).
Wang, Y.;, Li, H.-T., Fan, Z., Scamans, G., 2013. Characterisation of oxide films in al-Mg alloy melts. Mater. Sci. Forum 765 (2013), 220–224 (Trans Tech Publications, Switzerland).
Warrington, D., McCartney, D.G., 1989. Cast Metals 2 (3), 134–143.
Way, L.D., 2001. Mater. Sci. Technol. 17 (10), 1175–1190.
Weertman, J., 1996. Dislocation Based Fracture Mechanics. World Scientific Publishing Co, Singapore.
Weiberg, C., Quere, Y., 1987. Materials Sci. Forum 15–18, 943–948.
Wen, X., Wen, W., Zhang, Y., Xu, B., Zeng, Q., Liu, Y., Tong, L., Zhai, T., Li, Z., 2016. Effect of Fe content on recrystallisation texture evolution… and earing of cold rolled continuous cast AA5052 alloy sheets. Metall. Mater. Trans. A 47A (April), 1865–1880.
Wiese, J.F., Dantzig, J.A., 1990. Modelling stress development during the solidification of gray iron castings. Metall. Trans. A 21A, 489–497.

Willis, J.R., 1967. A comparison of the fracture criteria of Griffith and Barenblatt. J. Mech. Phys. Solid. 15, 151–162.
Wittenberger, E.J., Rhines, F.N., 1952. J. Metals 4 (4), 409–420 and Trans AIME 194 409-420.
Wu, R., Sandstrom, R., 1995. Creep cavity nucleation and growth in 12Cr–Mo–V steel. Mater. Sci. Technol. 11 (6), 579–588.
Xue, Y., Burton, C.L., Horstmeyer, M.F., McDowell, D.L., Berry, J.T., 2007. Multistage fatigue modelling of A356-T6 and A380-F. Metall. Mater. Trans. B 38B (4), 601–606.
Yamakov, V., 2011. Personal Communication.
Ye, Z.-H., Cai, W.-X., Li, J.-F., Chen, X.-R., Zhang, R.-F., et al., 2018. Intergranular corrosion behaviour of Al-Cu-Li alloy 2050. Metall. Mater. Trans. A 49A (June), 2471–2486.
Yonezawa, T., Watanabe, M., Hashimoto, A., 2015. Primary water stress corrosion cracking in 690 alloy. Metall. Mater. Trans. A 46A (June), 2768–2780.
Yousefian, P., Tiryakioglu, M., 2018. Pore formation during solidification of Al. MMTA 49A (Feb), 563–575.
Yu, L.X., Sun, Y.R., Sun, W.R., Zhang, W.H., Liu, F., Xin, X., Qi, F., Jia, D., Sun, X.F., Guo, S.R., Hu, Z.Q., 2013. Effect of B on … superalloy. Material Sci. Technol. 29 (12), 1470–1477.
Yuan, L., 2019. Solidification defects in additive manufactured materials. J. Met. 71 (9), 3221–3222.
Zhang, Y.N., Cao, X., Wanjara, P., Medraj, M., 2013. Oxide films in laser additive manufactured Inconel 718. Acta Mater. 61, 6562–6576.
Zhang, Z., 2007. Triplex Lloyd Group and Birmingham University. Personal communication, UK.
Zhonqiu, Z., 1990. An investigation on effect of hydrogen on ductility in heavy section steel 13Cr 4Ni 0.6Mo. Amer. Found. Soc. Trans. 98, 235–239.
Zhou, Y., Li, D., Zeng, X., Ding, W., 2013. The effect of particles on microstructure and mechanical behaviour of Mg-10Gd-3Y-0.4Zr alloy processed by ECAP. Mater. Sci. Forum 765, 444–448, 2013. (Trans Tech Publications, Switzerland).
Zurecki, Z., 1996. In: Saha, D. (Ed.), Gas Interactions in Nonferrous Metals Processing. TMS, pp. 79–93.

Index

Note: 'Page numbers followed by "f" indicate figures and "t" indicates tables.'

D

E

F

G

H

I

J

K

FSC
www.fsc.org
MIX
Papier | Fördert
gute Waldnutzung
FSC® C083411

Kolibri 360 GmbH
Ferdinand-Jühlke-Straße 7
99095 Erfurt, Deutschland
produktsicherheit@kolibri360.de